Concepts of Simultaneity

Concepts of Simultaneity

From Antiquity to Einstein and Beyond

Max Jammer

The Johns Hopkins University Press
Baltimore

Printed in the United States of America on acid-free paper
9 8 7 6 5 4 3 2 1

The Johns Hopkins University Press
2715 North Charles Street
Baltimore, Maryland 21218-4363
www.press.jhu.edu

Library of Congress Cataloging-in-Publication Data
Jammer, Max.
Concepts of simultaneity : from antiquity to Einstein and beyond / Max Jammer.
p. cm.
Includes bibliographical references and index.
ISBN 0-8018-8422-5
1. Simultaneity (Physics) 2. Physics—Philosophy. 3. Relativity (Physics) 4. Space and time. 5. Time—Philosophy. I. Title.

QC173.5.S56J36 2006
530.11—dc22 2006048564

A catalog record for this book is available from the British Library.

We have to bear in mind that all our propositions involving time are always propositions about simultaneous events.

Albert Einstein, 1905

Contents

	Preface	ix
	Introduction	1
1	Terminological Preliminaries	8
2	The Concept of Simultaneity in Antiquity	16
3	Medieval Conceptions of Simultaneity	47
4	The Concept of Simultaneity in the Sixteenth and Seventeenth Centuries	59
5	The Concept of Simultaneity in Classical Physics	68
6	The Transition to the Relativistic Conception of Simultaneity	95
7	Simultaneity in the Special Theory of Relativity	106
8	The Reception of the Relativistic Conception of Simultaneity	148
9	The Conventionality Thesis	171
10	The Promulgation of the Conventionality Thesis	192
11	Symmetry and Transitivity of Simultaneity	201
12	Arguments against the Conventionality Thesis	220
13	Clock Transport Synchrony	240
14	Recent Debates on the Conventionality of Simultaneity	251
15	Simultaneity in General Relativity and in Quantum Mechanics	271
	Epilogue	295
	Index	301

Preface

Quite a few notions in the physical sciences, such as "force" or "mass," are used in everyday language before becoming rigorously defined scientific concepts or technical terms. Only one concept of this kind, however, played a critical role in initiating a new physical theory that has fundamentally changed all our conceptions of physical reality and for which the question of whether it has a factual or only a conventional status in this theory remains a matter of dispute.

This unique notion is the concept of simultaneity of spatially separated events, and the theory that it initiated is Albert Einstein's special theory of relativity. In fact, the first paragraph of the seminal paper of this theory, published by Einstein in 1905 and hailed as "possibly the most important scientific paper . . . written in the twentieth century,"[1] carries the heading "*§ 1. Definition of Simultaneity.*"[2]

This monograph presents a comprehensive, coherent, critical, and completely documented analysis of the conceptual development of the notion of simultaneity from its earliest use in remote antiquity until its present status in modern physics.

Some sections of the text are based on papers that I have read at scientific meetings or on lectures given at various times. They include (1) "Some Fundamental Problems in the Special Theory of Relativity,"[3] a lecture presented in June 1978 at the International School of Physics Enrico Fermi in Varenna, Italy; (2) a series of lectures given in 1981 at the University of Otago in Dunedin, New Zealand; (3) "The Concept of Time,"[4] an address delivered in August 1983 at Gakushuin University in Tokyo, Japan; (4) "The History of

[1] R. W. Clark, *Einstein: The Life and Times* (New York: Avon Books, 1971), p. 116.

[2] A. Einstein, "Zur Elektrodynamik bewegter Körper," *Annalen der Physik* **17,** 891–921 (1905).

[3] In Toraldo di Francia, ed., *Problems in the Foundations of Physics, Course 72* (Amsterdam: North-Holland, 1979), pp. 202–236.

[4] In *Gakushuin Daigaku* **2,** 1–35 (1984), in Japanese.

the Concept of Distant Simultaneity,"[5] a talk given in May 1985 at the University of Rome; (5) a lecture on the concept of simultaneity, delivered in October 1987 at the Department of Physics of the University of Bari, Italy; (6) a series of lectures on the history and philosophy of the concept of time, presented in June 1989 at the University of Konstanz, Germany; (7) and a seminar conducted on the philosophy of time in the winter semester 1989/90 at the University of Western Ontario in London, Ontario, Canada.

I wish to thank Dr. Trevor Lipscombe, editor-in-chief of the Johns Hopkins University Press, for fruitful cooperation, and I thank an anonymous reader of the manuscript for useful comments. I also thank Nancy S. Wachter for her careful copyediting of the manuscript.

[5]In *Rendiconti della Accademia delle Scienze* **9,** 169–184 (1985), in Italian.

Concepts of Simultaneity

Introduction

Modern physics, as is well known, led to a radical revision of the fundamental concepts of classical physics, such as the concepts of space, time, matter, energy, and causality. The foundations of modern physics are the quantum theory and the theory of relativity, both of which originated in the early years of the twentieth century. Historians of physics generally agree with Arnold Sommerfeld[1] and Max von Laue[2] in dating Friday, 14 December 1900, as the "birthday" of the quantum theory. On that day, Max Planck[3] announced for

[1] "Die Quantentheorie ist ein Kind des 20. Jahrhunderts. Ihr Geburtstag ist der 14. December 1900." A. Sommerfeld, *Atombau und Spektrallinien* (Braunschweig: Vieweg, 1919, 1931), p. 39. "The quantum theory is the product of the twentieth century. It came to life on 14th December 1900." *Atomic Structure and Spectral Lines* (London: Methuen, 1923), p. 36.

[2] "Planck konnte am 14. Dezember 1900 die theoretische Ableitung des Strahlungsgesetzes vorlegen. Das war die Geburtsstunde der Quantentheorie." M. von Laue, *Traueransprache* [Memorial Address, delivered at Planck's funeral at the Albani Church in Göttingen on October 7, 1947], in M. Planck, *Physikalische Abhandlungen und Vorträge* (Braunschweig: Vieweg, 1958), vol. 3, p. 419.

[3] M. Planck, "Zur Theorie des Gesetzes der Energieverteilung im Normalspektrum," *Verhandlungen der Deutschen Physikalischen Gesellschaft* **2,** 237–245 (1900); "Über das Gesetz der Energieverteilung im Normalspektrum," *Annalen der Physik* **4,** 553–563 (1901); *Physikalische Abhandlungen und Vorträge* (Braunschweig: Vieweg, 1958), vol. 1, pp. 698–706, 717–727.

the first time what he then called "the natural constant *h*" and what later, under the name of "Planck's constant," became the "trademark" of quantum mechanics. For a similar reason Friday, 30 June 1905, may be called the "birthday" of the theory of relativity, for on that day Albert Einstein's seminal paper[4] on the special theory of relativity was received by the editorial board of the *Annalen der Physik.*

Priority questions are generally, and rightly, regarded as nugatory and not worth mentioning. In the present case, however, the following priority claim deserves our attention. The claim that, despite the chronological precedence of the birth of the quantum theory, it was Einstein's 1905 relativity paper that initiated the conceptual revolution of modern physics was made most eloquently not, as may be expected, by a relativist but by a most prominent quantum physicist. It was made in fact by one of the fathers of quantum mechanics, Werner Heisenberg. In his Gifford Lecture, delivered at the University of St Andrews in the winter semester 1955/56, Heisenberg declared: "Within the field of modern physics the theory of relativity has played a very important role. It was in this theory that the necessity for a change in the fundamental principles of physics was recognized *for the first time.*"[5]

A similar statement had already been made by Heisenberg in 1934 when he said: "The fundamental presuppositions of classical physics, which led to the scientific picture of the 19th century, had been challenged *for the first time* by Einstein's special relativity." Then, specifying exactly the premise of classical physics that gave rise to this challenge, he continued: "It was the assumption that it is meaningful without further consideration to call two events simultaneous in the case they do not occur at the same place."[6] Heisenberg's statement, that Einstein's 1905 analysis of the

[4] A. Einstein, "Zur Elektrodynamik bewegter Körper," *Annalen der Physik* **17,** 891–921 (1905); reprinted in *The Collected Papers of Albert Einstein* (Princeton, NJ: Princeton University Press, 1989), vol. 2, pp. 275–306; "On the electrodynamics of moving bodies," *Princeton Translation Project,* pp. 140–171. Also in A. Einstein, H. A. Lorentz, H. Minkowski, and H. Weyl, *The Principle of Relativity* (New York: Dover Publications, 1953), pp. 35–65.

[5] W. Heisenberg, *Physics and Philosophy* (New York: Harper & Row, 1958), p. 110 (emphasis added).

[6] W. Heisenberg, *Wandlungen in den Grundlagen der exakten Naturwissenschaft in jüngster Zeit* (Lecture delivered in Hannover on 17 September 1934), in *Die Naturwissenschaften* **22,** 669–675 (1934); reprinted in W. Blum, H. P. Dürr, H. Rechenberg, eds., *Werner Heisenberg—Gesammelte Werke* (Munich: Piper, 1984), vol. 1, pp. 96–101 (emphasis added).

concept of distant simultaneity (i.e., of spatially separated events) inaugurated the modern physical world picture, can be confirmed by the fact that Einstein himself once admitted: "By means of a revision of the concept of simultaneity in a shapable form I arrived at the special relativity theory."[7]

A magniloquent formulation of Heisenberg's claim was given more recently by the cosmologist Julian B. Barbour: "Einstein's definition of simultaneity opened the door into a world as unexpected as the one inadvertently discovered by Copernicus when trying to save uniformity in the heavens."[8]

Heisenberg, who had studied physics in Munich under Sommerfeld from 1920 to 1923 and knew, of course, his teacher's statement concerning the birth of the quantum theory, could nevertheless make this priority claim because he realized that the philosophical implications of a new physical theory did not necessarily need to be recognized at the birth of that theory. Indeed, as we know from documentary evidence, prior to 1906 Planck and his colleagues thought it was possible to "fit" the constant h into the conceptual framework of classical physics.[9] This was confirmed also by Fritz Reiche, a student of Planck from 1902 until 1907, when he recalled: "I would not say that he, Planck, had the feelings: 'I give you here something very new which is very funny and a very complete break.'"[10] In contrast, Einstein's 1905 relativity page immediately left no doubt that the classical notions of space and time could no longer be maintained. It was precisely because of its revolutionary innovations that the more conservative members of the editorial board of the *Annalen* considered the paper, with its seemingly bizarre notions of time dilation, length contraction, and relativity of simultaneity, written by a clerk of a patent office, more as a piece of "sci-

[7] Quoted by A. Fölsing, *Albert Einstein: a Biography* (New York: Viking, 1997), p. 176; *Albert Einstein: eine Biographie* (Frankfurt am Main: Suhrkamp, 1993), p. 201.

[8] J. B. Barbour, *Absolute or Relative Motion?* (Cambridge: Cambridge University Press, 1989), vol. 1, p. 676.

[9] Cf. M. Jammer, *The Conceptual Development of Quantum Mechanics* (New York: McGraw-Hill, 1966), p. 22; enlarged edition published as vol. 12 in the series *The History of Modern Physics* (New York: The American Institute of Physics, 1989), p. 17. See also T. S. Kuhn, *Black-body Theory and the Quantum Discontinuities* (Oxford: Clarendon Press, 1978), pp. 125–126.

[10] F. Reiche, Interview on 30 March 1957 in *Archives for the History of Quantum Physics* (Copenhagen: Niels Bohr Institute), p. 6. See also M. von Laue, *Aufsätze und Vorträge* (Braunschweig: Vieweg, 1961), p. xvii.

ence fiction" than as a serious scientific work. Thus, what later was called "possibly the most important paper that has been written in the twentieth century"[11] might have been returned to its author as unfit for publication had it not been for Planck, who as the representative of the German Physical Society was the chairman of the editorial board and who immediately understood the paper's importance.

Although it is certainly true that, as Heisenberg contended, Einstein's 1905 analysis of the concept of distant simultaneity inaugurated the conceptual revolution of modern physics, it would be wrong to assume that this notion became the subject of critical attention only with the advent of the theory of relativity. True, before 1905 or so, physicists did not think that this notion deserved much attention, but philosophers did. The fact that many philosophers, including such prominent thinkers as Aristotle, Leibniz, and Kant, thought that this notion required closer analysis seemed not to be well known, certainly not among physicists.

As we will see in the sequel, Einstein's treatment of this concept has a noteworthy prehistory that can be traced back to antiquity. It also has an equally important posthistory. Suffice it here to point out that the logical empiricists, and especially members of the Vienna Circle, like Alfred J. Ayer, Rudolf Carnap, Herbert Feigl, and Friedrich Waismann, repeatedly admitted to have been profoundly influenced by Einstein's treatment of space and time and, in particular, by his analysis of the concept of distant simultaneity. As Graham Nerlich, a philosopher of the University of Adelaide, rightly remarked in 1982: "It is hard to overestimate the impact of Einstein's definition of distant simultaneity on philosophy in this century, set, as the words were, in the context of a highly successful theory of physics."[12]

To show that such a statement is justified, at least as far as the philosophy of physics is concerned, it suffices to recall that, in accordance with the theory of relativity, even such an elementary concept as the length of a line segment or of a rod, moving relative to an inertial system, involves the concept of distant simultaneity. That not only temporal but also spatial measurements depend on the notion of simultaneity

[11] R. W. Clark, *Einstein—The Life and Time* (New York: Avon, 1972), p. 116.

[12] G. Nerlich, "Simultaneity and convention in special relativity," in R. McLaughlin (ed.), *What? Where? When? Why?* (Dordrecht: Reidel, 1982), p. 130.

follows from the simple fact that "the length of a moving line-segment is the distance between simultaneous positions of its endpoints," as Hans Reichenbach, in the chapter entitled "The Dependence of Spatial Measurements on the Definition of Simultaneity" in his influential book *The Philosophy of Space and Time*[13] convincingly demonstrated. Having shown that "space measurements are reducible to time measurements" he concluded that "time is therefore logically prior to space." Since, in turn, the notion of *time*, as Einstein demonstrated in 1905, presupposes a definition of *simultaneity*, it is clear that, indeed, the importance of the concept of simultaneity for kinematics, and therefore for physics in general, can hardly be exaggerated.

Of course, this holds especially for the theory of relativity. For instance, P. F. Browne rightly pointed out that *all* relativistic effects are ultimately "direct consequences of the relativity of simultaneity."[14] Or as Ernan McMullin wrote: "The 'relativity' of the new theory—one of the most solidly verified theories in the entire range of physics—is chiefly, therefore, a relativity of simultaneity."[15]

One of the major problems debated by philosophers of science in our time is the controversial question of whether the concept of distant simultaneity denotes something factual, empirically testable, or at least unambiguously definable, or whether it refers to merely an object of a convention, that is to an arbitrary stipulation without any factual content, as to which events are to be called simultaneous. This "problem of the conventionality thesis concerning the concept of distant simultaneity" seems to have far-reaching philosophical implications. If, as mentioned above, the concept of distant simultaneity is a fundamental ingredient in the logical structure of the theory of relativity but is in reality nothing but a convention, the question naturally arises of whether this does not imply that the whole theory of relativity and with it a major part of modern physics are merely fictions devoid of any actual content. A positive answer to this question would have disastrous consequences for the philosophical un-

[13] H. Reichenbach, *Philosophie der Raum-Zeit-Lehre* (Berlin: W. de Gruyter, 1928); *The Philosophy of Space and Time* (New York: Dover Publications, 1958), chapter 3.

[14] P. F. Browne, "Relativity of rotation," *Journal of Physics* **A10,** 727–744 (1977). Quotation on p. 731.

[15] E. McMullin, "Simultaneity," in the *New Catholic Encyclopedia* (New York: McGraw-Hill, 1967), vol. 13, p. 234.

derstanding and epistemological status of physics and with it of the whole of modern science.

The current debate about the conventionality thesis concerning the concept of distant simultaneity will be discussed in great detail in the final chapters of this book.

The concept of simultaneity, however, is of importance not only for issues related to the theory of relativity, but it also plays a significant role in other branches of physics, in particular, in quantum mechanics. A well-known example is the quantum-mechanical entanglement such as the one exhibited in the Einstein-Podolsky-Rosen type experiments, where the outcome of spatially separated measurements are instantaneously correlated. Indeed, "the uneasiness of fit between relativity and quantum mechanics regarding the treatment of measurements hinges on the concept of simultaneity."[16]

As we will see in the course of our study the notion of simultaneity also plays an important role in classical physics. Unfortunately, textbooks in the physical sciences, unless dealing exclusively with the theory of relativity in general, completely ignore the notion of simultaneity. A noteworthy exception is Herbert Goldstein's excellent text on classical mechanics which states explicitly at the beginning of its first chapter that "basic to any presentation of mechanics are a number of fundamental physical concepts, such as space, time, *simultaneity*, mass and force."[17]

The preceding remarks on the concept of simultaneity may mislead the reader to believing that only modern physicists and philosophers recognize the crucial importance of this notion. One of the major objectives of this treatise is to show that this concept has occupied the attention of philosophers and scientists throughout the whole history of human thought and played an important role in the writings of such intellectual giants as Aristotle, St. Augustine, Leibniz, and Kant.

It would be a serious mistake to associate the concept of simultaneity exclusively with philosophical or scientific reasoning. In fact, it was at the level

[16] H. Chang and N. Cartwright, "Causality and realism in the EPR experiment," *Erkenntnis* **38,** 169–190 (1993).

[17] H. Goldstein, *Classical Mechanics* (Cambridge, MA: Addison-Wesley, 1951), p. 1 (italics added).

of prescientific apprehension, a fundamental ingredient in the process of human apperception and conception of time. As Gerald James Whitrow rightly pointed out, "our conscious appreciation of the fact that one event follows another is of a different kind from our awareness of either event separately. If two events are to be represented as occurring in succession, then—paradoxically—they must also be thought of simultaneously."[18]

[18] G. J. Whitrow, *The Natural Philosophy of Time* (London: Thomas Nelson, 1961), p. 75.

CHAPTER ONE

Terminological Preliminaries

For the sake of verbal consistency and the prevention of possible misinterpretations it is desirable, if not necessary, to begin our study with some terminological comments. The topic of the present chapter is therefore primarily not the *concept* of simultaneity but rather the *word* or the verbal expression that denotes this concept. As far as semantic considerations are involved, without which a meaningful discussion of the differences between the terms under discussion would be impossible, it suffices at this stage to define the term "simultaneity," as understood by common sense, as the "temporal coincidence of events." We will ignore the philosophical and physical problems involved with this definition for now.

As far as we know, the earliest recorded term that has been interpreted as denoting simultaneity is the Egyptian hieroglyph ,[1] which is transliterated by "hȝw" and interpreted as a "term, that denotes the simultaneity

[1] See, for example, A. H. Gardiner, *Egyptian Grammar* (London: Oxford University Press, 1957), p. 579. For later or less frequently used hieroglyphic expressions of a similar meaning see E. A. Wallis Budge, *First Steps in Egyptian* (London: Kegan Paul, Trench, Trubner & Co., 1923), pp. 43, 81, 116, 250, or R. Lambert, *Lexique Hiéroglyphique* (Paris: Geuthner, 1925), p. 142.

of events."[2] As the well-known egyptologist Eberhard Otto showed, however, the original meaning of this term was not a temporal but rather a spatial relation denoting "local proximity or neighborhood."[3] It is therefore probably also the earliest known example of the metonymical use (the use of one word for another) of spatial terms to denote temporal relations that is frequently encountered both in ancient and in modern languages.[4] Today we still speak of a "short" or "long" interval of time; we say "thereafter" instead of "thenafter," or "always" instead of "at all times." In fact, we will soon see that the word "simultaneity" itself is such a metonymy.

Let us first point out that in his statement that the hieroglyph denotes "simultaneity of events," Otto did not use "event" in the sense in which it is used in modern physics. The term "event," derived from the Latin "e-venire" (to come out), was used at the time of William Shakespeare[5] to denote an occurrence, process, or phenomenon of indeterminate temporal duration, just like its German equivalent "Ereignis."[6] In the terminology of modern physics, however, the word "event," just like "Ereignis," became a technical term to denote "an occurrence of negligible spatial extension and temporal duration."[7] The use of the term in this sense gained general currency, especially with the advent of Einstein's 1905 relativity paper, in which the definition of simultaneity is followed by the statement "a system of values *x, y, z, t* . . . completely defines the place and time of an event."[8] In his 1907 review paper[9] Einstein called such a system a "Punktereignis" (point-event). Hermann Minkowski, in

[2] " . . . ein Terminus, der die Gleichzeitigkeit von Ereignissen angibt." E. Otto, "Altägyptische Zeitvorstellungen und Zeitbegriffe," in H. E. Stier and F. Ernst (eds.), *Die Welt als Geschichte* (Stuttgart: Kohlhammer, 1954), vol. 14, pp. 135–148.

[3] " . . . ein Wort, das eigentlich die räumliche Nähe, die Nachbarschaft bezeichnet." Ibid., p. 146.

[4] For examples of such space–time metonymies in Sumerian and ancient Hebrew, see M. Jammer, *Concepts of Space* (Cambridge, Massachusetts: Harvard University Press, 1954; enlarged edition, New York: Dover Publications, 1993), pp. 3–4.

[5] Cf. J. Bartlett, *A Complete Concordence of Shakespeare* (London: Macmillan, 1966), p. 455.

[6] The word "Ereignis" derives from "ir-ougen" (to appear to the eye), related to the German word "Auge" (eye). See G. Wahrig (ed.), *Brockhaus-Wahrig: Deutsches Wörterbuch* (Wiesbaden: Brockhaus, 1981), vol. 2, p. 548. From the etymological point of view the German word "Ereignis" resembles the word "phenomenon," which derives from the Greek "φαίνειν" (to make visible).

[7] Ibid., p. 548.

[8] " . . . Wertsystem *x, y, z, t,* welches Ort und Zeit eines Ereignisses . . . vollkommen bestimmt."

[9] A Einstein, "Über das Relativitätsprinzip und die aus demselben gezogenen Folgerungen," *Jahrbuch der Radioaktivität und Elektronik* **4,** 411–462 (1907); quotation on p. 415.

his famous 1905 lecture on space and time, called it a "Weltpunkt" (world-point).[10] Finally, let us quote from the introduction of a recent treatise on relativity: "We shall adopt the point of view that the basic problem of science in general is the description of 'events' which occur in the physical universe and the analysis of the relationship between these events. We use the term 'event,' however, in the idealized sense of a 'point-event,' that is, a physical occurrence which has no spatial extension and no duration in time."[11]

Using henceforth the term "event" in this sense we can rephrase our preliminary definition of simultaneity as follows: event $e = (x, y, z, t)$ and event $e' = (x', y', z', t')$ are simultaneous if and only if $t = t'$. This thus defined simultaneity will also be called "event simultaneity" to distinguish it from what we call "interval simultaneity," which refers to continuous sequences of events, called "processes" or "occurrences," and which we define as follows: process p, beginning at time t_1 and ending at time t_2, and process p', beginning at time t_1' and ending at time t_2', are simultaneous if and only if $t_1 = t_1'$ and $t_2 = t_2'$. If the distance between two simultaneous events is negligibly small we speak of a "local simultaneity," if not, we speak of a "distant simultaneity."

Returning now to the hieroglyph mentioned earlier, we see that its interpretation as "simultaneity of events," as suggested by Otto, does not agree with our definition of "event." For having always been used in combinations like "at the time of the reign of king . . . ," where "at the time" was expressed by that hieroglyph, it obviously refers not to events but rather to what we call "processes." But even its interpretation as "interval simultaneity" would be inaccurate; for the intervals or processes under consideration, say p and p', have temporal durations $\Delta t = t_2 - t_1$ and $\Delta t' = t_2' - t_1'$, which, in general, do not satisfy the condition that $t_1 = t_2'$ and $t_2' = t_2'$, but only the weaker condition that Δt and $\Delta t'$ overlap, that is, are not disjoint (symbolically, $\Delta t \cap \Delta t' \neq 0$). Such processes will be called "contemporaneous" in agreement with the etymology of the term "contemporaneity,"[12] which derives from the Latin "cum" (together, in common) and "tempus" (time). In our terminology the above-quoted hieroglyph denotes therefore not "simultaneity of events" but rather "contemporaneity of processes."

[10] H. Minkowski, "Raum und Zeit," *Physikalische Zeitschrift* **10,** 104–111 (1909).

[11] G. L. Naber, *The Geometry of Minkowski Spacetime—An Introduction to the Mathematics of the Special Theory of Relativity* (New York: Springer-Verlag, 1992), p. 1. For other modern usages of the term "event" see Q. Smith, *Language and Time* (Oxford: Oxford University Press, 1993), p. 24.

[12] For variations of this term, like "contemporariness" etc. and historical details see *The Oxford English Dictionary* (Oxford: Clarendon Press, 1933, 1961), vol. 2, pp. 894–895.

The interchange of the roles of space and time in the concept of "distant simultaneity" leads to the notion of events that occur at different times at the same location. Note that in contrast to "simultaneity," which plays an important role in the philosophy of space and time, this space–time-transposed analogue has never been found worthy of any philosophical reflection. In fact, it has never been given even a term of its own unless we accept the term "local recurrence" as a terminus technicus. The term "contemporaneity," however, has a space–time transposal, namely "collocation," derived from the Latin "cum" and "locus" ("place"), but it has been used since the seventeenth century primarily with respect to the arrangement of words in the field of literature. From the purely etymological point of view this terminological asymmetry between spatial and temporal terms has no justification, because the etymology of the term "simultaneity" assigns to it no preferential temporal connotation. Its etymological root is, of course, the Latin "simul," which in turn derives from the Sanskrit "sem" (or "sema"), meaning "together," both in the sense "together in space" and "together in time." It still survives in the German words "zus*am*men," "S*am*mlung," as well as in the Nordic (Danish, Swedish, Norwegian) expressions "*sam*tidig" ("simultaneous"), "*sam*tidighet" ("simultaneity"), and so on. In the Greek language "sema" became "ἅμα" (hama),[13] which was used by Aristotle, for example, in the sense of "together" not only in the temporal sense. Although the German language retained, as we have seen, the Sanskrit root "sem" in several of its words, it never used it to form a term denoting "simultaneity." Instead, it combined the words "gleich" (equal) and "Zeit" (time) to form "Gleichzeitigkeit" (simultaneity), a term listed previously, and probably for the first time, in Justus Georg Schottel's dictionary of the German language in 1662.[14]

To avoid equivocations it is imperative to distinguish sharply between certain terms that are often, and in many languages, employed as synonymous with "simultaneity." Because the terminology involved is almost the same in all languages we confine our discussion to the English language. The most important of such terms are "synchronism" and "isochronism" and their respective adjectives "synchronous" and "isochronous." To substantiate this contention we quote from the authoritative *Oxford English Dictionary* which

[13] Recall that in Greek the letter *s* is often replaced by the aspirate *h* as, e.g., in "ἅλς" (hals), which is the Latin "sal" and the English "salt" (cf., halogen = salt-forming).

[14] For further details see J. Grimm and W. Grimm, *Deutsches Wörterbuch* (Leipzig: Hirzel, 1912, 1967), vol. 41, pp. 8277–8281.

defines "simultaneous" as "existing, happening, occurring, operating, etc., at the same time; coincident in time." It defines "synchronous" as "existing at the same time; coincident in time; belonging to the same period, or occurring at the same moment of time; contemporary, simultaneous." Finally, it defines "isochronous" as "taking place in or occupying equal times; . . . equal in duration, or in intervals of occurrence . . . taking place in the same time, or at the same time intervals of time, as something else; equal in duration."[15] Webster's popular *New Collegiate Dictionary* identifies "synchronous" with "happening at the same time; concurrent in time; simultaneous."[16] The *Oxford English Reference Dictionary* states explicitly that both "simultaneous" and "synchronous" denote "occurring at the same time."[17]

This apparently incorrect treatment of different, albeit related, terms as synonyms should not be regarded as an error, for the compilers of the dictionaries report not only in what sense these terms are being used at the present but also how they have been used in the literature of the past. When these terms were introduced into the English language or given currency in it, they were usually interpreted in accordance with their etymology. Consider, for example, the term "synchronous" (or its variants like "synchronal," "synchronic," or "synchronical"). It derives from the Greek "σύν" (together) and "χρόνος" (time). By combining these two words some authors introduced the term in the sense of what we now call "simultaneous," others in the sense of what we now call "synchronous," for both interpretations agree with the notion of "together in time." Similarly, the term "isochronism," derived from the Greek "ἴσος" (the same) and "χρόνος," was interpreted by some as denoting "simultaneity," by others as what we now call "isochronism," when we speak, for example, of the isochronism of vibrations. When Galileo described the isochronism of pendulums, he did not yet have at his disposal the term "isocronismo" and had to say that they "vibrate in exactly the same time."[18] In 1786, when John Bonnycastle, the author of several influential

[15] *The Oxford English Dictionary* (Oxford: Clarendon Press, 1931, 1961), s.v.

[16] *Webster's New Collegiate Dictionary* (Springfield, Massachusetts: Merriam, 1960).

[17] J. Pearsell and B. Trumble (eds.), *The Oxford English Reference Dictionary* (New York: Oxford University Press, 1996), p. 1352. "Synchronous" and "isochronous" are also treated as synonyms in F. C. Graham's *The Basic Dictionary of Science* (New York: Macmillan, 1966), pp. 467 and 209.

[18] " . . . sotto tempi precisamento equali." G. Galilei, *Discorsi e dimostrazioni matematiche intorno a due nuove scienze* (Leiden: Elsevier, 1638), p. 107; *Le Opere di Galileo Galilei* (Firenze: Barbēra, 1933), vol. 8, p. 139; *Dialogues concerning Two New Sciences* (New York: Macmillan, 1914; Dover Publications, 1954), p. 95.

texts on mathematics and astronomy, mentioned this discovery by Galileo he wrote "Galileo . . . is said to have discovered the isochronism of the pendulum," using the term "isochronism" probably for the first time.[19]

In accordance with the modern usage of the terms "simultaneity," "synchronism," and "isochronism" we distinguish one from the other and explain this difference in terms of their most important application, the use of clocks. As stated in the beginning of this chapter, we define "simultaneity" as the "temporal coincidence of events." Assuming we have two clocks we regard the coincidence of their hands with certain numbers on their dials as an event. Then we can say that the clocks are "synchronized at a certain moment of time" if and only if at that moment of time the hands of the two clocks are in the same position, that is, they indicate the same time. If this condition is also satisfied at an arbitrary later moment of time we say that the clocks are "synchronized during an interval of time" or simply "synchronized." If the positions of the hands of the two clocks are the same at a certain moment of time they constitute two, generally separated, simultaneous events. It is therefore clear that the notion of "synchronism" involves or presupposes the concept of "simultaneity." It is also clear that "synchronism" always refers to two (or more) clocks. In contrast, the term "isochronism" refers to one and only one clock, for we say that a clock is "isochronous," if and only if it "runs" at a constant (or uniform) rate, that is, if the periods between consecutive "ticks" are equal. Here we face the problem of how to verify that two consecutive time intervals (or, more generally, two temporally separated time intervals) are equal. We cannot shift the later time interval into the position of the preceding (or earlier) time interval. The generally accepted method of calibrating time intervals with the duration of the rotation of the Earth, which defines the length of the mean solar day, is based on the assumption of a uniform rotational velocity of the Earth, but this assumption is not strictly correct because the tidal frictions of the oceans, for example, decelerate the rotation of the Earth. As a closer analysis shows,[20] even an appeal to the law of inertia, according to which a particle, not subjected to any interaction, moves with constant velocity, that is, moves through equal distances in equal intervals of time, would be of no avail.

[19] J. Bonnycastle, *An Introduction to Astronomy* (London: Johnson, 1781), p. 97.

[20] See, for example, H. Reichenbach, *Philosophie der Raum-Zeit-Lehre* (Berlin: Walter de Gruyter, 1928); reprinted in *Gesammelte Werke* (Braunschweig: Vieweg, 1977), vol. 2; *The Philosophy of Space and Time* (New York: Dover Publications, 1958), chapter 2, § 17.

Another temporal term, intimately related to "simultaneity" but not identical with "simultaneously," is the adverb "now." The English term "now," like the Latin "nunc" and the Greek "νῦν," derives from the Sanskrit "nu."[21] It deserves our attention not only because it is an "indexical"[22] term, related to simultaneity, but also because, as we will see in due course, it probably expressed the earliest human awareness of the conception of simultaneity and it played an important role in Aristotelian and medieval philosophy. In modern times Hans Reichenbach classified it as a "token-reflexive word, comparable in this respect with 'I', for it means the same as 'the time at which this token is uttered.' "[23] Ludwig Wittgenstein, in his discussion of time-language,[24] declared that "the function of the word 'now' is entirely different from that of a specification of time" and has a logic of its own. Einstein, as Rudolf Carnap recalls from a conversation with him, "said that the problem of the Now worried him seriously. He explained that the experience of the Now meant something special for man, essentially different from the past and the future, but that this important difference does not and cannot occur in physics. That this experience cannot be grasped by science seemed to him a matter of painful but inevitable resignation."[25]

The relation between "now" and "simultaneity" has scarcely ever been discussed. A noteworthy exception is Eugen Fink's statement: "The fundamental meaning of the Now is that of a universal simultaneity . . . it contains the whole world-wide extent of the simultaneous,"[26] a statement that perhaps explains Einstein's worry, because his theory of relativity denies the existence of a universal simultaneity. A comprehensive discussion of "simultaneity" cannot ignore the relation between this concept and the "now," even if it is

[21] According to the *Oxford English Dictionary* (Oxford: Clarendon Press, 1970), vol. 7, p. 245, "now" denotes "at this time; at the time spoken of or referred to." The German "nun" (now), also derived from the Sanscrit "nu," has a more frequently used synonym "jetzt," corresponding to the adjective "jetzig," which is a derivation of the medieval "iezuo." See J. Grimm and W. Grimm, *Deutsches Wörterbuch* (Leipzig: Hirzel, 1912, 1967), vol. 4, part 2, pp. 2317–2322.

[22] So called by Rudolf Carnap, because it "points" to the person who utters the word ("index" in Latin means "forefinger," "indicare" means "to point at"). For indexical terms see Y. Bar-Hillel, "Indexical expressions," *Mind* **63,** 359–379 (1954).

[23] H. Reichenbach, *The Elements of Symbolic Logic* (New York: The Free Press, 1947), p. 284.

[24] L. Wittgenstein, *The Blue and the Brown Books* (Oxford: Blackwell, 1958), pp. 107–108.

[25] R. Carnap, "Intellectual autobiography" in P. A. Schilpp, *The Philosophy of Rudolf Carnap* (La Salle, IL: Open Court [The Library of Living Philosophers, vol. XI], 1963), p. 37.

[26] E. Fink, *Zur Ontologischen Frühgeschichte von Raum-Zeit-Bewegung* (Den Haag: Nijhoff, 1957), p. 138.

true that such a relation holds only, as some philosophers of language contend, in what they call a "tenseless language."[27]

In classical physics, based on the Newtonian conception of absolute time, which "flows equably without relation to anything external,"[28] and on the assumption of the existence of instantaneous actions at a distance, none of the terms "simultaneity," "synchronism," "isochronism," or "now" raised any conceptual difficulties.

Even a notion like "spatial recurrence" in the meaning of repeated presence at the same local position, whether of an object, event, or process, had a clear-cut meaning in Newtonian physics, according to which "absolute space, in its own nature, without relation to anything external, remains always similar and immovable."[29] In modern physics, however, which denies the existence of absolute time, absolute space, and instantaneous actions, all these terms or notions involve serious problems as we will see in due course.

Finally, note that, in conformance with our terminology, spatially separated events can be defined as simultaneous if synchronized clocks, located in their immediate vicinities, indicate the same readings at the occurrences of these events. Clock synchronization and simultaneity are therefore intimately related concepts, at least insofar as the operational establishment of one of them assures the existence of the other.

[27] See, for example, Q. Smith, *Language and Time* (New York: Oxford University Press, 1993), chapter 1.

[28] "Tempus absolutum, verum et mathematicum, in se et natura sua absque relatione ad externum quodvis, aequabiliter fluit." I. Newton, *Philosophiae Naturalis Principia Mathematica* (London: Streater, 1687), p. 5.

[29] "Spatium absolutum natura sua absque relatione ad externum quodvis semper manet simile et immobile." Ibid., p. 5.

CHAPTER TWO

The Concept of Simultaneity in Antiquity

It is often said that the language of science is an extension or refinement of the language of ordinary life, because scientific concepts, no matter how sophisticated they may be, must ultimately be explainable by means of concepts used in the ordinary experiences of daily life. This is certainly true for the scientific concept of simultaneity because the term "simultaneous" was used in ordinary language long before it became the object of philosophical or scientific inquiry. Moreover, the notion of simultaneity must have been in the mind of humankind even before its conscious articulation, for when at the dawn of civilization prehistoric man observed the stars in the sky and thought that they were where he saw them, he conceived the idea of an all-pervasive "now." In this conception his mind implicitly applied the notion of distant simultaneity as a necessary component in the mental process of distinguishing his self from the world that surrounds him.

This does not mean, however, that he already possessed a distinctive verbal expression for this abstract notion at this early stage. The question of how he performed the transition to such an articulated expression is related, of course, to the general problem of the origin and development of language.

Discussing this issue, which is still a matter of dispute, would lead us too far from our topic. We focus, therefore, on only a few metalinguistic remarks that are relevant to the notion of simultaneity. Particularly relevant to our subject is the claim made by Edward Sapir and his disciple Benjamin Lee Whorf that "all observers are not led by the same physical evidence to the same picture of the universe, unless their linguistic backgrounds are similar."[1] In other words, the concepts and categories used by an observer to describe physical reality depend on the structure of his language, a thesis these authors call the "linguistic principle of relativity." They claim that this principle applies not only to an advanced stage of intellectual activity but also, and most importantly according to Whorf, to the rudimentary phase of concept formation even in primitive civilizations.

Whorf arrived at this conclusion as a result of analyzing various languages and, in particular, the Hopi language spoken by the westernmost group of Pueblo Indians of the Navajo Reservation in northeastern Arizona. This study showed him that "various grand generalizations of the Western world, such as time, velocity, and matter, are not essential to the construction of a consistent picture of the universe."[2] This applies also to the notion of simultaneity. According to Whorf "Hopi may be called a timeless language. It recognizes psychological time, which is like Bergson's 'duration,' but this 'time' is quite unlike the mathematical time, *t*, used by the physicists. Among the peculiar properties of Hopi time are that it varies with each observer and *does not permit of simultaneity*."[3] This constraint imposed on the expression of simultaneity relations is caused, in part at least, because Hopi verbs have no real tenses but are modified instead according to the length of time an event lasts or whether an action is completed or not. Because it lacks, in particular, the present tense by which the speaker of a statement implies the simultaneity of the contents of the statement with the act of uttering it, Hopi foregoes the most frequent application of simultaneity.

Hopi is not the only tenseless language. The ancient Semitic languages, such as the Hebrew of the Bible, have only two modifications of the verb: the "perfect," implying that the action referred to by the verb is completed,

[1] B. L. Whorf, "Science and linguistics," *Technical Review* **42,** 229–233, 247–248; reprinted in J. B. Carroll (ed.), *Language, Thought, and Reality—Selected Writings of Benjamin Lee Whorf* (Cambridge, Massachusetts: MIT Press; New York: Wiley, 1956), pp. 207–219.

[2] Ibid., p. 216.

[3] Ibid., p. 216 (emphasis added).

and the "imperfect," implying that the action is still going on. "Hebrew, unlike Greek and most other languages, possesses no form specifically to indicate date."[4] That this particularity makes it unwieldy, just as in Hopi, to express simultaneities can be shown by textual examples, in which the temporal relation of simultaneity is described in a roundabout way in terms of parallel statements.[5] It also increases the syntactic-logistic difficulties in presenting a description of simultaneous historical events. This feature was studied in detail by Shemaryahu Talmon in his essay "The Presentation of Synchroneity and Simultaneity in Biblical Narrative," which opens with the statement: " . . . the biblical author . . . found himself in a predicament when he faced the logistic problem of how to intelligibly present two episodes which synchronously occurred . . . in different geographical settings. . . . Simply to string them one to the other would result in the impression that they came about in a chronological sequence, and not concurrently, and thus would distort the picture."[6] Of course, modern Hebrew has a specific word to express "simultaneity," namely "bosmaniut" (בוזמניות), a word coined after the Aramaic adverb "be-simna" which means "at the same time."

Needless to say, according to the terminology suggested in the preceding chapter, the term "simultaneity," as used by Whorf and Talmon, should be replaced by the term "contemporaneity. The same holds for an essay by the Polish philologist Thaddaeus Zielinski, which is entitled "The Treatment of Simultaneous Events in the Antique Epic."[7] Note that, although Greek is certainly not a tenseless language, ancient Greek authors, as Zielinski shows, faced considerable difficulties in describing even only contemporaneous processes.

The absence of specific terms for the concept of simultaneity in ancient languages should come as no surprise. Distant simultaneity as we understand it today, especially in physics, is a notion that was hardly needed when time was measured by sundials, hourglasses, or clepsydras and when communication was transmitted by messengers or bonfires. Moreover, those clocks were used primarily to tell *how long* a process lasted rather than *when* it occurred. Indeed, reviewing the whole history of science in antiquity, we recognize that

[4] S. R. Driver, *A Treatise on the Use of the Tenses in Hebrew* (Oxford: Clarendon Press, 1892), p. 3.

[5] See, for example, *Genesis*, chapter 45, verse 14.

[6] *Scripta Hierosolymitana* **24,** 9–26 (1978).

[7] Th. Zielinski, "Die Behandlung gleichzeitiger Ereignisse im antiken Epos," *Philologus*, **8,** Suppl., 407–449 (1899–1901).

hardly any need existed to employ the notion of simultaneity in an operationally significant way. The only exceptions were certain astronomical measurements performed by the Alexandrian astronomers, such as those described by Aristarchos of Samos in his treatise "On the Sizes and Distances of the Sun and Moon" or the famous determination of the Earth's circumference by Eratosthenes of Cyrene. Eratosthenes's method consisted of performing simultaneously two observations: in Meroe (Alexandria) one had to measure the length of the shadow cast by the local obelisk at precisely the moment when the bottom of a deep well in Syene (Assuan) was seen to be fully illuminated by the rays of the sun, both places being supposed to lie on the same meridian. To ensure that both observations were performed really simultaneously—the Greek term used was "τῄ αὐτῆ ὥρα" and the Latin term was "eodem tempore"[8]—they had to occur at the summer solstice. Having thus obtained the difference in latitude between the two places one had only to measure directly the distance between them to obtain the circumference of the Earth.

Another noteworthy exception, not only in antiquity, was the need to establish distant simultaneity in genethialogy, the casting of nativities or horoscopes, an art that flourished in the Hellenistic period but originated in the astrological theories and practices of the Chaldeans and Babylonians of the second millennium B.C. To cast a horoscope the astrologer had to know the position of the heavenly bodies at the exact time of the birth of the person, for only then could he predict his future and advise him on the course of actions and decisions to be taken. How under such circumstances simultaneity was assured was vividly reported by Sextus Empiricus, a philosopher and physician of the third century A.D.: "by night, the Chaldean sat on a high peak watching the stars, while another man sat beside the woman in labour till she would be delivered, and when she had been delivered he signified the fact immediately (ευθύς) to the man on the peak by means of a gong; and he, when he heard it, noted the rising sign as that of the horoscope."[9] It was important to observe just the "rising sign," because the infant was believed to be submitted to the influence of the constellation that was being "born"

[8] Cleomedes, *De Motu Circulari Corporum Caelestium Libri Duo* (Leipzig: Teubner, 1891). H. Berber, *Die geographischen Fragmente des Eratosthenes* (Amsterdam: Meridian, 1964), pp. 122–123. For the precision of this measurement see B. R. Goldstein, "Eratosthenes on the 'measurement' of the earth," *Historia Mathematica* **11**, 411–416 (1984).

[9] Sextus Empiricus, *Against the Professors*, Book 5, 27–28 (London: Heinemann, 1961), pp. 334–335.

simultaneously with him. For "at the very instant in which he gives forth his first cry, all the astrological influences converge on his cradle and blend to develop his destiny."[10] Let us point out parenthetically that this approbation of what may be called "astrological simultaneity" was not confined to antiquity. Even a scientifically inclined person like Johannes Wolfgang von Goethe, the great German author and poet, described in his autobiography the moment of his birth as coinciding with an extremely "lucky" constellation of the stars:

> I was born at Frankfort-on-the-Main on the 28th of August 1749, at midnight, on the stroke of twelve. The position of the stars was favorable; the sun was in the sign of Virgo, and had reached the zenith; Jupiter and Venus were friendly. Mercury not in opposition; the moon alone, just full, exerted her power of reflection all the more as she had entered her planetary hour. She was therefore in opposition to my birth, which could not come about until this hour was past. These good aspects, which astrologers later on knew how to estimate very highly for me, may well have been the cause of my preservation, for, owing to the unskilfulness of the midwife, I came into the world as though dead, and the fact that I saw the light was only brought about with great trouble.[11]

Let us return to Sextus Empiricus to show that he played a very important role in the history of the concept of simultaneity not only because of his account of how the Chaldeans applied this concept for astrological purposes. By profession a physician of the so-called Empirical School (hence his name), he was also a skeptical philosopher who severely criticized the trustworthiness of scientific methods. It was in this spirit that he reprehended the Chaldean method of casting a horoscope. We will not discuss his mainly biological objections, such that the deliverance is not an instantaneous but usually time-extended process, that we do not know when, exactly, the life of the newborn child begins, that children are not always born at night, and so on. Nor will we mention his philosophical arguments against genethliac predictions. We are interested in his physical arguments against the method by which those astrologers claimed to have solved the problem of instantaneous transmission to ensure simultaneity. Thus, Sextus Empiricus pointed out that the sound of the gong needs a certain time to reach the peak of the moun-

[10] M. Gauquelin, *The Cosmic Clock* (London: Regnery, 1967), p. 37.
[11] *Goethes Werke* (Hamburg: Wegner, 1957), vol. 9, p. 10.

tain from the place in the valley. Although Sextus did not yet know the velocity of sound, which was measured for the first time, as far as we know, by Marin Mersenne in about 1630, he pointed out that, when one observes the felling of a tree from a distance, one sees the strike of the axe at the tree much earlier than one hears the blow. Concluding his comments he declared: "We have proved that exact simultaneity between the birth and the taking of the horoscope cannot be obtained."[12] Sextus Empiricus must therefore be credited with having been the first who criticized the method of signal transmission as a means for the solution of the problem of how to ensure simultaneity.

Note, however, that within the conceptual framework of classical physics such an acoustical signal transmission could in principle be used to establish distant simultaneity; one had only to know the velocity of sound, of the air (wind), and of course the distance involved. One could then easily calculate the time that the sound requires to propagate from the gong to the peak and compensate for this delay. The question of whether such a method is also acceptable in modern physics, which is based on the theory of relativity, will be discussed in due course.

In any case, Sextus Empiricus, the last and boldest of the Skeptics, believed that distant simultaneity was unattainable, at least by the method used by the Chaldeans, and he challenged therefore the legitimacy of astrology, which enjoyed the status of a highly respected science in Alexandria where he lived.

Incidentally, our translation into English of Sextus's statement (in note 12) is not a literal translation of his words, although it is a truthful rendering of what he had in mind. Using our terminology as outlined in the preceding chapter, we may say that what he had in mind was "event simultaneity" and not merely "contemporaneity." Because the term "hama" denoted still at his time "together," both in the spatial and in the temporal sense, he did not make use of it; and because he did not have at his disposal a noun denoting "simultaneity," he preferred the expression "at the same time."

Turn now from the Chaldeans and their astrology to the cradle of Greek civilization and study the use of the concept of simultaneity in the very first monuments of Greek literature, in Homer's *Iliad* and *Odyssey*.

In view of what was said in the beginning of this chapter we are not surprised that the notion of simultaneity did not play an important role in these

[12] Op. cit. (note 9), V. 99 (p. 367).

works, which were probably written in the ninth century B.C. True, the term "hama" appeared quite frequently but almost always in the nontemporal meaning of "together."[13] The only exceptions were those cases in which "hama" was used to denote the simultaneity of an action with a certain hour of the day.[14] Even in these cases, the temporal sense of "hama" was induced because it denotes a "togetherness" with a temporal date. Because it is certainly true that, as a German philologist pointed out, in Homer's epics "in general only living beings or what is thought to be alive are connected by 'hama',"[15] the infrequency of the use of "hama" in the temporal sense finds its natural explanation. In a few cases simultaneity was expressed by the use of the term "now," as for example in the phrase "mortals that now are on the earth."[16] Occasionally, however, νῦν was used, mostly in combination with δέ, also in a nontemporal sense to express the reality of a state of affairs in contrast to the previously expressed irreality of it.[17]

To understand how Homer also used "νῦν" in such a nontemporal sense we need only recall that we, too, use the term "now" in such a sense when at the end of a logical or mathematical deduction we say, "it *now* follows that . . . " Clearly, the emphasis here is not so much on "presently" as on "finally" or "in conclusion." Let us call this usage of "νῦν" its use in the "logical sense" in contrast to its use in the "temporal sense." That the term "hama" (ἅμα)—which, as we recall, derived from the Sanskrit "sam" (together)—apart from its spatial and temporal senses likewise has a "logical sense" will become clear when we discuss the notion of simultaneity in Aristotle's writings.

We would have to go into too much detail to study how Greek writers, such as the poets or dramatists Pindar, Aischylos, and Sophocles of the sixth or fifth century B.C., used the notion of simultaneity. Besides, our topic is the concept of simultaneity in physics and philosophy, not in literature. Let

[13] See, for example, *Iliad*, Book 1, 495: "πάντες ἅμα" ("all in one company"); Book 8, 64: "ἅμ' οἰμωγή τε καὶ εὐχωλή" ("groaning together with the cry of triumph"). *Odyssey*, Book 1, 98: "ἐπ' ἀπείρονα γαῖαν ἅμα πνοιῇς ἀνίμοιο" ("both over the waters of the sea and over the boundless land"); 11, 371: "οἵ τοι ἅμ' αὐτῷ Ἴλιον εἰς ἅμ . . . " ("who went to Ilium together with you . . . ").

[14] See, for example, *Iliad*, Book 9, 682: "ἅμ' ἠοῖ φαινομένηφι" ("at break of day").

[15] "Im allgemeinen finden sich nur lebende Wesen oder belebt Gedachtes mit ἅμα verbunden." C. Capelle, *Vollständiges Wörterbuch über die Gedichte des Homer und der Homeriden* (Leipzig: Hahn, 1889), p. 41.

[16] "οἱ νῦν βροτοί εἰσιν" ("mortals of our day"), *Iliad*, Book 1, 272.

[17] See, for example, *Iliad*, Book 1, 417: "νῦν δ' ἅμα τ' ὠκύμορος" (but now you are doomed to a speedy death).

us turn, therefore, to the earliest philosophers in ancient Greece, the Pre-Socratists or "*physicists*," as Aristotle called them in distinction from their predecessors, the theologians, because they tried to explain nature by principles and causes. Their philosophy of time focused on the controversy between *becoming* or *change* versus *being* or *permanence*. The problem of whether the flux of time is real or merely an illusion engaged their philosophical interest, a problem that in no way involved any temporal relations between separated events. It is not surprising, therefore, that the Pre-Socratists scarcely made any explicit reference to the concept of simultaneity. A careful study, for example, of the writings of Democritus of Abdera, who was the most learned of the Ionian physicists and head of the atomistic school and who died between 380 and 370 B.C., shows that he only once made use of the notion of simultaneity, and this was in the context of his ethical and not physical writings. He wrote, "One should, as far as possible, divide out one's property among one's children, at the same time (hama) watching over them to see that they do nothing foolish when they have it in their hands."[18] Clearly, "hama" was used here in the purely temporal sense of "simultaneity" or, perhaps more accurately, of "contemporaneity," because a distribution of property is usually a temporally extended process.

We speak of an "implicit use of the concept of simultaneity" in a statement if the term "simultaneity," or any of its equivalents, is not explicitly stated, but the concept of simultaneity is necessary to make the statement meaningful. Statements of this kind are frequently encountered but generally devoid of any philosophical or scientific interest. We therefore confine our discussion to only those cases that are of outstanding philosophical or scientific importance.

The earliest examples of this kind are probably the famous four "paradoxes" or arguments against the possibility of motion raised by Zeno of Elea who flourished about 460 B.C. Bertrand Russell, for example, emphasized their importance when he declared that they "have afforded grounds for all theories of space and time and infinity which have been constructed from his day to our own."[19]

[18] H. Diels, *Die Fragmente der Vorsokratiker* (Berlin: Weidmannsche Verlagsbuchhandlung, 1956), vol. 2, p. 203; K. Freeman, *Ancilla to the Pre-Socratic Philosophers* (Oxford: Blackwell, 1956), p. 117.

[19] B. Russell, *Our Knowledge of the External World* (London: George Allen & Unwin, 1926), p. 183.

Not surprisingly, therefore, Zeno's four arguments became the subject of numerous critical commentaries,[20] but their implicit, yet crucial dependence on the notion of simultaneity has hardly, if ever, been recognized or noted by their commentators. The only source of our knowledge of them are some statements by Aristotle in *The Physics*. Although they were very brief and rather polemical, Aristotle provided sufficient ground to demonstrate that Zeno's four arguments, with the possible exception of the first, make implicit use of the concept of simultaneity.

The first argument, usually called "the Dichotomy" or "the Race Course," according to Aristotle, says "that motion is impossible because, however near the mobile is to any given point, it will always have to cover the half, and then the half of that, and so on without limit before it gets there"[21]; but this is impossible because "an illimitable process . . . cannot be accomplished in a limited time."[22] Because this argument deals with only one single moving object and with events occurring at different instants, namely the temporally successive arrivals of the mobile at different points, it cannot involve the notion of simultaneity.[23]

Zeno's second argument, which Aristotle called "the Achilles" and later commentators called "Achilles and the Tortoise," refers, in contrast to the former, to two moving bodies. "It purports to show that the slowest [the Tortoise] will never be overtaken in its course by the swiftest [Achilles], inasmuch as, reckoning from any given instant, the pursuer, before he can catch the pursued, must reach the point from which the pursued started at that instant, and so the slower will always be some distance in advance of the swifter."[24] Obviously, this Aristotelian quotation of the argument makes no explicit reference to the notion of simultaneity. Nor does Simplicius's much more detailed formulation of the argument include any reference to simultaneity.[25]

[20] See, for example, the article by G. Vlastos in *The Encyclopedia of Philosophy* (New York: MacMillan, 1972), vol. 8, pp. 369–379, or by K. von Fritz in *Paulys Realencyclopädie der Classischen Altertumswissenschaft* (München: Druckenmüller, 1972), vol. 19, pp. 58–83; A. Grünbaum, *Modern Science and Zeno's Paradoxes* (Middletown, Connecticut: 1967); W. C. Salmon (ed.), *Zeno's Paradoxes* (Indianapolis, Indiana: Bobbs-Merrill, 1967).

[21] Aristotle, *The Physics* 239 b 11–13.

[22] Ibid., 233 a 21.

[23] Unless, of course, one adopts the very general point of view that "to be present is to be simultaneous." See R. G. Gale, "Has the present any duration?", *Nous* **5**, pp. 39–42 (1971).

[24] Aristotle, *The Physics* 239 b 14–19.

[25] Simplicius, In *Aristotelis Physicorum Libros Commentaria*, 1013.31.

Figure 2.1

That this notion nevertheless plays a central role in this argument can be seen as follows. Let *A* denote the initial position of Achilles and *B* the position of the tortoise (fig. 2.1). It is tacitly assumed that Achilles and the tortoise start the race at the same instant, an assumption that clearly involves the concept of distant simultaneity. Furthermore, at the moment when Achilles arrives at *B* the tortoise is supposed to arrive at *C*, a statement that again implies the notion of distant simultaneity. At the moment when Achilles reaches *C* the tortoise arrives at *D*, and so on ad infinitum. Clearly, the concept of distant simultaneity is tacitly applied an infinite number of times. This conclusion remains valid even if we do not assume that Achilles and the tortoise start moving at the same instant, provided, of course, Achilles does not reach the starting point of the tortoise prior to its departure.

Aristotle claimed that this Achilles argument "is the same as the former one which depends on bisection, with the difference that the division of the magnitudes we successively take is not a division into halves" (but according to any ratio we like to assume between the two speeds).[26] This reduction of the "Achilles" to "the Dichotomy" has been challenged by the majority of modern commentators as logically faulty. One of the few who have disagreed with the majority view is Jonathan Barnes, who vindicated Aristotle as follows. A closer inspection of the "Achilles," he contended, shows only that "they [Achilles and the Tortoise] do not meet before they meet," but not that they never meet. To prove that they never meet the Achilles argument has to incorporate the Dichotomy argument.[27] Barnes's argumentation for the reliance of the "Achilles" on the "Dichotomy" did not exclude the need of the simultaneity concept, because it left the "Achilles" essentially intact. The following, albeit anachronistic, stratagem may do the trick, though. Introduce a reference system S in which—say, at its origin—the tortoise is always at rest. Thus S moves relative to the racecourse to the right with the velocity that the tortoise had originally (fig. 2.1). In S Achilles now always approaches the tortoise from the left and according to the gist of the Dichotomy argument

[26] Aristotle, *The Physics* 239 b 18–20, 239 b 25–26.

[27] J. Barnes, *The Presocratic Philosophers* (London: Routledge & Kegan Paul, 1982), pp. 274–275.

will never reach the tortoise. That, as Aristotle pointed out, "the division of the magnitudes we successively take is not a division into halves" does not impair the conclusion that in S only Achilles moves and the tortoise remains at rest. But the fact that in S only one body (Achilles) is in motion excludes even an implicit use of the notion of simultaneity, at least as far as we confine our attention to events within S. However, that this strategem completely eliminates the notion of simultaneity may be questioned on the grounds that S is assumed to move simultaneously with the tortoise.

Zeno's third argument, known as "the Arrow," claimed that motion is impossible because, as Aristotle phrased it, "everything is either at rest or in motion, but nothing is in motion when it occupies a space equal to itself, and what is in flight is always at a given instant occupying a space equal to itself, hence the flying arrow is motionless."[28] An equivalent but more elegant formulation was given by Diogenes Laertius in the third century A.D.: "That what moves, moves neither in the place in which it is, nor in the place in which it is not."[29]

Like the Dichotomy, this argument deals with only a single moving object. One may therefore assume that it does not imply any notion of simultaneity. According to Kurt von Fritz, however, this assumption is not strictly correct. For, if fully explicated, the argument runs as follows: "No object can *simultaneously* be at two places, it is therefore always there where it is. Yet when it is at one place it does not move. But at a place other than that, at which it is, it can not move."[30] Obviously, in von Fritz's formulation this argument does imply the notion of distant simultaneity.

Aristotle invalidated Zeno's third argument on the ground that "it rests on the assumption that time is made up of 'nows,' and if this be not granted the inference fails."[31] But "time is not composed of atomic 'nows' ("ἐκ τῶν νῦν"), any more than any other magnitude is made up of atomic elements."[32] The continuity of time, the fact that it is not composed of indivisible instants, invalidates, according to Aristotle, Zeno's third argument against motion. Let us assume, however, that contrary to Aristotle but in agreement with

[28] Aristotle, *The Physics* 239 b 5–9.

[29] Diogenes Laertius, *Vitae Philosophorum* (Oxford: Clarendon Press, 1964), vol. 2, p. 475.

[30] "Kein Körper kann *gleichzeitig* an zwei Orten sein, ist also immer dort wo er ist." (emphasis added). K. von Fritz, *Schriften zur griechischen Logik* (Stuttgart: Fromann-Holzboog, 1978), vol. 1, p. 71.

[31] Aristotle, *The Physics* 239 b 31–33.

[32] Ibid., 239 b 7–9.

the medieval Muslim philosophy of the Kalam[33] time is composed of indivisible "nows" or instants and motion is composed of "leaps." Also under such "cinematographic" assumptions the notion of simultaneity would be involved, provided the "now" is interpreted in the above-mentioned sense of "universal simultaneity."[34]

The last of Zeno's four arguments against motion, the so-called "Stadium" or the "Moving Rows," assumed, according to Aristotle,

> a number of objects all equal with each other in dimensions, forming two equal rows and arranged so that one row stretches from one end of the racecourse to the middle of it, and the other from the middle to the other end. Then if you let the two rows, moving in opposite directions but at the same rate, pass each other, Zeno concluded that half of the time they occupy in passing each other is equal to the whole of it. . . . This is his demonstration. Let there be a number of objects AAAA, equal in number and bulk to those that compose the two rows but stationary in the middle of the stadium. Then let the objects BBBB, in number and dimension equal to the A's, form one of the rows stretching from the middle of the A's in one direction; and from the inner end of the B's let CCCC stretch in the opposite direction, being equal in number, dimension, and rate of movement to the B's. Then when they cross, the first B and the first C will *simultaneously* (*"hama"*) reach the extreme A's in contrary directions. During this process the first C has passed all the B's, whereas the first B has only passed half the A's, and therefore only taken half the time; for it takes an equal time for the C to pass one B as for the B to pass one A. But during this same half-time the first B has also passed all the C's because measured by their progress through the A's the B's and C's have had *the same time* in which to cross each other, (for the first C and the first B arrive at the opposite ends *simultaneously*).[35]

That Zeno, as quoted by Aristotle, made explicit use of the concepts of distant simultaneity and of interval simultaneity is indicated by the terms printed in italics. That he also made implicit use of distant simultaneity fol-

[33] See, e.g., Abu Ishaq Ibrahim Al-Nazzam, *Kitab fi al-haraka* (Treatise on Motion).

[34] See chapter 1, note 26.

[35] Aristotle, *The Physics* 239 b 33–240 a 18. The Greek text is philologically obscure and allows therefore slightly different interpretations. The quoted translation agrees essentially with that given by P. H. Wicksted and F. M. Cornford in *Aristotle, The Physics* (London: Heinemann, 1934), vol. 2, pp. 185–197 (emphasis added).

$[A_1]$ $[A_2]$ $[A_3]$ $[A_4]$

$[B_4]$ $[B_3]$ $[B_2]$ $[B_1]$ ⟶

⟵ $[C_1]$ $[C_2]$ $[C_3]$ $[C_4]$

Figure 2.2

lows from his statement that "the first B (B_1) and the first C (C_1) will *simultaneously* reach the extreme A's (A_1 and A_4) in contrary directions" (fig. 2.2). For having moved through equal distances with equal speed, they must have started their motion at their spatially different starting points simultaneously. Of course, if it is assumed that the initial positions of the row of the B's and the row of the C's are not symmetrical with respect to the rows of the A's, as Pierre Bayle[36] and Jacques Lechalier[37] assumed, our last conclusion becomes invalid.

Zeno was a disciple of Parmenides, the head of the Eleatic School. Parmenides taught that the notions of plurality, motion, change, and hence also of the flux of time (in contrast to Heraclitus) contradict reason and are therefore merely illusions. In reality only one unchanging and indivisible Being exists, sometimes called simply the "One." Zeno contrived his paradoxes to defend his teacher's doctrine. How, then, was it permissible for Zeno to make use of temporal concepts like the plurality of objects or of simultaneity, as he did, for example, in the "Stadium," without violating the basic tenet of the very doctrine that he intended to defend? The answer to this question can be found in Plato's dialogue *Parmenides*[38] (though only with respect to Zeno's argument against plurality). Zeno was not inconsistent because he applied, expressed in modern terms, the logical form of a *reductio ad absurdum*: he proved a proposition, for example, that of the impossibility of motion, by deducing a contradiction from the negation of this same proposition, or with respect to our topic, he made use of the temporal notion of simultaneity to prove its untenability.

[36] P. Bayle, *Dictionnaire Historique et Critique* [originally published 1695–1697], (Geneva: Slotkin Reprints, 1969), vol. 15, pp. 40–41.

[37] J. Lechalier, "Sur les deux derniers arguments de Zénon d'Élée contra l'existence de mouvement," *Revue de Métaphysique* **18,** 345–355 (1910) (see diagram on p. 348).

[38] In the *Parmenides* (128 D) Zeno is reported to have said: "My answer [argument] is addressed to the partisans of the many, whose attack I return with interest by retorting upon them that their hypothesis of the being of many, if carried out, appears to be still more ridiculous than the hypothesis of the being of one." B. Jowett, *The Dialogues of Plato* (New York: Random House, 1937), vol. 2, p. 89. It is probably no exaggeration to say that Zeno invented the *reductio ad absurdum,* even though Parmenides is often credited with having been the first to base his metaphysics on logic.

Parmenides does not seem to have made explicit use of the notion of simultaneity in his philosophical writings. In any case, the extant fragments of his famous philosophical poem, "The Way of Truth," the most ancient monument of metaphysical speculation among the Greeks, do not contain the term "hama." However, Parmenides did use the adverb "homou" (ὁμοῦ), which is akin to "hama,"[39] and the adverb "*νῦν*" (now). And he did so in what is generally regarded as the cardinal statement of his ontological theory: "It never was, nor will it be, because it is now, a whole, altogether."[40] Modern scholars differ on how to interpret almost every single word in this statement. Jonathan Barnes[41] discussed four different interpretations; Richard Sorabji[42] presented twice as many. We confine our attention to only those interpretations that suggest the possibility that the statement involves at least an implicit use of the notion of simultaneity. Guilym E. L. Owen, for example, interprets "homou" as "continuous" in time and claims that Parmenides made the following implicit assumption: "Times of which exactly the same things are true (at which the same states of affairs obtain, and which are not distinguished by their antecedents or sequels) are the same time."[43]

If the subject of Parmenides's sentence refers to "Being," which according to the Eleatics never changes, then this interpretation implies that all phases of "Being" occur "at the same time," that is, occur simultaneously; but "simultaneity" in this context must then be interpreted in accordance with its etymological root as "together" and not as "neither earlier nor later" in a tensed language. Such an interpretation seems to suggest that Parmenides anticipated the well-known conception of the static "B-series" proposed by the Cambridge philosopher John McTaggart Ellis McTaggart, who like Parmenides believed that "nothing that exists can be temporal, and that therefore time is unreal."[44] A similar interpretation was proposed by Malcom Schofield when he declared, referring to Parmenides's statement: "What it posits is tanta-

[39] See, for example, G. R. Berry (ed.), *The Classic Greek Dictionary* (Chicago: Follet, 1949), p. 488.

[40] H. Diels, op. cit. (note 18), vol. 1, p. 235.

[41] J. Barnes, op. cit. (note 27), pp. 192–193.

[42] R. Sorabji, *Time, Creation and the Continuum* (Ithaca, New York: Cornell University Press, 1983), pp. 99–108.

[43] G. E. L. Owen, "Plato and Parmenides on the timeless present," *The Monist* **50,** 317–340 (1966). See also his essay "Eleatic questions," *Classical Quarterly* **10,** 84–102 (1960).

[44] J. M. E. McTaggert, "The unreality of time," *The Monist* **17,** 457–474 (1908); *The Nature of Existence* (Cambridge: Cambridge University Press, 1921, 1927, 1968), book 5, chapter 32.

mount to what is envisaged when an omniscient being is conceived as comprehending to him all time *simultaneously*, laid before him like a spatial continuum."[45] We would have to go into too much detail to discuss the relation between Parmenides's statement and the notion of "eternity," which is a major subject in Sorabji's analysis of the eight different interpretations. Nor will we comment on his scholarly historical treatment of those authors who argued that the terms "now" and "is" as used by Parmenides are devoid of any temporal connotation.

Let us, instead, conclude our discussion on whether Parmenides made implicit use of the notion of simultaneity with the following remark. Julian Barbour, in his book *The End of Time*, which intended to show that modern physics vindicates the Parmenidian denial of time, quoted a definition of time, which is said to have been "much loved by John Wheeler": "Time is nature's way of preventing everything from happening all at once."[46] If Parmenides, who was mentioned on the first page of this book, had accepted this definition then his denial of time would imply that everything happens "all at once" or "simultaneously," provided, of course, the term "simultaneously" is being used in a tenseless sense, just as the term "hama" had been used originally to denote "together" or "altogether," the last word in Parmenides's statement.

Plato was undoubtedly acquainted with Parmenides's writings but did not accept his doctrine of the exclusive existence of the timeless "One." Plato's ontology acknowledged not only the existence of a world of immutable and timeless Forms or Ideas, as they exist for example in mathematics, each of which has the characteristics of Parmenides' "Being," but also a world of impermanent sensible things that are patterned after their Forms. Thus, time is defined as the image of eternity. The question of how things partake of their ideas was a central topic in Plato's dialogue *Parmenides*. On the question of whether each participant object partakes of the whole idea or only part of it or whether the whole idea, being one, is in each of the many participants, Parmenides is said to have answered: "While it is one and the same, the whole of it would be in many separate individuals simultaneously ("hama"), and thus it would itself be separate from itself." "No," is the objection, "for it [the

[45] M. Schofield, "Did Parmenides discover eternity?" *Archiv für Geschichte der Philosophie* **52,** 113–135 (1970) (emphasis added).

[46] J. Barbour, *The End of Time* (New York: Oxford University Press, 2000), pp. 44–45.

idea] may be like day, which is one and the same, is in many places simultaneously ("hama"), and yet not separated from itself; so each idea, though one and the same, might be in all its participants simultaneously ("hama")."[47] Numerous other passages may be quoted in *Parmenides* and in other dialogues in which Plato unreservedly makes use of the term "hama." It may be questioned, however, whether this term was really used, for instance, in the passage just quoted, in its temporal meaning. That the answer is positive can be seen by inspecting a related passage in *Parmenides*, where it is asked whether the one can have "come into being" contrary to its own nature, and where it is said that "the end comes into being last ("ὕστατον")."[48]

Although these issues, in the context of which Plato applies the notion of simultaneity, are beset with profound philosophical difficulties, his usage of this notion as such was, at least from the logical point of view, completely unproblematic. The only statement, in which his application of the concept of simultaneity raised a logical problem, was a passage in his dialogue *Timaeus*, where it was said that "time and the heaven [the world] came into being together in order that, having been created simultaneously ("hama"), if ever there was to be a dissolution of them, they may be dissolved simultaneously ("hama")."[49]

This statement by Plato on the simultaneity of the creation of time with the creation of the material world engaged the attention of philosophers, theologians, and scientists throughout the ages, from Aristotle[50] through St. Augustine[51] and Kant[52] to modern cosmologists like Hawking.[53] To understand why the statement may pose a logical problem it is important to verify at first that the term "hama" in it really has the meaning of temporal "simultaneity" and not that of the etymological nontemporal term of "together." This clarification is imperative because, as we will see, even Aristotle still used "hama" both in the nontemporal and in the temporal sense. To convince ourselves that Plato did indeed use it in the temporal sense it suffices to turn

[47] Plato, *Parmenides* 131 B.

[48] Ibid., 153 C.

[49] "ἅμα γεννηθέντες ἅμα λυθῶσιν" Plato, *Timaeus* 38 B.

[50] Aristotle, *The Physics* 251 b 11–19; *On the Heavens (De Caelo)* 300 b 16–21.

[51] "Non est mundus factum in tempore, sed cum tempore." St, Augustine, *De Civitate Dei*, book 11, chapter 5.

[52] I. Kant, *Kritik der Reinen Vernunft (Critique of Pure Reason)*, (First Antinomy), book 2, section 2.

[53] S. Hawking, *A Brief History of Time* (Toronto: Bantam Books, 1988).

our attention to a passage preceding this statement in which Plato declared: "There were no days and nights and months and years before ("πρίν") the heaven was created."[54] Because the term "πρίν" (the old Latin "pris" and German "früh") had at least at Plato's time only a temporal meaning and because this passage has almost the same meaning as the statement under discussion, it is clear that the term "hama" in the statement denotes temporal simultaneity. But, because "simultaneously" means "neither earlier nor later," it may be asked whether the statement "time and the heaven . . . have been created simultaneously" is not self-contradictory, or, in other words, whether it does not presuppose the existence of time by assigning to the act of creation a moment of time and hence contradicts itself. Or finally, the problem may be formulated as follows. If the creation of time really occurred it must have occurred in time.

This formulation suggests resolving the problem by assuming the existence of two categories of time, T and T', where T is the time created and T' is the time in which this creation occurs. This assumption may be supported by the fact that Plato repeatedly describes using temporal terms, the state of the world as it was before the creation of time.[55] If this assumption is accepted then it may be said that the term "hama" in its temporal sense of "simultaneity" refers to T', and the term "time" in the expression "time and the heaven came into being" as quoted above refers to T. The objection that such a resolution of the problem shifts only the problem from T to T' and may lead to an infinite regress can be met by the remark that T' has to be interpreted as disorderly time in contrast to T, which is an orderly time due to the (simultaneous) creation of the heavenly spheres (the stars, planets, sun, and moon), which by their rotations determine the course of time. Because in a disorderly time the temporal relation "earlier" and "later" are not defined, to be "simultaneous" cannot be interpreted in T' as "being not earlier nor later" as it could in T. The idea that Plato tacitly assumed the existence of two different categories of time, one kind of orderly time, created by the Demiurge, and another kind of chaotic time, which preceded the former, had been proposed in antiquity by Gaius Velleius, a contemporary of Cicero, and by Plutarch and Atticus in their commentaries on Plato's *Timaeus*.

[54] Plato, *Timaeus* 37 C.

[55] Plato, *Timaeus* 28 B 2, 39 E. For further details see R. Sorabji, op. cit. (note 42), pp. 272–276.

In contrast to this interpretation, which ascribes to Plato the use of the term "time" in two different connotations of temporality, interpretations of the *Timaeus* exist which contend that the term "time" as used in this dialogue has no temporal meaning at all. Thus, Richard D. Mohr, for example, declares: "I will suggest that when Plato says that the Demiurge makes time, he means that the Demiurge makes a clock, nothing more, nothing less."[56] And later on he adds: "Plato would not repeatedly say that the ordered heavens and time came into being simultaneously (37 D 5–6, 38 B 6), if he in fact meant that they were one and the same thing, thus rendering otiose any claim about their simultaneity."[57] According to this interpretation Plato, when using, for example, in 37 E, the terms "day", "night," "month," and "year," did not refer to what is temporally measurable, but rather to the moving planets and other celestial bodies whose motions constitute measured time. A similar interpretation had been proposed earlier by W. K. C. Guthrie[58] and by William C. Kneale.[59] In fact, as Simplicius reports in his sixth-century commentary on Aristotle's *Physics*, already "Eudemus [a pupil of Aristotle's] got the idea that Plato said that time was the revolution of the heaven" and that this idea "does not involve Plato in the absurdity, as Alexander [of Aphrodisias, head of the Lyceum at about 200 B.C.] argued, that there was time before time."[60] According to such a nontemporal definition of time the statement that "time and heaven came into being simultaneously" would not be self-contradictory because of the identity between the concepts of "time" and (rotating) "heaven." But the snag is that the term "simultaneously" would not only be "otiose," but its Greek original "hama" would have to be interpreted in the nontemporal sense as "together," which would contradict the term "πρίν," as explained previously.

So far we have confined our discussion of Plato's use of the notion of simultaneity to his account of the sensible world, the created copy of the world of Forms or Ideas. In this created world there exist motion, change, and time,

[56] R. D. Mohr, *The Platonic Cosmology* (Leiden: Brill, 1985), p. 54.

[57] Ibid., p. 70.

[58] "For Plato *chronos* itself is a clock, not mere succession or duration . . . " W. K. C. Guthrie, *A History of Greek Philosophy* (Cambridge: Cambridge University Press, 1978, 1986), vol. 5, p. 300.

[59] W. C. Kneale, "Eternity" in P. Edwards (ed.), *The Encyclopedia of Philosophy* (New York: Macmillan, 1967), vol. 3, p. 64.

[60] Simplicius, *In Aristotelis Libros Quattuor Priores Commentaria* (Berlin: Reimer, 1882); J. O. Urmson (ed.), *Simplicius, On Aristotle's Physics* **4,** 1–5, 10–14 (Ithaca, New York: Cornell University Press, 1992), p. 11.

which allowed us to interpret the term "simultaneously" as denoting "not earlier nor later." This raises the question of whether such an interpretation is permissible also with respect to Plato's world of Ideas to which he assigned a higher degree of reality and of which the created world is only a corruptible image. Or, more generally, we may ask whether the concept of simultaneity could have been used by Plato with respect to the eternal world of Ideas at all. To answer this question let us quote the following passage:

> When the father and creator saw the creature he had made moving and living . . . he rejoiced,[61] and in his joy determined to make the copy still more like the original; and as this was eternal, he sought to make the universe eternal, as far as might be. Now the nature of the ideal being was everlasting, but to bestow this attribute in its fulness upon a creature was impossible. Wherefore he resolved to have a moving image of eternity, and when he set in order the heaven, he made this image eternal but moving according to number, while eternity itself rests in unity; and this image we call time. . . . The past and future are created species of time, which we unconsciously but wrongly transfer to the eternal essence; for we say that he "was," he "is," he "will be," but the truth is that "is" alone is properly attributed to him, and that "was" and "will be" are only to be spoken of becoming in time.[62]

Clearly, this "is" should be understood only in a nontemporal or tenseless sense, for time exists only in the world of change and corruption. It follows logically, therefore, that the notion of simultaneity in the sense of "not earlier nor later" also has no raison d'être in the world of Ideas. However, if we admit a nontemporal usage of the term "simultaneously" (in accordance with the etymology of "hama") and accept the above-quoted aphorism that "time is nature's way of preventing everything from happening all at once," then we may also say that in Plato's world of Ideas, where time does not exist, everything happens simultaneously.

In the conclusion of our discussion of the role of "simultaneity" in Plato's *Timaeus* note that Plato himself warns us that he may "not be able to give

[61] Cf. *Genesis* 1, 31, where it is said: "And God saw everything that he had made, and behold, it was very good."

[62] Plato, *Timaeus* 37 D. This point has been stressed by Bertrand Russell who wondered why *Timaeus* has been more influential throughout the history of philosophy than Plato's other works, although "it contains more that is simply silly than is found in his other writings." B. Russell, *History of Western Philosophy* (London: George Allen and Unwin, 1948), pp. 105–106.

notions which are altogether and in every respect exact and consistent with one another."[63] Furthermore, to assist his readers Plato makes frequent use of myths and allegories. But these remarks seem to have little relevance to the subject of our discussion.

Plato's most famous disciple and rival, Aristotle, rejected his teacher's doctrine of two worlds, separated from each other, the world of eternal and immutable Ideas and the world of created and changing matter. According to Aristotle only one uncreated world exists in which the Ideas or Forms are inherent in matter. Because time is inseparably connected with the motion of matter, it too has never been created. Aristotle, therefore, did not have to face the logical problem involved in the notion of a simultaneous creation of time and the material world. Nevertheless, the concepts of simultaneity and the "now" play an important role in his philosophical study of time, which is undoubtedly the most exhaustive and penetrating treatment of this subject the ancient Greeks were able to produce. We will refer to it only as far as it deals with the notion of simultaneity ("hama") and the related concept of the "now."

Aristotle uses the term "hama" in four different connotations, each of which is a possible interpretation of "together," the etymological root of "hama" (see chapter 1): (1) logical togetherness (ἅμα δὲ φανερόν), (2) natural togetherness (ἅμα τῇ φύσει), (3) spatial togetherness (ἅμα κατὰ τόπον), and (4) temporal togetherness (ἅμα κατὰ χρόνον).

In the *Categories* Aristotle begins his treatment of "hama" with the question of how this term is usually used in ordinary language. He answers as follows: "The term 'simultaneous' [ἅμα] is primarily and most appropriately applied to those things the genesis of the one of which is simultaneous with that of the other; for in such cases neither is prior or posterior to the other. Such things are said to be simultaneous in point of time."[64]

Although Aristotle assigned priority to the temporal use of "hama," which of course is also our main concern, we defer our discussion of it to after some brief comments on the other three modes of using this term.

1. "Hama" was often used by Aristotle, mostly in combination with the conjunctive particle "de" (δέ), to express logical opposition as in the English expressions "at the same time" or "while," which, although containing a tem-

[63]Plato, *Timaeus* 29 D.
[64]*Categories* 14 b 23–28.

poral term, have the purely logical nontemporal meaning of "although." Examples of this logical use can be found in *The Physics* 185 a 14, in *Metaphysics* 1008 a 30, in *Nicomachean Ethics* 1163 b 22, and in *Politics* 1294 a 32. A further example, which is interesting because it is explicitly classified by Aristotle as a "natural togetherness" but may be judged by the modern reader as belonging to the class of "logical togetherness," is found in *Categories* and reads: "Correlatives are thought to come into existence simultaneously . . . as in the case of the double and the half. The existence of the half necessitates the existence of that of which it is a half."[65] Using the same example, a later passage reads: "Those things, again, are 'simultaneous' in point of nature, the being of each of which involves that of the other, while at the same time neither is the case of the other's being. This is the case with regard to the double and the half, for these are reciprocally dependent since, if there is a double, there is also a half, and if there is a half, there is also a double, while at the same time neither is the cause of the other."[66]

2. An example of "hama" denoting "natural togetherness" follows immediately thereafter:

> Again, those species which are distinguished one from another and opposed one to another within the same genus are said to be "simultaneous" in nature. I mean those species which are distinguished each from each by one and the same method of division. Thus the "winged" species is simultaneous with the "terrestrial" and the "water" species. These are distinguished within the same genus, and are opposed each to each, for the genus "animal" has the "winged", the "terrestrial", and the "water" species, and no one of these is prior or posterior to another, on the contrary, all such things appear to be "simultaneous" in nature. Each of these also, the terrestrial, the winged, and the water species, can be divided again into subspecies. Those species, then, also will be "simultaneous" in point of nature, which, belonging to the same genus, are distinguished each from each by one and the same method of differentiation.[67]

3. The meaning of "hama" as "spatial togetherness" is given by a formal definition: "Things are said to be 'spatially together' when the immediate and proper place of each is identical with that of the other, and 'apart' when this

[65] *Categories* 7 b 15–18.
[66] *Categories* 14 b 27–32.
[67] *Categories* 14 b 33–15 a 4.

is not the case."[68] This definition seems to contradict the generally accepted axiom that two bodies cannot occupy the same place. To resolve this contradiction some commentators claim that this definition refers only to unextended things like points "which occupy no space at all [but] have proper places, defined unequivocally by position."[69] A different resolution of this contradiction suggests itself if we consider another passage in *The Physics* where "hama" is used in the meaning of "spatial togetherness" and where it is said that in the case of a movement or change the initiator of the movement "must be in direct contact" (hama) with the thing it moves.[70] True, immediately before that definition the terms "together," "apart," "touching," "between," "next in succession," "contiguous," and "continuous" are listed, apparently as distinct from one another. Still, it seems possible that, when formulating the quoted definition, Aristotle referred primarily to a mover moving an object. Moreover, the definition is followed by the statement that two things "touch each other when their extremes are in this sense [the one mentioned in the definition] 'together.' " In short, "hama" in its spatial connotation means "touching each other."

4. Turning now to our main topic, Aristotle's conception of "temporal togetherness" or "simultaneity," we quoted earlier the statement: "The term 'simultaneous' [hama] is primarily and most appropriately applied to those things the genesis of the one of which is simultaneous with that of the other; for in such cases neither is prior or posterior to the other."[71] It can hardly be called a definition of temporal simultaneity. For the main part of the statement—ignoring the additional last twelve words—is based on a *petitio principii:* the *definiens* is identical with the *definiendum,* both being the term "simultaneous." The addition "for in such cases . . . " gives only the reason why the preceding statement is formulated as it is. Only if we ignore this fact can we say that this statement defines "simultaneity" in terms of "prior" and "posterior."

This statement has been quoted from the *Categories,* which in general is regarded as one of the earliest writings of Aristotle. Some experts doubt whether the extant text of it is really genuine. Because Aristotle expounds his

[68] *The Physics* 226 b 22–24.

[69] See footnotes on pp. 34–35 in P. H. Wicksteed and F. M. Cornford (eds.), *Aristotle—The Physics* (London: Heinemann, 1934), vol. 2.

[70] *The Physics* 243 a 5–6.

[71] See note 64.

theory of time in the last four chapters of book IV of *The Physics*, the extant text of which (at least until 222 b 29) is undoubtedly genuine, we may expect to find in *The Physics* perhaps a more acceptable definition of the temporal notion of simultaneity. The very last sentence of the undoubtedly genuine part, however, the sentence in which all the temporal notions that have been defined are listed, mentions only the terms "now," "sometime," "but now," "already," "some time ago," and "suddenly."[72] "Simultaneously" is not mentioned.

Still, there is one sentence in (the undoubtedly genuine part of) *The Physics* that seems to be a definition of "simultaneity." It says: "Further, if simultaneity in time (ἅμα κατὰ χρόνον), and not being before or after, means coinciding and being in the very 'now' wherein they coincide, then. . . . "[73] The first part of it, ending with the word "after," reminds us of the above-quoted statement from the *Categories*. It alone can hardly be interpreted as being a definition of "simultaneity," because to serve as a definition of this term the word "and" (καί) in the expression "and not being" would have to be replaced by "which means" or at least by "or," because "and" denotes a relation of addition and not of logical equivalence or synonymity. Considering the whole sentence preceding the word "then" we note that it begins with the conjunction "if" (εἰ), which suggests that it is a conditional proposition and not a categorical proposition as it should be if it was a definition.

The last-mentioned difficulty may be overcome by pointing out that the use of conditionals is a typical feature of Aristotle's style and may often be ignored. Accordingly Aristotle's definition of simultaneity may be reformulated as follows: "Things are 'temporally together' when the immediate and proper time of the occurrence of each is identical with that of the other (in the sense that they occur in the very 'now' wherein they coincide)." This definition without the parentheses would then be completely analogous to the definition of "spatial togetherness" quoted in mode 3. Moreover, the omission of the parentheses would have two advantages: (a) It would not confine the "time of occurrence" to the temporal duration of the "now," which raises serious problems because the "now" may well have no temporal duration at all; and it would make the definition applicable also for what we have called "interval simultaneity." (b) It would make the definition immune to the

[72] *The Physics* 222 b 27–29.
[73] Ibid., 218 a 26–28.

charge of being merely a *petitio principii*, a charge leveled by those who, like Edward Hussey, claim that "to define simultaneity in terms of 'being in the same now' looks like a begging of the question."[74]

On the other hand, Aristotle's association of the concept of simultaneity with that of the "now" has been used by modern scholars to draw conclusions concerning the logical properties of the simultaneity relation as conceived by Aristotle. It would be worthwhile to discuss one example in some detail. In his unique book *Relativity and Geometry* Roberto Torretti comments on Aristotle's definition of simultaneity from the modern point of view. After stating that "the so-called flow or flight of time is nothing but the ceaseless transit of events across the now, from the future to the past," Torretti declares: "If an event takes some time, while it happens, the now so to speak cuts through it, dividing that part of it which is already gone from that which is still to come. Two events which are thus cleaved by the same now are said to be *simultaneous*. Simultaneity, defined in this way, is evidently reflexive and symmetric, but it is not transitive. . . . However, if we conceive simultaneity as a relation between (idealized) *durationless* events we automatically ensure that it is transitive and hence an equivalence."[75] Thus, in figure 2.3, which represents three temporally extended events *a, b,* and *c*, events *a* and *b* are simultaneous because the instant or "now" n_1 cuts through both and so are the events *a* and *c* because the "now" n_2 cuts through both, but events *b* and *c* are not simultaneous because there is no "now" that cuts through both.

Torretti distinguished between the simultaneity of temporally extended events and the simultaneity of durationless events. For Aristotle the notion of a durationless event was unacceptable, because, as he explained in his *Metaphysics*, whatever happens is a movement or transition from the potential to the actual. True, he occasionally employed the term "suddenly" (ἐξαίφνης), especially in his theory of optical phenomena, but he also insisted that this term denoted "a *passage of time* [only] too short to be perceptible" (222 b 16). But the "now," which in so far as it is an "instant" divides the past from the future and in so far as it denotes the "present" links the past with the future, is indivisible. Hence, as Aristotle proved by a *reductio ad absurdum*, nothing

[74] E. Hussey, *Aristotle's Physics—Books III and IV* (Oxford: Clarendon Press, 1983), p. 141.

[75] R. Torretti, *Relativity and Geometry* (Oxford: Pergamon Press, 1983), p. 220. A relation *R* on a set (*a, b, c, d,* . . .) is reflexive if *x R x*, symmetric if *x R y* implies *y R x*, and transitive if *x R y* and *y R z* imply *x R z* for every *x, y, z* of the set.

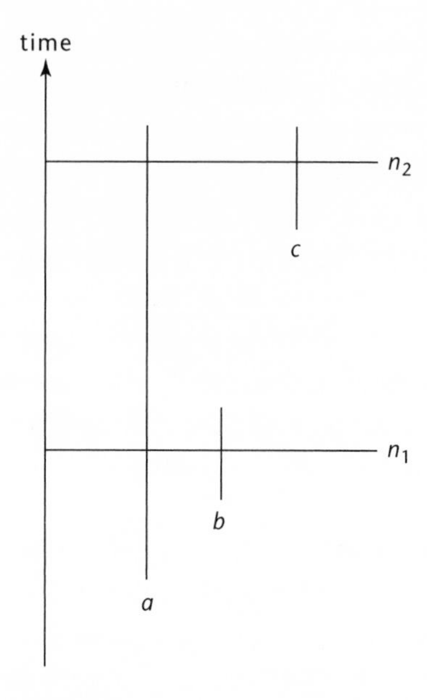

Figure 2.3

can move "in it" nor could anything be said to be at rest "in it" (ἐν τῷ νῦν 234 b 24–34).

Clearly the three statements, (1) Aristotle's definition of simultaneity as the coincidence of occurrences *in* the "now," (2) the thesis of the indivisibility of the "now," and (3) the denial of a possible instantaneity of occurrences, are, if taken together, logically incompatible. Clearly, it suffices to renounce only one of these three statements to obtain a logically consistent theory. Reasons for renouncing statements 2 or 3 have been adduced by Eduard Zeller[76] and Simon Moser.[77] Aristotle's definition of simultaneity was renounced by David Bostock because in the Aristotelian theory of time "the concept of simultaneity is the primitive temporal concept that is used and not further explained."[78]

In addition to the two passages quoted (*Categories* 14 b 23–28 and *The Physics* 218 a 26–28), in which Aristotle discussed "hama" in its temporal con-

[76] E. Zeller, *Die Philosophie der Griechen in ihrer geschichtlichen Entwicklung* (Leipzig: Fues, 1879; Hildesheim: Olms, 1963), vol. 2, part 2, p. 399.

[77] S. Moser, *Grundbegriffe der Naturphilosophie bei Wilhelm von Ockham* (Insbruck: Rauch, 1932), p. 116.

[78] D. Bostock, "Aristotle's account of time," *Phronesis* **25,** 148–169 (1980). Quotation on p. 164.

notation but without stating that he intended them to serve as definitions of the concept of simultaneity, only one more statement in all his writings deals with the meaning of this term as temporal simultaneity, though mainly in the sense of "local simultaneity." It is found in his treatise *Sense and its Objects (De Sensu)*, which deals with the physiological and psychological problem of whether, as he states it, "it is possible or not to perceive several things simultaneously."[79] "By simultaneity," Aristotle declared, "I mean in a time which, for the various things relatively to each other, is one and atomic."[80] Clearly, the phrase "I mean" (λέγω[81]) indicates that Aristotle intends this statement to serve as a definition. Comparison with the two above-quoted statements, in which he discussed the notion of simultaneity, seems to suggest that Aristotle used the expression "in a time . . . which is one and atomic" to denote what he usually called "now." But, as Aristotle repeatedly declared in *The Physics*, "the 'now' is not a part of time at all."[82] And what is "not a part time" cannot be "a *time* which is one and atomic." We must therefore conclude that the only statement that Aristotle explicitly declared to be a definition of the concept of simultaneity is hardly compatible with his general philosophy of time.

In any case, whether it is intended to serve as a definition, Aristotle's statement that temporal simultaneity means "being in the same now" has no operational content. It does not tell us how to find out, or how to verify, whether spatially separated occurrences happen in the same "now." Even Aristotle's affirmations that "at any given moment time is the same" (218 b 11), "time is identical everywhere simultaneously" (220 b 6), and "time is the same everywhere" (223 b 12), which collectively are called "the postulate of simultaneity,"[83] will be of no avail. Nor can Aristotle's much debated definition of time as the number of motion (219 b 2 and elsewhere) provide an operational definition of the simultaneity of motions, because it could, at best, only serve to determine whether the motions are of equal duration.

That a rigorous definition of simultaneity cannot be obtained without the specification of a physical operational procedure was never recognized in an-

[79] *De Sensu* 448 b 20.

[80] "τὸ δ' ἅμα λέγω ἐν ἑνὶ καὶ ἀτόμῳ χρόνῳ πρὸς ἄλληλα." Ibid., 448 b 22.

[81] The Greek word λέγειν denotes not only "to read" but frequently also "to mean."

[82] *The Physics* 218 a 7.

[83] See, for example, M. Inwood, "Aristotle on the reality of time," in L. Judson (ed.), *Aristotle's Physics* (Oxford: Clarendon Press, 1995), pp. 151–178, especially pp. 168–169.

tiquity. One reason, though not the principal one, is probably that such a procedure would be inconceivable in Aristotle's philosophically interesting applications of the notion of simultaneity. The philosophically most important and frequently quoted example is certainly the statement in *Metaphysics*, in which Aristotle explained his anti-Platonic thesis concerning the relation between Ideas or forms and matter: "The form of a bronze sphere exists simultaneously ("hama") with the bronze sphere" (1070 a 23). Obviously, in such an example an operational verification would be impossible.[84]

Of some importance in this respect is also chapter 7 of Aristotle's treatise *On the Senses* which deals with the problem: "Can we perceive two things in the same moment of time?"[85] His answer is this: "By a single sense we cannot perceive two objects simultaneously unless they combine with each other. For the combination requires to be something unitary, and of a unitary object the perception is single and a single sensation is one possessing internal simultaneity" (447 b 11–14). "But certainly, when objects of the same sense, if dual, cannot be simultaneously perceived, it is clear that still less will this be possible in the case of objects of two different senses, *e.g.* white and sweet" (447 b 21–26). Nowhere else in his writings did Aristotle use the term "hama" in the meaning of local temporal togetherness more frequently than in these and the following sentences where it appears more than a dozen times, and in none of these is it accessible to an operational definition. What most effectively prevented Aristotle, his commentators, and later generations from recognizing the need for an operational definition of the concept of simultaneity, however, is Aristotle's theory of vision and light.

Aristotle declares that in contrast to sound, which "seems to be a motion of something which travels" (446 b 34), "light is not a motion" (446 a 30). It does not "penetrate the medium first before reaching us"; rather, the medium is altered "all at once (ἀθρόον[86]) just as water freezes all at one time" ("ἅμα πᾶν" 447 a 3). In other words, according to Aristotle light does not need time to propagate from one place to another or, expressed in the language of physics, the velocity of light is infinite.

[84] "Hama" is here used in the sense of both temporal and spatial togetherness.

[85] Aristotle, *De Sensu* 447 a 14.

[86] The adverb ἀθρόον derives, just like ἅμα, from the Sanskrit where it was spelled "sadhriyãnc," meaning literally "collected together"; it is a generalization of "hama" in the sense that it refers to several or many individuals whereas "hama" usually refers to only two.

The instantaneousness of the process of vision, a logical consequence of Aristotle's theory of light, seemed of course to tally with ordinary experience. Let us call the assumption that in the act of visually observing a distant event, such as a flash of a lightning, the event and its perception in the eye of the observer are simultaneous, "the visual simultaneity thesis." This thesis was common to all optical theories in antiquity with the possible exception of that of Empedocles. It persisted also throughout the Middle Ages, mainly because of Aristotle's influence. Its validity was questioned for the first time at the end of the fourteenth century by Blasius of Perma but was disproved only in 1676, when Olaf Roemer, on the basis of his astronomical observations of the eclipses of Jupiter's satellites, discovered that the velocity of light is finite. Clearly, the visual simultaneity thesis, even if not explicitly affirmed and only subconsciously accepted, as it undoubtedly has been (and still is) in general, was not conducive to a quest for an operational definition of simultaneity. Moreover, because of this thesis everybody took the notion of simultaneity for granted.

Aristotle must therefore be credited with having been the first to realize the importance of the concept of simultaneity for the philosophy of time even though his definition of it may be criticized and his adherence to the visual simultaneity thesis may have inhibited, because of his enormous influence on the course of philosophy, any operational definition of this concept.

In general, Aristotle's successors at the Lyceum and his later commentators accepted his philosophy of time and, in particular, his treatment of the notion of simultaneity without serious reservations. One of the few exceptions was Strato of Lampsacus, who succeeded Theophrastus in 288 B.C. as head of the Lyceum in Athens. Strato criticized Aristotle's definition of time as the number of motion[87] on the grounds that any number is a definite quantity, whereas time is a continuum and, in his opinion therefore, an indefinite quantity or relation and as such could not be counted. Furthermore, and most importantly, the parts of any number are always simultaneous, whereas the parts of time, being always in succession, can never be simultaneous. As Simplicius, the famous sixth-century commentator of Aristotle's *Physics* reports in his *Corollary on Time*, Strato denied that time is the number of motion because a number is a discrete quantity whereas time is continuous and the continuous is not denumerable. In addition, all parts of a

[87] Aristotle, *The Physics* 220 a 25.

number must necessarily exist, for there could not be a triad without three integers. But this is impossible in the case of time, because the earlier and the later would have to exist simultaneously, which would entail that "a time of time for time" would exist.[88] What is of interest for us is not so much Strato's conception of number but rather his argument that a simultaneity of the past with the future would entail the existence of an additional kind of time in which this simultaneity exists.

Strato's argumentation for the existence of a hypertime can also be applied to frequently used expressions such as "the past precedes the future," for it may be asked in what time does this act of preceding take place. A spatial analogue exists in the following syllogism: (a) Everything, that exists, exists in space; (b) space exists; ergo (c) there exists a space (or hyperspace) in which space exists.

The main problem that engaged the attention of Aristotle's commentators was his treatment of the problem of whether time exists at all. In chapter 10 of *Physics* Aristotle discusses the conceptual difficulties one has to face in contending that time exists.[89] As Simplicius reports, this problem and especially the so-called paradox of the parts of time,[90] has also greatly intrigued his teacher Damascius. According to Simplicius, the solution that Damascius proposed is this: time exists, but the whole of it does not exist simultaneously in reality.

We conclude our discussion of the role of simultaneity in ancient philosophy with some remarks on an issue that plays a dominant role in modern theories of time but was almost completely ignored in the ancient philosophy of time: the relation between causality and time. Aristotle's definition of time as the number of movement, combined with his conception of movement or change as a transition from potentiality to actuality, suppressed any

[88] *Corollarium de Tempore* in Simplicius, *In Aristotelis Physicorum Libros Quattuor Priores Commentaria* (Berlin: Reimer, 1992), pp. 771–800. Quotation on p. 789; Simplicius, *Corollaries on Place and Time* (London: Duckworth, 1992), p. 108.

[89] Aristotle, *The Physics* 217 b 29–218 b 20.

[90] Aristotle's formulation of the paradox of the parts of time is this: "Some of it is past and no longer exists, and the rest is future and does not yet exist; and all time, whether in its limitless totality or any given length of time we take, is entirely made up of the no-longer and the not-yet; and how can we conceive of that which is composed of non-existents sharing in existence in any way?" (217 b 36–218 a 4). (Recall that the "now" as "the present" is according to Aristotle not a part of time.) This problem has been dealt with in a similar manner many times in the history of philosophy, cf., for example, Aurelius Augustinus (St. Augustine), *Confessions*, chapter 14.

attempt of relating time with causality even though he admitted that "time causes things to age."[91] It was only when the attainment of certain knowledge and undeniable truth was renounced by the so-called Sceptics, as for instance by Pyrrho of Elis, a contemporary of Aristotle, that the notion of causality and its relation to time became the subject of critical reflection.

Photius, who at the end of the ninth century A.D. became bishop of Constantinople, reported in *Bibliotheca*,[92] a collection of excerpts from philosophical writings, that Aenesidemus of Cnossos, who taught at the end of the first century A.D. in Alexandria, was the first to criticize the notion of causality in connection with the concept of time. Sextus Empiricus, the last and boldest of the Greek Sceptics, who lived two centuries after Aenesidemus, was deeply influenced by the *Pyrrhonean Discourses* in which Aenesidemus rejected the concept of causality. According to Sextus, Aenesidemus argued as follows: "If anything is the cause of anything, either the simultaneous is the cause of the simultaneous ("hama"), or the prior of the posterior, or the posterior of the prior; but the simultaneous is not the cause of the simultaneous, nor the prior of the posterior, nor the posterior of the prior, as we shall establish. Therefore there does not exist any cause."[93]

To show how Aenesidemus proved these contentions Sextus continues as follows:

> The simultaneous cannot be the cause of the simultaneous owing to the coexistence of both and the fact that this one is no more capable of generating that one than is that one of this one, since both are equal in point of existence. Nor will the prior be capable of producing that which comes into being later; for if, when the cause exists, that whereof it is cause is not yet existent, neither is the former any longer a cause, as it has not that whereof it is the cause, nor is the latter any longer an effect, since that whereof it is the effect does not coexist with it. For each of these is a relative thing, and relatives must necessarily co-exist with each other, instead of one preceding and the other following. It only remains for us, then, to say that the posterior is the cause of the prior; but this is a most absurd notion, worthy of men who turn things topsy-turvey; for we shall have to say that the effect is older than that what produced it, and con-

[91] Aristotle, *De Coelo (On the Heavens)* 279 a 19.

[92] Photius, *Bibliotheca (Myrobiblion)* (Berlin: Reimer, 1824), Codex 212, pp. 169–171. Photius erroneously called Aenesidemus of Cnossos "Aenesidemus of Agae."

[93] Sextus Empiricus, *Adversus Physicos* I, 232–233 (London: Heinemann, 1960), p. 117.

> sequently is not an effect at all since it is without that whereof it is the effect. So just as it is foolish to say that the son is older than his father, or that the harvest is earlier in date that the sowing, so it is silly to maintain that what is as yet non-existent is the cause of what already exists. But if the simultaneous is not the cause of the simultaneous, nor the prior of the posterior, nor the posterior of the prior, and besides these there is no other possibility, no cause will exist.[94]

We have purposely quoted this passage *in extenso* because it touches on issues that are of great importance in the modern history of the concept of simultaneity as we will see in due course. Suffice it at present only to point out that Aenesidemus used the notion of simultaneity to disprove causality; in modern philosophy and in a certain sense also in modern physics, the notion of causality is used to define simultaneity.

[94] Ibid., I, 233–236.

CHAPTER THREE

Medieval Conceptions of Simultaneity

In the writings of the early medieval philosophers metaphysical speculations were inextricably intermingled with theological ideas. This fusion of the conclusions of reason with the facts of revelation was characteristic for the patristic philosophers, in particular, who were thought to have recognized similarities between Plato and the Judeo-Christian tradition as, for example, that Plato, in contrast to Aristotle, taught that the world was created by God. Some of these philosophers even claimed that Plato had drawn from the writings of the Old Testament. It is not surprising, therefore, that when dealing with the problem of time they devoted much attention to the concept of eternity, which according to Plato's *Timaeus* served the Demiurge as the prototype for creating time as its moving image.[1]

Some logical difficulties had to be resolved, however. For if God or the Demiurge created the world simultaneously with time, as Plato had taught—or if, as the Bible says, "in the beginning God created heaven and earth"[2]—

[1] See chapter 2, notes 49, 54, and 55.
[2] *Genesis* 1, 1.

there should have been before creation a period of inaction on the part of the Creator, whose eternity was, of course, never doubted. St. Augustine (Aurelius Augustinus) discussed this problem *in extenso* in *Confessions*, which he wrote at the end of the fourth century. He answered the question, "What was God doing before He made heaven and earth?," by saying: "It is not in time that you [God] precede times; all your 'years' subsist in simultaneity because they do not change; your 'years' are 'one day' and your 'today' is 'eternity.' "[3]

Meditating about the term "before" St. Augustine of course rejected the "merrily" given answer that "He was preparing hell for pryers into mysteries" as "eluding the pressure of the question." Instead he argued that "if before heaven and earth there was no time, why is it demanded, what Thou then didst? For there was no 'then,' when there was no time." But if there was no time before creation, how could he claim that all of God's "years subsist in simultaneity?" In other words, did he use the term "simul" in this context in a temporal sense, in which case he would have contradicted his own statement that prior to creation there was no time? If we study his writings we find that he frequently applied the term "simul" in a definitely temporal sense. The most explicit example can be found in *De civitate Dei*, where he wrote: "Who does not see that time would not have existed had not some creature been made, which by some motion would bring about change, and that since the various parts of this motion and change cannot exist simultaneously (*simul*), when one passes away and another succeeds it in shorter or longer intervals of duration, time would be the result?"[4] On the other hand, if we recall that St. Augustine frequently[5] quoted the Latin version "*creavit omnia simul*" of *Ecclesiasticus* 18, 1, which in the *Septuagint* reads "ἔκτισεν τὰ πάντα κοινῇ," where the term "κοινῇ" (common) has the original not necessarily temporal meaning of the Sanskrit "sem" (together), we cannot exclude the possibility that St. Augustine used *simul* in "*anni tui omnes simul stant*"[6] in the nontemporal meaning of simply "together."

This nontemporal use of *simul* was, of course, an exception. In general, Augustine used this term in its temporal sense as synonymous with "at the

[3] " . . . anni tui omnes simul stant." *St. Augustine's Confessions* (London: Heinemann, 1979), vol. 2, p. 236.

[4] St. Augustine, *De Civitate Dei*, XI, 6. See also ibid., XII, 16 and his *De Genesi ad Litteram Liber Imperfectus* VI, 11, 19, VII, 28, 42 and *De Vera Religione* 7, 13.

[5] See, for example, *De Civitate Dei*, VI, 6.

[6] His choice of "*anni*" (years) was probably influenced by *Psalm* 90, 4 where it is said: "For a thousand years in they sight are but as yesterday when it is past."

same time." An interesting example of this temporal usage can be found in book 7, chapter 6, of *Confessions,* in which he explains the reason for his disbelief in astrological divinations. He tells us the story of a rich woman and a poor maid servant who became pregnant at the same time. To cast the horoscopes of their expected babies, they had to know the exact times of the births. Their husbands therefore "most carefully reckoned the days, hours, and even the lesser divisions of the hours until both women gave birth, as it happened at the same instant." (*Atque ita factum esse, ut cum iste coniugis, ille autem ancillae dies et horas minutioresque horarum articulos cautissima observatione numerant, enixe essent ambae simul.*) "Hence, both of them were constrained to allow the very same horoscope, even to the very smallest points. As soon as the women began to be in labor, they both gave notice to one another . . . and had messengers ready to send to one another as soon as each had notice of the child's birth. Thus, then, the messengers sent from one to the other met in such equal distance from either house that neither of the calculators could observe any other position of the stars than had the other." And yet the son of the rich woman "throve well in riches, raised himself to honor, whereas that little servant . . . continued to serve his masters."

Augustine thus claims to have refuted astrology by recording the story of two persons who, although born simultaneously, fared quite differently in their lives.

What is of interest for us in this story is, of course, not so much Augustine's refutation of astrology but rather the *method* that he applied to refute astrology as a heretical doctrine. It is the method of dispatching two messengers, supposed to be running with equal velocities and meeting at "equal distance from either house," that is, at the midpoint of their initial distance, to verify the simultaneity of two spatially separated events, namely the two parturitions.

The method described by Augustine may well be regarded as probably the earliest recorded example of an *operational* definition or verification of distant simultaneity, even though it still needed some elaboration. In fact, it may be regarded as anticipating Einstein's earliest operational definition of distant simultaneity, though not his first published one, as we shall see in chapter 7.

The analogy between Augustine's operational verification and Einstein's operational definition of distant simultaneity needs some qualification. Whereas Augustine did not inform us on what grounds it is justified to assume that the two messengers were running equally fast, an assumption that of course is logically necessary for the validity of his method, this problem did not arise in the context of Einstein's definition of simultaneity, as we will see in due course.

The most important successor of St. Augustine was Anicius Manlius Torquatus Severinus Boethius who studied at Rome and Athens and was influenced not only by Plato and the Neo-Platonists but also by Aristotle, part of whose writings he translated into Latin. Following Aristotle, he believed in the eternity of the world in the sense that it had no beginning and would have no end, but he distinguished this kind of eternity from the eternity of God. In *Consolations of Philosophy*, written during his long captivity before his death in 525, Boethius defined divine eternity as "the complete simultaneous and perfect possession of interminable life" or, as he expressed it in Latin, "aeternitas est interminabilis vitae tota simul et perfecta possessio."[7] To understand the exact meaning of "simul" in this definition of eternity we have to refer to his short tractate *Trinitas Unus Deus ac Non Tres Dei*, usually called *De Trinitate*, in which he distinguished eternity from sempiternity, the state of lasting for ever.[8] He declared that the now that abides unmoved, the *nunc stans*, constitutes eternity, whereas the now that flows in time makes sempiternity.[9] This seems to suggest that Boethius has used the term *simul* in his definition of divine eternity not in the sense of a temporal relation between events but rather in the original meaning of the Greek term "hama" as denoting a timeless "together" in the *nunc stans*.

This conclusion can be corroborated by a study of what seems to have been the source of Boethius's definition of eternity, a statement made by Plotinus of Lycopolis, one of the founders of Neo-Platonism in the third century. His writings, edited and published by his disciple Porphyry in six *Enneads* or series of nine essays each, contained in the third *Ennead* a definition of eternity as "the life, which belongs to that which exists and is in being, all together and full, completely without extension or interval."[10] The similarity of this definition to that of Boethius is too striking to be dismissed as being

[7] Boethius, *Philosophiae Consolationes*, book V, chapter 6, pp. 9–10; *The Theological Tractates and the Consolation of Philosophy* (London: Heinemann, 1973), p. 422.

[8] The Latin adverb *semper* means "always" or "at all times."

[9] Boethius, *De Trinitate*, chapter 4, pp. 67–70. *The Theological Tractates* (op. cit.), pp. 20–23.

[10] Plotinus, *Enneads* III, 7 (London: Heinemann, 1980), vol. 3, p. 305.

merely coincidental. Both define eternity as some kind of "life" and qualify it as "full" in a certain sense. If we compare these definitions word by word we find that the term "all together" (ὁμοῦ πᾶσα) in Plotinus's formulation corresponds to Boethius's expression "simul." Now, the word "homou" (ὁμοῦ), akin to "hama" in its original meaning, denotes spatial togetherness or at least unqualified "togetherness." In fact, "homou pan" (ὁμοῦ πᾶν) was used previously by Parmenides in his description of the unchanging and hence timeless Being.[11] We can therefore conclude that Boethius also used in his definition of eternity the term *simul* in the nontemporal sense of "together." Furthermore, according to Plato's dialogue *Timaeus*, which undoubtedly was known to Boethius, "time" is the "moving image of eternity," whereas "eternity" itself is unmoving and hence timeless, so that a temporal concept cannot be associated with it.

Boethius's definition of eternity and, in particular, his use of the term *simul* have become the subject of an animated debate among scholars, which was prompted by the publication of an article by Eleonore Stump and Norman Kretzmann[12] in 1981. Their article was inspired by Einstein's conception of simultaneity as a concept whose validity depends on the choice of the reference frame relative to which it is applied. Stump and Kretzmann agree that Boethius used *simul* in his definition of eternity as a nontemporal notion because it is an ingredient of "eternity," a mode of existence which differs fundamentally from the mode of existence called "time," neither of which is reducible to the other. These two modes of existence, however, cannot be unrelated to each other because God, although an eternal being, created everything that is in time, and human beings, existing in time, are supposed to be interacting with God, for example, in prayer. These authors tried, therefore, to establish some relationship between these two modes of existence, disparate as they are. Because eternity, as Boethius rightly emphasized, is also characterized by life (vitae possessio), presentness, it is claimed, must be a feature of eternity in contrast to temporal priority or posteriority, which implies temporal succession. Because the ordinary concept of simultaneity, denoting the existence of beings or the occurrence of events *at the same time*, is obviously a purely temporal notion, it cannot serve to form a relation be-

[11] "It never was, nor Will Be, because it Is now, a Whole all together, One, continuous." Parmenides, Fragment B 8, 5. K. Freeman, *Ancilla to the Pre-Socratic Philosophers* (Oxford: Blackwell, 1952), p. 43.

[12] L. Stump and N. Kretzmann, "Eternity," *The Journal of Philosophy* **78,** 429–458 (1981).

tween the two modes of existence. But by relativizing the concept of simultaneity to different reference frames, these being the temporal and the eternal modes of existence, Stump and Kretzmann claimed to have found a relationship between eternity and time.

Calling it "ET simultaneity," in which E stands for "eternity" and T for "time," they defined it as a relation that satisfies the following three necessary and sufficient conditions, in which *x* and *y* denote entities or events: (1) either *x* is eternal and *y* is temporal, or vice versa; (2) for some observer, A, in the unique eternal reference frame, *x* and *y* are both present, that is, either *x* is eternally present and *y* is observed as temporally present, or vice versa; (3) for some observer, B, in one of the infinitely many temporal reference frames, *x* and *y* are both present, that is, either *x* is observed as eternally present and *y* is temporally present, or vice versa. It is easy to see that ET simultaneity, unlike ordinary simultaneity, is a symmetric but not reflexive or transitive relation. Each of the two modes of existence has, of course, a simultaneity relation of its own. The existence of entities or occurrence of events at one and the same time is called "T simultaneity" and at one and the same eternal present "E simultaneity."

In the recent philosophical or theological literature hardly another essay has prompted so many responses, for or against, as the Stump and Kretzmann essay "Eternity." It would lead us too far astray to discuss these responses even only summarily.[13] Stump and Kretzmann's essay prompted this interest because Boethius's definition of eternity had been accepted by Thomas Aquinas in *Summa Theologica*, which was written in the years 1265–1272 and since then has formed the basis of the dogmatic teachings of the Catholic Church.

In their writings on the philosophy of time the scholastics of the thirteenth and fourteenth centuries focused their attention primarily on the problems of the essence and the existence of time. Their discussion of the former problem (*quid est*) usually started with the Aristotelian definition of time as the number of motion with respect to the "before" and "after" (*tempus est numerus motus secundum prius et posterius*), and they expended great effort in defending this definition against the objection of circularity by arguing, of course, that the "before" and "after" in this definition should not be inter-

[13] For a list of responses to the Stump–Kretzmann paper see M. Jammer, *Einstein and Religion* (Princeton, New Jersey: Princeton University Press, 1999), p. 175.

preted in a temporal sense.[14] Their discussions of the existence of time (*an sit*) centered on the question of whether time exists objectively or only subjectively (*in anima*).[15] Clearly, such issues could hardly provide any opportunity to discuss the notion of simultaneity. Some scholastics, however, discussed a third topic that could not only involve the notion of simultaneity but even anticipate its relativity as conceived in modern physics. This problem dealt with the question of the uniqueness or rather the singleness of time, that is, the question of whether there exists only one time. If several times existed, because "simultaneously" means "at the same time," every time would have its own simultaneity relation and two events simultaneous in one time would not necessarily be simultaneous with respect to another time.

Of course, the scholastics were aware that Aristotle himself had already raised the question concerning the singleness of time, but the argumentation for his positive answer did not satisfy them. To understand why the scholastics rejected Aristotle's arguments we must first recall his arguments. After having pointed out that time is not the number of any particular movement but of all movements Aristotle asked: "Can each of them have a different time, and must there be more than one time running simultaneously? Surely not. For a time that is both equal and simultaneous is one and the same time, and even those that are not simultaneous are one in kind; for if there were dogs and horses, and seven of each, the number would be the same."[16] In an earlier passage Aristotle distinguished sharply between two meanings of "number," "for we speak of the 'numbers' that are counted in the thing in question, and also of the 'numbers' by which we count them and in which we calculate; we are to note that time is the countable thing that we are counting, not the numbers we count in—which two things are different."[17] Aristotle's emphasis that time as *numerus motus* is not the mathematical number, the *numerus quo numeranus*, the number we use in counting, but rather the *numerus numeratus*; the quantitative measure inherent in motion contra-

[14] Cf., for example, "Dicendum est quod prius et posterius ponuntur in definitione temporis secundum quod causantur in motu ex magnitudine et non secundum quod mensurantur ex tempore." Thomas Aquinas, *In Octo Liberos de Physico Audito sive Physicorum Aristotelis Commentaria* (Paris, 1268), IV, lectio 17.

[15] An extremely subjective interpretation of time had already been given by Aurelius Augustinus (St. Augustine) (354–430) when he declared in his *Confessions* (chapter 27): "It is in you, O mind of mine, that I measure the periods of time" (In te, anime meus, tempora metior).

[16] Aristotle, *The Physics* 223 b 2–6.

[17] Ibid., 219 b 6–9.

dicts, therefore, his statement that the use of the notion of number as applied to time is the same as the use of the notion of number when speaking of seven dogs and seven horses. The scholastics realized therefore that Aristotle could only infer by his reasoning that the duration of different time intervals can be equal but not that the time of different simultaneous occurrences is numerically the same.

Turning now to the question of whether there can be a plurality of times the scholastics referred to Aristotle's affirmation of the plurality of motions and to his statement: "It may be asked to what kind of motion time does pertain. We may answer, 'It does not matter.' For things begin and cease to be, and grow, and change their qualities and their places 'in time', so far, then, as change can be regarded as movement, so far time must be a numerator of every kind of movement. We conclude therefore that time is simply the number of continuous movement, not of any particular kind of it."[18] The scholastics epitomized Aristotle's conclusion with the phrase "tot tempora quot motus" (so many times as there are motions). But such a statement can hardly be reconciled with Aristotle's declaration that "time is identical everywhere in earth, and sea and sky simultaneously."[19]

Before discussing how the scholastics attempted to overcome this discrepancy let us digress by making some anachronistic remarks on the "simultaneity of time" and its logical implications, which unfortunately seem never to have been drawn by the scholastics.

If we assume in accordance with the phrase "tot tempora quot motus" that several different times exist, we cannot say that they proceed "simultaneously." First, "simultaneity" is a relation between events (or intervals) and times are not events (or intervals). Second, "simultaneity" is a temporal relation and as such presupposes a given time. To say that, for example, two times are simultaneous would require the existence of a third time, some hypertime, whose existence is more than questionable. Furthermore, as stated previously, two events that are simultaneous with reference to one time need not be simultaneous with reference to another kind of time, provided such a time exists.

Alexander of Aphrodisias, one of the most influential commentators on Aristotle, in his treatise "De Tempore," written at the end of the second cen-

[18] Ibid., 223 a 30–223 b 1.

[19] Ibid., 223 a 18. See also 218 b 14, 219 b 11, 220 b 6, and 223 b 7–12.

tury A.D. and extant only in Arabic and Latin translations, stressed the need to reformulate the Aristotelian definition of time in such a way that it excludes the possibility of a plurality of times. Probably prompted by the passage in Aristotle's *Physics*, in which a uniform rotation is called "the best standard because it is easiest to count,"[20] Alexander proposed defining time as the measure not of motion in general but of only the motion of the outmost celestial sphere. In addition to excluding a plurality of times the choice of such a universal standard, Alexander added, is advantageous "because no faster motion exists and any magnitude can be counted or measured only by what is smaller than it."[21]

From a modern point of view, but still recalling what we earlier called "the visual simultaneity thesis," we may say that Alexander was the first to define a worldwide standard time and to impart thereby the notion of distant simultaneity with a spatially unbounded meaning. But of course the horology of that age was not yet sufficiently advanced to make any operational use of this possibility. Still, by promoting the *primum mobile* to what may be called the standard clock, which by the number of its rotational motions defines a univocal time, Alexander gave meaning to Aristotle's statement that "time is identical everywhere." In other words, and anachronistically expressed, the convention that Alexander introduced bestows semantic legitimacy on the concept of distant simultaneity. The famous Persian philosopher and physician Ibn Sina, also known by his Latin name as Avicenna, did not content himself with accepting the idea of a universal time or simultaneity as being based merely on a convention, but he claimed to explain it as an effect of causality. In *Kitab Assifa* (Book of Healing), which is actually a philosophic encyclopedia of numerous volumes, dealing with Logic, Physics, Mathematics, and Metaphysics, written about the year 1000, Ibn Sina, influenced by Aristotle and no less by Neo-Platonists, presents a theory of emanations. These effluences originate in God and propagate from the celestial spheres down to our Earth. This process, although lasting eternally, occurs in such a way that cause and effect in it are simultaneous. The cause—and not only the measure—of all motions in the world is the circular motion of the heavens.[22] In contrast to Alexander, who considered only the geometric-kinematic aspects

[20] Aristotle, *The Physics* 223 b 20.

[21] Alexander of Aphrodisias, *De Tempore* 94, 16. See also R. W. Sharples, "Alexander of Aphrodisias, On Time," *Phronesis* **27,** 58–81 (1982).

[22] Ibn Sina, *Kitab Assifa* (Venice: 1508), book 2, chapter 13.

of the celestial motion for his definition of time, Ibn Sina introduced dynamical features in his theory of time without specifying their mathematical structure.

Ibn Roshd, also known as Averroes or simply "the Commentator" as the Scholastics used to call him, because of his detailed commentaries on the whole works of Aristotle, agreed with Ibn Sina's theory according to which the motion of the *primum mobile* is the cause and driving force of all other motions and therefore defines time. As Ibn Roshd pointed out in his commentary to Aristotle's *Physics*, however, this theory leads to the recognition of time only *per accidens*. Something else is required to make time recognizable everywhere. Ibn Roshd also recalled that Plato's identification of time with celestial motion has been criticized because people, who, on a cloudy night, cannot see the sky, for example, should not have the feeling that time passes by, which is, of course, not the case. He concluded therefore that, apart from our perception of the existence of time from the motion of the celestial sphere, we experience the presence of time directly in ourselves, through our own mutability, in our own "esse transmutabile." In short, according to Ibn Roshd it is in ourselves, in our physically and psychologically changing constitution, that we become aware of the flow of time.[23]

Ibn Roshd's somewhat subjectivistic theory of time, which acknowledges that the whole universe is, so to speak, a single clockwork, whose rhythm is regulated by means of an instantaneously operating causal efficiency, which originates in the diurnal motion of the heaven, but which the human organism can perceive without looking at the sky, was accepted by Albertus Magnus[24] and his disciple Thomas Aquinas. Although the concept of simultaneity was not mentioned explicitly in any of these theories of time, it played an important role in them. This becomes apparent in the reformulation of Ibn Roshd's conception of time by the nominalist William of Ockham, who is known for having rejected the doctrine that abstract or general terms, the so-called universals, have an objective reality and exist in several things at once; for the same thing, Ockham contended, cannot exist simultaneously in several different things. But it is not in this context that Ockham introduced the notion of simultaneity in his interpretation of Ibn Roshd's theory.

[23] Ibn Roshd, *Aristotelis de Physico Auditu Libri Octo cum Averrois Cordubensis Commentariis* (Venice: 1562); reprinted (Frankfurt a.M.: Minerva, 1962), vol. 4, p. 282.

[24] Albertus Magnus, *Opera Omnia* (Paris: Vives, 1890), vol. 3, p. 313.

In *Tractatus de Successivis*, written in the early fourteenth century, Ockham explained Ibn Roshd's theory of time as follows:

> The Commentator means that anyone who senses any movement can sense time or, in other words, can understand that time exists—but not directly and essentially. In fact, here is the process followed for this. A man sees a celestial body move or perceives some external movement or even imagines a movement. That done, he can imagine that he coexists with some body moved by continuous and uniform movement (*potest imaginari se coexistere alicui uniformiter et continue moto*); consequently, he can understand this proposition: I coexist with a body moved by a continuous and uniform movement.[25]

Ockham repeated this idea elsewhere where he said: "Someone blind from birth does not know the proposition, heaven moves, for he has never seen the movement of heaven; he can, however, grasp the movement of heaven by a composite concept. It suffices for him to grasp the proposition, I coexist with a certain body moving continuously and uniformly."[26] It is reasonable to assume that Ockham used the term "coexistence" and not "contemporaneity" or "simultaneity" purposely to avoid an explicitly stated logical circle, for the latter two terms clearly presuppose the notion of time. The last mentioned theories of time have a modern analogue in a statement made by a contemporary well-known philosopher of science who declared: "It makes sense to speak of the 'temporality' of the world as an internal property, whereby our experience of time has its cause in the fact that the human being participates unescapably as a physical system at the changing processes that occur in his environment."[27]

All these scholastic discussions[28] concerning the nature of time intended to show that the conception of a simultaneous existence of a plurality of times (plura tempora simul existentia), although not circular, is inadmissible.

[25] William of Ockham, *Tractatus de Successivis*, chapter 2, p. 105. Quoted after Pierre Duhem, *Medieval Cosmology* (Chicago: The University of Chicago Press, 1985), p. 316.

[26] William of Ockham, *Questiones super librum Physicorum*, quaestio XLV. Duhem, op. cit. p. 319.

[27] B. Kanitscheider, *Philosophie und Moderne Physik* (Darmstadt: Wissenschaftliche Buchgesellschaft, 1979), p. 11.

[28] For a detailed exposition of these and other discussions, including those by Bonaventura, Duns Scotus, Aegidius Romanus, Roger Bacon, Hervaeus Natalis, Richard of Mediavilla, Petrus Aureoli, William of Ockham, and Walter Burley, see A. Maier, "Scholastische Diskussionen über die Wesensbestimmung der Zeit," *Scholastik* **26,** 520–556.

In all these discussions it was also tacitly assumed that the application of the scale of time, given by the celestial motions, to the terrestrial processes or events involves no temporal delay but proceeds simultaneously. The denial of "tot tempora quot motus" implied, of course, the impossibility of a plurality not only of time but also of the "now" (nunc) at a given instant.

In our account of these deliberations, which may sound strange today, we should not forget that the concept of "motion" had ontological priority over the notion of "time" for the scholastic philosophers and theologians. The reversal of this priority relation occurred only in the Italian natural philosophy of Bernardino Telesio and Francesco Patrizi in the middle of the sixteenth century. Just like space, which in the Aristotelian and scholastic philosophy had no independent existence prior to material objects, now gained ontological priority over the objects that it contains (for it now became legitimate to assume the existence of space without any objects contained therein), so also time now became the matrix or substratum of all occurrences irrespective of whether they are related with motion.[29]

[29] B. Telesio, *De Rerum Natura* (Naples: 1586), 11, 22.

CHAPTER FOUR

The Concept of Simultaneity in the Sixteenth and Seventeenth Centuries

Three important innovations in the sixteenth and the seventeenth centuries led to a far-reaching revision of the concept of time and with it of the concept of simultaneity: (1) the philosophical abandonment of the Aristotelian association of the concept of time with that of motion, (2) the widespread technical use of mechanical clocks, and (3) the scientific discovery of the finite velocity of light. As mentioned at the end of chapter 3, the philosophical change was initiated by the Italian natural philosophers Telesio and Patrizi who maintained that Aristotle was right in stating that time cannot be measured without motion but wrong in stating that it cannot exist without motion. Their point of view made it possible to resolve the problem of the multiplicity of time (tot tempora quot motus) and to reduce it to a duplicity of time, in which one kind of time, usually defined as subjective or spiritual time, as previously conceived by Augustine, serves, so to speak, as a pacemaker or regulator of the time measured by motion. The theory of time proposed in 1597 by the Spanish Jesuit Franciscus Suarez in *Disputationes*

Metaphysicae[1] illustrates this point and shows its relation to the notion of simultaneity. He distinguishes between a time measured by motion, which he calls "intrinsic successive duration" (duratio successiva intrinsica) and a mental time, "imaginary continuous succession" (successio continua imaginaria). All times of the former kind, measured by motions of physical objects, are brought into coalescence by being immersed into the latter kind of time. Such a coordination introduces the concept of simultaneity either directly between the two kinds of time or indirectly through the intermediation of the moving body whose motion measures the "intrinsic duration." This latter alternative would have led to a novel conception of simultaneity, for it would correlate one event with two times and not, as usually, two events with one time. Yet Suarez avoids this complication by claiming that the "intrinsic duration" is "coexisting and, so to speak, filling ("replens") a part of that imaginary duration." That only duration (duratio), in contrast to succession (successio), constitutes the nature of time is the basic assumption in the theory of time proposed by the anti-Aristotelian physician and alchemist Jan Baptista van Helmont in his tractate *De Tempore*.[2] This work refers to the notion of simultaneity in so far as it describes a synchronization procedure of a pendulum with a sundial.

The next step toward the introduction of absolute time with its concomitant notion of simultaneity was the total abolishment of the "intrinsic duration" in favor of the exclusive survival of the "imaginary succession," which is unaffected by physical processes. This stage is displayed in the writings of Pierre Gassendi who at the end of his intellectual career professed an absolutist theory of time. "Time," he wrote, "is not something dependent upon motion or posterior to it, but is merely indicated as something which is subject to some measurement. For otherwise it would be impossible to know how much time we spend doing something or not doing it. Therefore we raise our eyes to the celestial motion and say that time fled in proportion to its quantity. And since the observation of this motion was commonly found to be difficult, the movements of readily familiar objects such as water, sand, wheels, or pins of sundials were adapted to the celestial motion so that since it was easy to glance at them, it was possible to take count of them and of

[1] F. Suarez, *Metaphysicarum Disputationum Tomi duo* (Maguatiae [Mainz]: 1600), Disputatio 30 ("De rerum duratione"), p. 955. Cf. M. Čapek, "The conflict between the absolutist and the relational theory before Newton," *Journal of the History of Ideas* **4,** 595–608 (1987).

[2] W. Pagel, "J. B. van Helmont: De Tempore and biological time," *Osiris* **8,** 346–417 (1945).

time. This is the reason why I said a short while ago that the heavens are a sort of general clock ("quoddam generale horologium") for they are inasmuch as all our clocks imitate them as closely as possible and are called upon to help us when we cannot see them." Although, as just mentioned, the heavens are a sort of general clock they are not time itself; for "time would continue its regular flow even if the heavens were not moving."[3] Gassendi criticized the proponents of the twofold theory of time, for "clearly, there is no other time except that which is called imaginary and which is necessary and which, as they admit, continued to flow alone when the heavens were standing still as long as Joshua was fighting the kings of Amorites."[4] Gassendi's contention that only one kind of time exists eliminates the duplicity of two kinds of simultaneity. This follows for him also as a logical consequence of the ubiquity of a moment of time, which he refers to in his comparison between space (or "place") and time by stating: "As space as a whole is unlimited, so time as a whole has neither the beginning nor the end, and, as any moment of time is the same in all places, so any place is in all times."[5]

Note that in the philosophies of time, like Gassendi's, that prepared the ground for the notion of absolute time as professed later by Isaac Barrow and Isaac Newton, the concept of distant simultaneity played only a minor role. It did so because it was rarely needed in practice. It became indispensable only with the widespread use of mechanical clocks.

As pointed out earlier, clocks had been used in antiquity, but almost exclusively to measure the length of time intervals. Nor had they been sufficiently accurate to serve as means for synchronizations. Thus, for example, the best device that antiquity produced for the measurement of time, the famous water clock of Ctesibius, who lived in the second century B.C. in Alexandria, was accurate only to rather large fractions of an hour. Still in the Middle Ages almost all clocks, usually employed by monks in monasteries to indicate at what time to pray, were water driven and rather inaccurate.[6] To produce an accurate clock some device was needed that provided a constant periodic motion to which a clock could be geared. Such a device was dis-

[3] P. Gassendi, *Syntagma Philosophicum* (Physica, liber 2, cap. 7) in *Opera Omnia*, vol. 1 (Lugduni [Lyon]: Anisson et Devenet, 1658; reprinted Stuttgart: Frommann, 1964), vol. 1, p. 198.

[4] Ibid., p. 199.

[5] "Quodlibet temporis momentum idem est in omnibus locis." Loc. cit.

[6] For more details on the history of ancient and medieval timekeeping see K. Lippincott, *The Story of Time* (London: Merrell Holberton, 2000), chapter 2.

covered by Galileo Galilei in 1581, when he attended service at the Cathedral of Pisa and watched a swinging chandelier. Using his pulsebeat, he found what we now call the law of the simple pendulum, namely that the time of the oscillation of a pendulum is (practically) independent of its amplitude. Because of this discovery and the fact that in his old age he made use of this "isocronismo"[7] for the design of a clock, he may be regarded as the inventor of the pendulum clock.

Only in 1657, more than a decade after Galileo's death, did Christiaan Huygens put Galileo's discovery into practice by devising an attachment at the fulcrum of the pendulum to make it swing in the proper arc and by coupling the mechanism with falling weights to counteract its loss of energy by friction and air resistance. About twenty years later Huygens proposed the use of a spiral spring to be added to the conventional balance, a proposal that led to the production of the first portable clock.

It was mainly due to the invention of portable clocks that the notions of distant simultaneity and synchronization gained practicability. The transport of a portable timekeeper made it possible to synchronize, for example, the turret clocks in a town or in different towns of a country. The concept of simultaneity played a significant role especially in cartography. With the rapid increase of maritime commerce between Europe and the Far East and the newly discovered American continent, cartography gained great importance. In particular, it proved useful to determine the geographical latitude and longitude of a far-away place or of the location of a ship at sea. At any place the determination of the geographical latitude, being equal to the altitude of the celestial pole, could easily be carried out by measuring, for example, the altitude of the Pole star. The determination of the geographical longitude, on the other hand, was a much more complicated task. The first to propose a convenient method by making use of the concept of distant simultaneity was the Flemish physician, mathematician, and founder of the Belgian school of cartography, Reinerus Gemma, also named Frisius because he was born in 1508 in Dokkum, East Friesland. In chapter 8, "De novo modo inveniendi longitudinem," of his book *De Usu Globi*,[8] published in Antwerp, Gemma sug-

[7] See chapter 1, note 18.

[8] R. Gemma, *De Usu Globi* (Antwerp: Gregorium et Bontium, 1530), chapter 8. For further details see F. van Ortroy, "Bio-bibliographie de Gemma Frisius fondateur de l'école belge de géographie," *Memoires de l'Académie Royale de Belgique (classe des lettres, deuxième série)* **11** (1911); and A. Pogo, "Gemma Frisius," *Isis* **22,** 469–485 (1935).

gested the following method to determine the geographical longitude of a ship at sea: synchronize a reliable portable clock with a master clock stationary at the point of departure, say with the clock at the Cathedral of Antwerp; take the synchronized clock on board the ship and when at high sea compare its reading with the ship's local apparent time, say its noon time observed from the sun's position; because a difference of one hour corresponds to 15 degrees of longitude, the time difference obtained by this comparison divided by 15 gives the longitude of the ship's location relative to the point of departure. This method is essentially still used today, the difference being only that instead of establishing distant simultaneity by means of clock transport, it is established by means of electromagnetic signals, the time signals broadcast by radio, and the prime meridian, by international convention, is that of Greenwich. Hence, the longitude is simply given by the difference between local time and Greenwich time *at the same instant.*

Note that, in turn, the quest for obtaining precise simultaneity contributed to the technical improvement of clocks. Still in the seventeenth century portable clocks were not sufficiently accurate to provide precise measurements of longitudes. The governments of Spain, France, and the Netherlands offered attractive awards for their successful construction. The first reliable marine chronometer was constructed in 1773 by the horologist John Harrison, who had dedicated his whole life to this project. His work greatly influenced the subsequent production of all precision watches.

The search for the improvement of longitude determinations is historically related also to the third of the above-mentioned innovations, the discovery of the finite velocity of light. It is well known that in 1610 Galileo discovered the existence of four of the moons of Jupiter with his newly invented telescope and soon realized that their frequent eclipses, providing a reliable method of determining standard time, could be used for the determination of terrestrial longitude. They became, therefore, an important subject of astronomical observations. In 1665 the astronomer Giovanni Domenico Cassini published a table of the motions of these satellites which made it possible to predict the precise times of the beginning and end of each of their eclipses by Jupiter as viewed from the Earth. Ole Roemer discovered in 1676 that these eclipses start about seven minutes earlier than predicted, when the Earth is between the sun and Jupiter, and about seven minutes later than predicted, when the Earth is beyond the sun. He explained these observations as an indication that the velocity of light or, as he called it, "the retardment of light," is a finite quantity.[9] When Roemer's conclusion gained general acceptance in the early eighteenth century, mainly because of James Bradley's 1728 discovery of stellar aberration, it became clear that events at different distances from the observer, seen as occurring or having occurred simultaneously, did not really occur simultaneously. In other words, Roemer's discovery put an end to what we have called "the visual simultaneity thesis."

Before the abandonment of this simultaneity thesis, when people were looking at the nocturnal sky and saw, for example, the seven bright stars of the constellation of the Great Bear (Ursa Major), they thought that these stars were precisely where they were seen, even if they were not equidistant from the observer. Today, knowing the distance and the proper velocity of each of these stars and the velocity of light, we can easily calculate the real simultaneous positions they occupy at the moment of observation. The result shows that reality differs considerably from its appearance. The simultaneous picture of the world, as seen by our eyes, is an illusion. But, of course, if distances and velocities are small, as they are in our ordinary life, such a difference between illusion and reality becomes negligible. We owe this far-reaching philosophical insight to Roemer's discovery of the finite velocity of light.

Note, however, that the idea of a finite velocity of light had been conceived long before Roemer, but merely as a speculation. Empedocles of Agrigentum had previously claimed that light needs time "to travel."[10] Other proponents of this idea were the Islamic scholars Alhazen and Avicenna of the early eleventh century and, a few decades before Roemer, the English philosopher Francis Bacon.

An early reference to the relation between the instant of seeing an object and its distance from the observer can be found in St. Augustine's *Epistola ad Deogratias*, written in the early fifth century. Commenting on the Biblical phrase "in ictu oculi" (in the twinkle of the eye)[11] and adhering still to the Greek (Euclid, Ptolemy) extramission theory of vision, according to which radiation is emitted from the observer's eye to the object that is seen, Augustine argued for the instantaneity of vision or light on the grounds that oth-

[9] O. Roemer, "A Demonstration Concerning the Motion of Light," *Philosophical Transactions* **XII,** 893–894 (1677). For details see I. B. Cohen, *Roemer and the First Determination of the Velocity of Light* (New York: Burndy Library, 1944). That, contrary to many reports, "Roemer did not give explicitly any value for the velocity of light" was stated by Andrzej Wróblewski in his paper "De Mora Luminis," *American Journal of Physics* **53,** 620–639 (1985).

[10] Cf. Aristotle, *De Anima*, 418 b 20–22.

[11] *1 Corinthians* XV, 52.

erwise "nearer objects would be seen sooner than distant objects."[12] A thousand years later the Italian philosopher and physicist Blasius of Parma, who adopted the introvision theory (Alhazen, Roger Bacon), according to which the visible object is the source of radiation that enters the eye, argued against Augustine's reasoning as follows. The simultaneous visibility of distant and near objects does not necessarily imply the instantaneity of light propagation because "before you direct your eye toward those objects, their species (rays) were diffused to the eye." If, however, Blasius continued, such objects "were suddenly and instantaneously presented to your eye at unequal distances at the instant when you directed your eye toward them, then you would see a castle long before the sun."[13] As the context clearly shows, when Blasius wrote that two objects were "suddenly and instantaneously" (subito et in instanti) presented to your eyes he meant that they were "simultaneously" presented to your eyes. This was simply because, at that time, the term "simultaneously" was not yet in common usage.

Paraphrasing "simultaneity" by other terms also was common at later times. An important example is the philosophical school called "Occasionalism." Its name, derived from the Latin word "occasio" (time of happening), indicates an intimate relation with the notion of simultaneity. Although its history can be traced to ancient and medieval philosophy,[14] it usually refers to those philosophers of the seventeenth century, who like Arnold Geulincx and Nicholas Malebranche tried to resolve a problem raised by the Cartesian bifurcation of reality into *res cognitans* (mind) and *res extensa* (body). The difficulty that Descartes's dualist metaphysics left unresolved was the question of how an unextended substance, like the mind, can exercise an influence on an extended substance, like the body, and, conversely, how can an extended substance affect an unextended substance. Geulincx proposed to resolve this problem by assuming the existence of a divine interference or, as he phrased it, a "concourse of God" in every single case of an apparent interaction between mind and body. God intervenes *on the occasion* of every volition and excites in our bodies the movement that our mind or soul of it-

[12] A. Augustinus, "Epistola 102 ad Deogratias" in J. P. Migne, *Patrologia Latina* (Paris: Montrouge, 1845), vol. 33, col. 372, question 1, n. 5.

[13] G. F. Vescovini (ed.), "Le questioni di 'Perspectiva' di Biagio Pelacani da Parma," *Rinascimento* **1,** 163–243 (1961). Quotation on p. 221.

[14] L. Stein, "Antike und mittelalterliche Vorläufer des Occasionalismus," *Archiv für Geschichte der Philosophie* **2,** 193–245 (1889); M. Fakhry, *Islamic Occasionalism* (London: Allen & Unwin, 1958).

self would be unable to perform. As Geulincx expressed it, "since a person does not do what he does not know how to do it, mind does not act on body, nor body on mind." Human volition is only the "causa occasionalis" and not the "causa efficiens" of our actions.[15]

Obviously, all such expressions like "on the occasion of" implied the concept of simultaneity. In fact, Geulincx himself, to illustrate the simultaneity between events in the realm of *res extensa* and events in the realm of *res cogitans*, repeatedly used the metaphor of synchronized clocks.[16] Gottfried Wilhelm Leibniz, independently of Geulincx,[17] applied the same metaphor in his theory of pre-established harmony according to which body and soul act simultaneously, but without interacting with each other, like two synchronized clocks.[18] The main difference between the two applications of this metaphor is this: according to Geulincx God establishes the simultaneity between events in those two realms at every moment separately like a watchmaker who constantly regulates one clock by another; according to Leibniz the synchronism of the two clocks has been pre-established by God simultaneously with his creation of the world.

The psychophysical synchronism as a solution of the mind-body problem has subsequently been discussed by many philosophers, but, in general, without emphasizing its relation to the notion of simultaneity. A noteworthy exception is Arthur Schopenhauer who in his *World as Will and Idea* wrote of man's will that "every true act of his will is also at once and inevitably a movement of his body; he cannot actually will the act without *at the same time* being aware that it appears as a movement of the body."[19]

Just the notion of simultaneity was used by the occasionalists to solve a philosophical problem, it was employed by scientists for the solution of an important astronomical problem, the determination of the interplanetary distances of the solar system. According to Johann Kepler's "third law," published 1619 in *Harmonice Mundi*, the square of the period of revolution of a

[15] A. Geulincx, *Annotata ad Ethicam* in *Opera Philosophica* (Den Haag: Nijhoff, 1893), vol. 3, pp. 205–207.

[16] Ibid., pp. 124, 140, 155.

[17] Cf. p. 237 in L. Stein's article mentioned in note 14.

[18] G. W. Leibniz, "Système nouveau de la nature et de la communication des substances, aussi bien que de l'union qu'il y a entre l'âme et le corps," *Journal des Savants* (June-July) 1695.

[19] A. Schopenhauer, *Die Welt als Wille und Vorstellung* (Leipzig: Brockhaus, 1819, 1922), p. 120; *The World as Will and Representation* (New York: Dover, 1969), vol. 1, p. 100 (emphasis added).

planet is proportional to the cube of its distance from the sun. Because the distance of the sun from the Earth was well known and the periods of revolution of the planets were easily observable, it sufficed to measure the distance of only one of the planets from the Earth in absolute units, to determine the distances of all planets. To this end Cassini proposed to measure the distance of the planet Mars from the Earth by the parallax method, that is, by measuring the angular difference of the directions in which the planet is seen from two locations of a known distance apart. The two locations of these historical observations in 1671 were Cassini's observatory in Paris and Cayenne in French Guiana, where Cassini's colleague Jean Richer was leading an astronomical expedition. Because of the Earth's own motion the parallax method could yield a reliable result only if the two observations could be made simultaneously. But, as mentioned earlier, before 1773 clocks for such a long-distance synchronization were not yet available. To resolve this simultaneity problem Cassini made use of his table of the motions of Jupiter's satellites (recorded in 1665) and thus succeeded in establishing the first precise determination of interplanetary distances in the history of astronomy.

CHAPTER FIVE

The Concept of Simultaneity in Classical Physics

The prehistory of the classical or Newtonian conception of time, and with it of simultaneity, did not start with Gassendi's revision of the Aristotelian theory of time. It can at least be traced back to the kinematical theories of the so-called Calculatores of the Merton School in Oxford, which flourished in the early fourteenth century. One may even claim that, strictly speaking, certain Greek kinematicians, like Autolycus of Pitane, who flourished about 310 B.C., or other ancient authors of purely kinematical treatises, who used the notion of time as logically prior to motion, anticipated its classical conception. In any case, by presenting time as the independent variable in their graphical representations of uniform or accelerated motions, as, for example, in the famous Merton theorem of uniform acceleration (or mean-speed theorem), the members of the Merton School, like Thomas Bradwardine, William Heytesbury, Richard Swineshead, and John Dumbleton, prepared the ground for Galileo's kinematics. In *Discorsi e Dimostrazioni Matematiche intorno à due Nuove Scienze* (1638), especially in the chapter entitled "Third Day" containing his kinematics, Galileo made frequent, though mostly implicit, use of the concept of simultaneity without defining it. He obviously assumed that the reader knows its meaning from his everyday language. The *Discorsi,* in turn, served as the source of Isaac Newton's kinematical theorems in his writings on mechanics.

Newton's conception of absolute time, however, was primarily influenced by Isaac Barrow's *Lectiones Geometricae* (1669), which he reportedly revised and edited in the same year in which Barrow resigned his chair as Lucasian professor at the University of Cambridge to Newton. Because Newton's conception of simultaneity is intricately connected with the notion of absolute time, and because his notion of absolute time is based on Barrow's philosophy of time, it is appropriate to discuss, in brief, Barrow's theory of absolute time.

Barrow's philosophy of time appears to have been strongly influenced by the philosophy of space of his colleague, the Cambridge Platonist Henry More, who was fifteen years his senior. In particular, More's conception of space as the omnipresence of God must have greatly appealed to Barrow, who had been a theologian before his appointment as professor of geometry in 1662. In addition, More's anti-Cartesian separation of space from matter and his argument for the reality of space from its measurability, as presented in his *Antidote against Atheism* (1653), must have greatly appealed to Barrow, who as a mathematician declared repeatedly that the object of science is *quantity*. It is not surprising, therefore, that just as More liberated space from its Cartesian bondage with matter, so Barrow disjoined time from its Aristotelian conjunction with motion. In *Lectiones Geometricae* Barrow wrote, repeating More's argument of the eternity of space, though without mentioning More's name: "Just as space existed before the world was created and even now there exists an infinite space beyond the world (with which God coexists) . . . so time exists before the world and simultaneously with the world (*prius mundo et simul cum mundo*)." He then asked whether the notion of time implies the concept of motion and answered: "Not at all as far as its absolute, intrinsic nature is concerned. . . . Whether things run or stand still, whether we sleep or wake, time flows in its even tenor (*aequo tenore tempus labitur*). Even if all the stars would have remained at the places where they had been created, nothing would have been lost to time (*nihil inde quicquam tempori decessisset*). The temporal relations of earlier, afterwards, and simultaneity, even in that tranquil state, would have had their proper existence (*prius, posterius, simul etiam in illo transquillo statu fuisset in se*)."[1] This passage, in which Barrow

[1] I. Barrow, *Lectiones Opticae & Geometricae* (London: Scott, 1674). *Lectiones Geometricae* (reprinted: Hildesheim: Georg Olms, 1976), p. 3.

speaks of the "absolute . . . nature of time," which "flows in its even tenor," has historical importance. It is probably the first attribution of the predicate "absolute" to the concept of time and reverberates in Newton's Scholium to his *Mathematical Principles of Natural Philosophy*, which reads: "Absolute, true, and mathematical time, of itself, and from its own nature, flows equably without relation to anything external."[2] Whether Newton's use of the term "absolute time" resulted from his study of Barrow's *Lectiones* or as a correlate to his notion of "absolute motion" and whether his often criticized definition of "absolute time" is physically meaningful are issues beyond our present interest. Nor will we discuss why the theory of relativity confuted the concept of "absolute time." Note, however, that it would have been logical, before the advent of the theory of relativity, to call the notion of simultaneity of classical physics "absolute simultaneity" because it denotes the temporal coincidence of events in absolute time. But such a term seems never to have been used in the era of classical physics.

Although the notions of time and simultaneity as defined and used in the theory of relativity will be discussed later, two issues in Barrow's philosophy of time deserve to be mentioned in the present context and to be confronted with modern physics.

The first issue involves the notion of simultaneity, which Barrow regarded, of course, as a temporal relation of coincidence in absolute time. The passage in question appears in the beginning of his treatment of time and reads: "Time absolutely is quantity, admitting in some manner the chief affections of quantity, equality, inequality, and proportion; nor do I believe there is anyone but allows that those things existed equal times, which rose and perished simultaneously."[3] In other words, processes, which begin simultaneously and end simultaneously, last for equal times. Although valid in classical physics, this statement loses its validity in relativistic physics because of the dilation of time, as illustrated by the well-known example of the fast-moving space traveler in the so-called twin paradox. The second issue refers to Barrow's remark that motion, although not an integral part of the absolute and intrinsic nature of time, must be used to measure the flow of time. Noting that for this purpose the motion of the stars, especially of the sun or

[2] "Tempus absolutum verum & Mathematicum, in se & natura sua absq. relatione ad externum quodvis, aequabiliter fluit." *Philosophiae Naturalis Principia Mathematica* (London: Streater, 1687), p. 5.

[3] I. Barrow, op. cit., p. 2.

moon, in general, are used, he asks: "How can we know that the sun moves with an equal motion, that the time of one day or year, for example, is exactly equal to that of another? . . . No one can pretend to assert as a certainty, that the life of Methusalem, who lived almost a thousand years, was really longer than that of a man, who dies before he arrives at a hundred."[4]

Barrow raised here, seemingly for the first time, the important problem of whether it is possible to test or verify the equality of the durations of temporally separated time intervals. His answer is positive, for he argues that the equality of the durations of celestrial motions can be tested by comparing them with the periodic motions of artificial timekeepers, such as clepsydras or sandglasses, because "the water or sand contained in them remains entirely the same as to quantity, figure, and force of descending, and the vessel that contains them, as likewise the little hole they run through, do not undergo any kind of mutation. . . . There is no reason whatever for us not to allow the times of every running out of the water or sand to be equal."[5] From our modern point of view, Barrow's appeal to what Leibniz later called "the principle of sufficient reason" is questionable, because the "force of descending," for example, or as we call it gravitation, may itself be a function of time if cosmological time intervals are involved. In fact, modern philosophers of science reject Barrow's argumentation. For example, Hans Reichenbach asked: "How can we test this assumption [concerning the time-independence of the behavior of certain physical mechanisms]?" And he answers: "There is basically no means to compare two successive periods of a clock, just as there is no means to compare two measuring rods when one lies behind the other."[6]

The conceptually similar problem of whether it is possible to test or verify whether two (or more) spatially separated events are simultaneous was never discussed by Barrow, although he explicitly stated that "there is a great affinity and analogy between space and time." This affinity, according to Barrow, found its most important expression in the analogy between time and a geometrical line. As a straight line can be thought of either as an aggregate of points or as the trace of a point in motion, so time may be regarded either as composed of instants or as the continuous flow of one instant. Barrow's insistence on the one-dimensionality of time and his conclusion "we

[4] Ibid., p. 5.

[5] Ibid., p. 5.

[6] H. Reichenbach, *Philosophie der Raum-Zeit-Lehre* (Berlin: W. de Gruyter, 1928; Braunschweig: Vieweg, 1977); *The Philosophy of Space and Time* (New York: Dover Publications, 1958), p. 116.

shall therefore always represent time by a right line"[7] have since held decisive importance for the study of time and simultaneity.

When discussing the notion of simultaneity in Newtonian physics we have to distinguish between Isaac Newton's conception of this notion, as far as it is retrievable from his own writings, and the meaning and role of this notion in Newtonian or classical physics. Despite their agreement from the physical point of view, these two notions differ significantly in their conceptual foundations and theoretical justifications.

Turning first to Newton's own ideas about simultaneity, note that Newton never analyzed this notion or proposed a criterion for events to be called simultaneous, although he used the term "simul" quite frequently, as for example in his description of astronomical observations with his newly invented sextant.[8]

The most detailed reference to the notion of simultaneity Newton ever made was a statement in *De Gravitatione et Aequipondio Fluridorum*, written between 1664 and 1668, in which he wrote: "There is a very different relationship between . . . space and duration. For we do not ascribe various durations to the different parts of space, but say that all endure together. The moment of duration is the same at Rome and at London, on the Earth and on the stars, and throughout the heavens. And just as we understand any moment of duration to be diffused throughout all spaces, according to its kind, without any thought of its parts, so it is no more contradictory that Mind also, according to its kind, can be diffused through space without any thought of its parts."[9] The ubiquity of a moment of time throughout space, which underlies Newton's conception of simultaneity, had been mentioned previously by Gassendi, who declared in *Syntagma Philosophicum* that "any instant of time is the same at all places."[10] Newton must have been acquainted with Gassendi's philosophy of time because he had read Walter Charleton's compendium of Gassendi. Newton's comparison of the "diffusion" of a moment of time or duration, as he calls it, throughout space with the omnipresence of "Mens," however, suggests that theological considerations may have motivated Newton's conception of simultaneity at that time. In fact,

[7] "Tempus itaque per rectam lineam semper designabimus." I. Barrow, op. cit., p. 7.

[8] A. R. Hall and M. B. Hall (eds.), *Unpublished Scientific Papers of Isaac Newton* (Cambridge: Cambridge University Press, 1962), p. 388.

[9] Ibid., p. 104.

[10] P. Gassendi, op. cit. (reference IV-3).

when he made this statement he also wrote "God is everywhere, created minds are somewhere, and body is in the space it occupies."[11]

It may be claimed that for Newton the notion of simultaneity, and of distant simultaneity in particular, did not pose any problem, for once he had conceived the existence of absolute time, "which flows uniformly without reference to anything external," the term "simultaneity" could be interpreted as meaning merely "at the same absolute time." This still does not resolve the problem, because to acknowledge that such a time can be conceived does not imply that it really exists and can be measured or be applied operationally. Newton himself admitted that "it is possible that there is no uniform motion by which time may have an exact measure."[12] If Newton nevertheless proclaimed the existence of absolute time, and with it by implication the existence of absolute simultaneity, he did so for two reasons. First, he was convinced that he had proved the existence of absolute motion, for instance, by his famous rotating-bucket experiment. But clearly, he argued, the existence of absolute motion implies the existence of absolute space and absolute time. The question of whether Newton's argumentation involves a *petitio principii* will be discussed in a later chapter.[13] Second, Newton's belief in the existence of absolute time is religiously motivated. Although the following statements by Newton were printed only in the second (1713) and third (1727) edition of the *Principia*, we know that they reflect Newton's lifelong religious convictions. Referring to God, Newton wrote: "He is eternal and infinite, omnipresent and omniscient. . . . He endures for ever, and is everywhere present, and, by existing always and everywhere, he constitutes duration and space. . . . Since each and every particle of space is *always* and each and every indivisible moment of duration is *everywhere*, certainly the maker and lord of all things will not be *never* or *nowhere* . . . God is one and the same God always and everywhere."[14] This conflation of spatial and temporal notions with divine attributes allows us to infer that, for Newton, God's omnipresence was also the ground and justification of the notion of distant simultaneity. If we ask ourselves why it is that human beings have no direct and immediate perception of the simultaneity of spatially separated events, we an-

[11] A. R. Hall and M. B. Hall, op. cit., p. 103.

[12] Note 2, p. 7.

[13] See also R. Rynasiewicz, "By their properties, causes and effects: Newton's scholium on time, space, place and motion," *Studies in History and Philosophy of Science* **26,** 133–153, 295–321 (1995).

[14] I. Newton, *The Principia* (edited by I. B. Cohen) (Berkeley, California: University of California Press, 1999), p. 941; *Sir Isaac Newton's Mathematical Principles of Natural Philosophy and his System of the World* (edited by F. Cajori) (Berkeley, California: University of California Press, 1947), p. 545.

swer it is because no human observer can be present at these events when they occur. A human observer can experience directly only what occurs in his immediate vicinity. According to Newton, this spatial restriction does not apply to God. True, Newton did not apply the term "observer"; but he used the term "Sensorium" or "Sensory" when he wrote: "Does it not appear from the Phaenomena that there is a Being incorporeal, living, intelligent, omnipresent, who in infinite Space, as it were in his Sensory, sees the things themselves intimately, and thoroughly perceives them, and comprehends them wholly by their immediate presence to himself."[15] In short, God's omnipresence warrants the actual existence of an absolute worldwide distant simultaneity. We call it an *absolute* distant simultaneity because Newton undoubtedly regarded divine ubiquity as existing in *absolute* time and not in what he called *"relative, apparent, and common time"* (tempus "relativum apparens & vulgare"). Newton would not have used the adjective *vulgaris,* which already at his time was often used in the sense of "profane," as an attribute of God's omnipresence. In this context, it is important to remember that Newton regarded religious arguments as an integral or at least congenerous part of his physics. That he had theological considerations in mind, in particular, when writing his *Principia* can be seen from his letter of 10 December 1692, to Richard Bentley, a theologian and member of the Royal Society, in which he confessed: "When I wrote my treatise about our system, I had an eye upon such principles as might work with considering men for the belief of a Deity; and nothing can rejoice me more than to find it useful for this purpose."[16] Because of its theological connotations, namely of being associated with God's omnipresence, Newton's conception of simultaneity, as stated above, differs from the notion of simultaneity of classical physics after Newton, in which theological considerations were no longer regarded as constitutive arguments in its theoretical systems.

Another distinction between Newton's notion of absolute distant simultaneity and the concept of distant simultaneity in post-Newtonian classical physics refers to the measurability of these notions. Although rarely, if ever,

[15] I. Newton, *Opticks* (4th edition, 1730) (London: 1730; New York: Dover Publications, 1952), p. 370. See also the first reply of Samuel Clarke, the spokesman of Newton, to Leibniz's first letter in *A Collection of Papers which passed between the late learned Mr. Leibnitz and Dr. Clarke* (London: Knapton, 1717), in H. G. Alexander (ed.), *The Leibniz-Clarke Correspondence* (New York: Philosophical Library, 1956), p. 13, where it is stated that "God sees all things, by his immediate presence to them; he being actually present to the things themselves, to all things in the universe."

[16] H. W. Turnbull (ed.), *The Correspondence of Isaac Newton* (Cambridge: Cambridge University Press, 1961), vol. 3, p. 233.

explicitly discussing this concept, classical physicists did not doubt that the concept of distant simultaneity is an operationally testable notion, for example, by the employment of synchronized clocks. In contrast, Newton's notion of absolute distant simultaneity, that is, simultaneity in *absolute* time, cannot be operationally testable, for absolute time, according to Newton, flows uniformly "without reference to anything external."[17] If absolute distant simultaneity were operationally testable, it would be a relation in absolute time that would be accessible to empirical observation in contradiction to the unrelatedness of absolute time to "anything external."

Three years after the publication of Newton's *Principia* John Locke published his magnum opus *An Essay concerning Human Understanding*,[18] in which he called the *Principia* a "never-enough-to-be-admired book."[19] Nevertheless, Locke did not accept Newton's conceptions of time and, by implication, of simultaneity. To understand Locke's theory of time we must recall the foundations of Locke's philosophical system. As every student of philosophy knows, Locke, in opposition to Plato and Descartes, denied the existence of innate ideas and claimed that all ideas come either from sensation or from reflection, which "might properly enough [also] be called 'internal sense.' "[20] He distinguished between simple ideas about external objects, which have their origin in sensation, and complex ideas, which are formed by the combination of simple ideas. Duration and time, according to Locke, are complex ideas that arise in our mind when we reflect on the appearance of several ideas one after another, producing the idea of succession. Clearly, such an approach to the problem of time did not admit the notions of absolute time and absolute distant simultaneity, but this does not imply that Locke could not use the notion of simultaneity with reference to duration or time as conceived by him. True, nowhere in his *Essay* did Locke define the concept of simultaneity nor did he make even explicit use of this term. He frequently applied it implicitly, however, when he used such terms as "whilst" or "co-existing." A typical example is his statement: "Whilst we are thinking, or whilst we receive successively several ideas in our minds, we know that we do exist; and so we call the existence or the continuation of the existence of our-

[17] "absq. relatione ad externum quodvis." See note 2.

[18] J. Locke, *An Essay concerning Humane* [sic] *Understanding* (London: Holt, 1690). *An Essay concerning Human Understanding* (second edition) (London: Churchil and Manship, 1694; 32th edition, 1860) (new edition, London: G. Routledge, 1894).

[19] Op. cit. (new edition), book 4, chapter 4, section 11, p. 511.

[20] Ibid., book 2, chapter 1, section 4, p. 60.

selves commensurate to the succession of any ideas in our minds, the duration of ourselves, or any such thing co-existing with our thinking."[21] Locke, it seems, was not aware that terms like "whilst" (the British equivalent of the conjunctive "while") or "co-existing" implicitly contain the notion of simultaneity. Because Locke applied these terms in his definitions of duration and of time, which he defined as the measure of duration,[22] it may be said that he unwittingly assigned to the notion of simultaneity logical priority over the notion of time, a procedure by which he anticipated Leibniz and Einstein. With respect to Locke's definition of time as measured duration we may also say that he anticipated Reichenbach's previously quoted statement,[23] because in his chapter on "duration" Locke emphatically declared that "no two parts of duration can be certainly known to be equal" and although "duration in itself is to be considered as going on in one constant, equal, uniform course . . . none of the measures of it which we make use of can be known to do so: nor can we be assured that their assigned parts are equal in duration one to another; for two successive lengths of duration, however measured, can never be demonstrated to be equal."[24] One may even say that Locke anticipated Menyhert (Melchior) Palagyi's psychological or Hermann Minkowski's physical concept of space–time. At the end of his discussion on "duration and expansion [space] considered together" Locke made the following statement: "Expansion and duration do mutually embrace and comprehend each other; every part of space being in every part of duration, and every part of duration in every part of expansion. Such a combination of two distinct ideas is, I suppose, scarce to be found in all that great variety we do or conceive, and may afford matter to farther speculation."[25] That such a "speculation" eventually became a well-founded theory will be apparent when we discuss the notion of simultaneity in the theory of relativity.

Locke's critique of Newton's absolute time was shared by George Berkeley, but not only on purely epistemological grounds. Although Berkeley's epistemology, epitomized by his famous statement "esse est percipi" (to exist is to be perceived), would have sufficed to reject the existence (esse) of Newton's conception of absolute time, which, as Newton himself admitted, cannot be perceived (percipi), Berkeley attacked Newton's absolute time by refuting the

[21] Ibid., book 2, chapter 14, section 3, p. 122.

[22] "Time is duration set out by measures." Ibid., book 2, chapter 14, section 17, p. 126.

[23] See note 6.

[24] J. Locke, op. cit., book 2, chapter 14, section 21, p. 129.

[25] Ibid., book 2, chapter 15, section 12, p. 140.

very argument that, as mentioned earlier, led Newton to this notion, namely Newton's interpretation of his rotating-bucket experiment. In *De Motu*,[26] published in 1721, Berkeley argued, similar to Ernst Mach in 1883, that Newton's inference of the existence of absolute motion is based on the implicit and unproven assumption that the experiment would yield the same result if performed in empty space.[27] Previously, in *Treatise concerning the Principles of Human Knowledge*, published in 1710, Berkeley declared: "It does not appear to me that there can be any motion other than *relative*; so to conceive motion there must be at least conceived two bodies, whereof the distance or position in regard to each other is varied. Hence, if there was one only body in being it could not possibly be moved."[28] With respect to time he wrote: "For my own part, when ever I attempt to frame a simple idea of *time*, abstracted from the succession of ideas in my mind, which flows uniformly and is participated by all beings, I am lost and embrangled in inextricable difficulties."[29] Time, according to Berkeley, consists merely of passing sensations in the mind of percipient beings, is entirely relative to them, and thus may differ from person to person. Berkeley denied therefore the existence not only of absolute time but also of a common time, a time "participated by all beings." In fact, in his *Commonplace Book*, written in 1705 at Trinity College at the age of twenty, he expressed these ideas by noting that "time is sensation; therefore onely in ye mind and the same τὸ νῦν not common to all intelligences."[30]

By rejecting absolute time Berkeley rejected also the concept of absolute simultaneity. And if time, according to Berkeley, is only an "abstraction from the succession of ideas in our minds" and the "now" (τὸ νῦν) "is not common to all intelligences," we must conclude that the notion of simultaneity, also with respect to Berkeley's "time," is for him an unacceptable concept. However, Berkeley was not only a philosopher but also a theologian and, in the last two decades of his life, the Anglican Bishop of Cloyne. He thought it imperative, therefore, to revise his juvenile interpretation of the "τὸ νῦν" as "not common to all in-

[26] G. Berkeley, *De Motu, sive de motus principio et natura et de causa communicationis mutuum* in A. G. Fraser (ed.), *The Works of George Berkeley* (Oxford: Clarendon Press, 1891), pp. 501–527.

[27] For details see M. Jammer, *Concepts of Space—The History of Theories of Space in Physics* (Cambridge, Massachusetts: Harvard University Press, 1954, 1969; New York: Dover Publications, 1993), chapter 4.

[28] G. Berkeley, *A Treatise concerning the Principles of Human Knowledge* (Chicago: Open Court, 1913), section 112, p. 97.

[29] Ibid., section 98, p. 87.

[30] G. Berkeley, *Commonplace Book, mathematical, ethical, physical, and metaphysical*, in A. C. Fraser (ed.), *The Works of George Berkeley* (Oxford: Clarendon Press, 1891), vol. 1, pp. 1–7.

telligences," which therefore excludes the existence of simultaneity. In a letter to Samuel Johnson, the first president of King's College, now called Columbia University in New York, he wrote in 1730: "By the τὸ νῦν I suppose to be implied that all things, past and to come, are actually present to the mind of God, and that there is in Him no change, variation, or succession."[31] If we recall that for Berkeley "time" was nothing but "the succession of ideas in our minds," it follows that for God, in whose mind, as Berkeley expressively stated, there is no succession, time as defined by Berkeley does not exist. Moreover, because past, present, and future are "actually present," in God's mind, He sees them as what human beings would call "simultaneously." That kind of simultaneity exists when we look at the representation of the past, present, and future in a diagram of what is now called the "block universe," a term that was coined in 1890 by William James and will be discussed subsequently.

The most famous and eloquent opponent of Newton's theory of absolute or, in modern parlance, substantival space and time was Gottfried Wilhelm Leibniz. The debate between Leibniz and Newton on the nature of space and time is well-known thanks to the publication of *The Leibniz–Clarke Correspondence.*[32] What is not so well-known is that Leibniz's arguments against Newton's theory of absolute time involved Leibniz's notion of simultaneity. According to Newton's theory of absolute time, moments of time exist in their own right, whereas according to Leibniz's relational theory of time, they are classes of events defined by means of the concept of simultaneity. To avoid a vicious circle Leibniz could not employ the term "simultaneity" in its usual sense of denoting simply the relation of occurring "at the same time," because his definition of "time" presupposed the notion of simultaneity. What, then, was Leibniz's interpretation of the notion of simultaneity within the context of his relational theory of time?

One of his earliest statements of his relational conception of space and time is a remark, made in 1695, that "extension or space and surfaces . . . are but sets of relations or order of coexistence."[33] In 1700 he wrote to Burcher de Volder of Leyden that "extension seems to me to be but a continuous or-

[31] For Berkeley's letter to Johnson, dated 24 March 1730, see A. A. Luce and T. E. Jessop (eds.), *The Works of George Berkeley Bishop of Cloyne* (London: Nelson, 1948), vol. 2, p. 293.

[32] See note 15.

[33] G. W. Leibniz, "Remarques sur les objections de M. Foucher" in I. C. Gerhardt (ed.), *Die Philosophischen Schriften von G. W. Leibniz* (Berlin: Weidmann, 1875; Hildesheim: Olms, 1960), vol. 4, p. 491.

der of coexistents and time a continuous order of successive existents."[34] And in another letter to de Volder he wrote in 1703 that "space is the order of possible coexistents and time the order of possible inconsistents."[35] In his correspondence with scholars he frequently repeated these definitions,[36] but he did not consider at that time the possible objection that they might be criticized because the term "coexistence" in the 1695 statement and the term "successive" in the 1700 statement have a temporal connotation, so that to define "time" as "a continuous order of successive existents" is a vicious circle. Leibniz himself may have criticized his 1700 definition of time when in 1703 he defined it as "the order of possible inconsistents," that is, in terms of a logical and not temporal relation. In the last years of his life Leibniz replaced even this logical relation by the causal relation and proposed what is generally regarded today as the prototype of the modern causal theories of space and time. He did this in his essay *Initum rerum Mathematicarum metaphysica*,[37] which was published in 1855 from a manuscript found in the library of Hanover. Opinions differ widely concerning the soundness and validity of Leibniz's deductive construction of time and space. Julius Baumann criticized it as "frail and flimsy,"[38] Bertrand Russell described it as a "fatally vicious circle,"[39] whereas Hans Reichenbach called it "a profound explanation of time and space"[40] and John Winnie, "simple and ingenious."[41] In any case, Leibniz's causal theory of time and space is of

[34] Ibid., vol. 2, p. 221.

[35] Ibid., vol. 2, p. 253; L. E. Loemker (ed.), *G. W. Leibniz: Philosophical Papers and Letters* (Chicago: Chicago University Press, 1956; Dordrecht: Reidel, 1969), p. 531.

[36] See, for example, Gerhardt, op. cit., vol. 4, p. 523; Loemker, op. cit., p. 496.

[37] G. W. Leibniz, *Initium rerum Mathematicarum metaphysica* in I. C. Gerhardt, *Leibnizens Mathematische Schriften* (Halle: Weidmann, 1855; Hildesheim: Olms, 1960), vol. 7, pp. 17–29; L. E. Loemker (ed.), *G. W. Leibniz: The Metaphysical Foundations of Mathematics* (Chicago: University of Chicago Press, 1956), vol. 2, pp. 1082–1094.

[38] " . . . die ganze Deduction ist hinfällig und nichtig." J. J. Baumann, *Die Lehren von Raum, Zeit und Mathematik in der neueren Philosophie* (Berlin: Reimer, 1869), p. 95.

[39] B. Russell, *A Critical Exposition of the Philosophy of Leibniz* (London: George Allen & Unwin, 1900, 1967), p. 130.

[40] "Eine ausserordentlich tiefgehende Erklärung von Zeit und Raum." H. Reichenbach, "Die Bewegungslehre bei Newton, Leibniz und Huyghens," *Kant-Studien* **29**, 416–438; "The theory of motion according to Newton, Leibniz, and Huyghens," in H. Reichenbach (ed.), *Modern Philosophy of Science—Selected Essays* (London: Routledge & Kegan Paul, 1959), pp. 46–66; reprinted in M. Reichenbach and R. S. Cohen (eds.), *H. Reichenbach: Selected Writings, 1909–1953* (Dordrecht: Reidel, 1987), vol. 2, pp. 48–68. Quotation on p. 52.

[41] J. A. Winnie, "The causal theory of space-time," in J. Earman, C. Glymour, and J. Stachel (eds.), *Foundations of Space-Time Theories* (Minneapolis: University of Minnesota Press, 1977), pp. 134–205. Quotation on p. 136.

great importance for us because of the special role it assigns to the concept of simultaneity. It reads as follows:

> If a plurality of states of things is assumed to exist which involves no opposition to each other, they are said to exist *simultaneously*. Thus we deny that what occurred last year and this year are simultaneous, for they involve incompatible states of the same thing.
>
> If one of two states which are not simultaneous involves a reason for the other, the former is held to be *prior*, the latter *posterior*. My earlier state involves a reason for the existence of my later state. And since my prior state, by reason of the connection between all things, involves the prior state of other things as well, it also involves a reason for the later state of these other things and is thus prior to them. Therefore whatever exists is either simultaneous with other existences or prior or posterior.
>
> *Time* is the order of existence of those things which are not simultaneous. Thus time is the universal order of changes when we do not take into consideration the particular kinds of change.
>
> *Duration* is magnitude of time. If the magnitude of time is diminished uniformly and continuously, time disappears into moment, whose magnitude is zero.
>
> *Space* is the order of coexisting things, or the order of existence for things which are simultaneous.[42]

When reviewing Leibniz's theory of time and space, note that it remarkably begins with a definition of simultaneity and thus ascribes to this notion logical priority over all spatiotemporal concepts. Before Leibniz published his theory, it had always been understood that simultaneity could be defined

[42] "Si plures ponantur existere rerum status, nihil oppositum involventes, dicentur existere simul. Itaque quae anno praeterito et praesente facta sunt negamus esse simul, involvunt enim oppositos ejusdem rei status. Si eorum quae non sunt simul unum rationem alterius involvat, illud prius, hoc posterius habetur. Status meus prior rationem involvit, ut posterior existat. Et cum status meus prior, ob omnium rerum connexionem, etiam statum aliarum rerum priorem involvat, hinc status meus prior etiam rationem involvit status posterioris aliarum rerum atque adeo et aliarum rerum statu est prior. Et ideo quicquid existit alteri existenti aut simul est aut prius aut posterius.

Tempus est ordo existendi eorum quae non sunt simul. Atque adeo est ordo mutationum generalis, ubi mutationum species non spectatur.

Duratio est temporis magnitudo. Si temporis magnitudo aequabiliter continue minuatur, tempus abit in Momentum, cujus magnitudo nulla est.

Spatium est ordo coexistendi seu ordo existendi inter ea quae sunt simul."

G. W. Leibniz, *Initium rerum Mathematicarum metaphysica*, I. C. Gerhardt (ed.), op. cit., p. 18.

only after the notion of time had been defined. In Leibniz's approach the logical order is reversed, and precedence and posteriority, hence time, are defined in terms of simultaneity or its negation. Leibniz may have conceived the idea of defining simultaneity without any reference to time from his study of Aristotle's writings. In the *Categories*,[43] for example, Aristotle declared that "nothing admits contrary predicates at one and the same time" (5 b 39), from which it follows that predicates, which are admitted simultaneously, cannot be contrary or in Leibniz's terminology cannot "involve opposition" (oppositum involventes). But, of course, the inference from "simultaneity implies nonopposition" to "nonopposition implies simultaneity" is logically illegitimate. Or, with the term "incompredicability," introduced by Christoph Sigwart,[44] the inference from the statement "incompredicability implies nonsimultaneity" that "nonincompredicability implies simultaneity" means to commit the fallacy of the denial of the antecedent. Nobody seems to have ever imputed this fallacy to Leibniz, but Baumann criticized Leibniz on a related charge. Baumann stated:

> The theorem of logic which was in Leibniz's mind said: predicates which include no contraries can simultaneously belong to a thing. Leibniz changed it to: states of several things, which did not include contraries, coexist if posited; the logical theorem refers to one and the same thing, whereas Leibniz transforms it into a relation of states of several things; but even so the proposition does not yet assert what Leibniz intends it to say unless we replace the words "if posited" by the words "if posited simultaneously." For states belonging to several things which do not include contraries can be posited in succession just as well as simultaneously or together.[45]

[43] See also, for example, Aristotle, *Metaphysics* 1005 b 19, where according to the Principle of Contradiction it is said that "the same predicate cannot at the same time belong and not belong to the same subject."

[44] C. Sigwart, *Logik* (Freiburg: Mohr, 1873, 1878, 1893, 1904); *Logic* (H. Dendy, transl.) (London: Sonnenschein, 1895), vol. 1 (1895), p. 132.

[45] "Ein Satz der Logik, an welchen Leibniz hier denkt, besagt: Prädicate, welche nichts Entgegengesetztes einschliessen, können einem Ding zugleich zukommen; Leibniz macht daraus Zustände mehrerer Dinge, welche nicht Entgegengesetztes einschliessen, existiren zugleich, falls sie gesetzt werden; der logische Satz geht auf die Zustände eines und des nämlichen Dinges, Leibniz überträgt ihn auf das Verhältnis von Zuständen mehrerer Dinge; aber auch so sagt der Satz noch gar nicht, was Leibniz will, wenn wir zu den Worten: falls sie gesetzt werden, nicht hinzufügen: falls sie zugleich gesetzt werden. Denn Zustände mehrerer Dinge, die nichts Entgegengesetztes einschliessen, können nach einander ebenso gut wie gleichzeitig oder zugleich gesetzt werden." J. J. Baumann, op. cit. (note 38), p. 95.

Indeed, within Leibniz's formalism, the difficulty of extending the relations defined as pertaining to one and the same thing to several things seemed insurmountable if the formalism was interpreted as a self-contained logical system. Because of this impasse and similar difficulties related to the ideality and continuity of time modern commentators, like Rescher,[46] McGuire,[47] and Arthur,[48] regard Leibniz's relational theory of time as part of his metaphysical doctrine of monads and pre-established harmony.

With these remarks we conclude our discussion on the role of the concept of simultaneity in Leibniz's causal theory of time. For further details, refer to the essay by Winnie[49] and to Bas C. van Fraassen's book on the philosophy of time and space.[50]

Leibniz's relational theory of time and space was accepted by Immanuel Kant in his early precritical writings on problems in the philosophy of nature. In his first published essay, *Thoughts on the True Estimation of Living Forces* (1747),[51] in which he tried to settle the controversy between the Cartesians and the Leibnizians about the measure of *vis viva*,[52] he endorsed the Leibnizian conceptions of space and time and still adhered to them in *Monadologia Physica* (1756).[53] But in the early 1760s, influenced by Leonhard Euler's *Reflexions sur l'Espace et le Temps* (1750),[54] Kant declared himself in favor of the Newtonian conceptions of absolute space and time and still took them for granted in *On the First Ground of the Distinctions in Space (1768)*.[55] Finally, in

[46] N. Rescher, *The Philosophy of Leibniz* (Englewood Cliffs, New Jersey; Prentice-Hall, 1967).

[47] J. E. McGuire, " 'Labyrinthus Continui': Leibniz on Substances, Activity and Matter" in K. Machamer and R. G. Turnbull (eds.), *Motion and Time, Space and Matter* (Columbus, Ohio: Ohio State University Press, 1976).

[48] R. T. W. Arthur, "Leibniz's theory of time" in K. Okruhlik and J. R. Brown (eds.), *The Natural Philosophy of Leibniz* (Dordrecht: Reidel, 1985), pp. 263–313. See, however, G. A. Hartz and J. A. Cover, "Space and time in the Leibnizian metaphysic," *Nous* **22,** 493–519 (1988).

[49] See note 41.

[50] B. C. van Fraassen, *An Introduction to the Philosophy of Time and Space* (New York: Random House, 1970), chapter 2.

[51] I. Kant, *Gedanken von der wahren Schätzung der lebendigen Kräfte*, Gesammelte Werke (Berlin: Reimer, 1910), vol. 1, pp. 1–181.

[52] For details see M. Jammer, *Concepts of Force* (Cambridge, Massachusetts: Harvard University Press, 1957; New York: Dover Publications, 1999), pp. 178–180.

[53] I. Kant, *Monadologia Physica*, Gesammelte Werke, op. cit., vol. 1, pp. 473–487.

[54] L. Euler, "Reflexions sur l'Espace et le Temps," in *Leonhardi Euleri Opera Omnia* (Geneva: Orell Füss, 1942), vol. 2, pp. 376–383.

[55] I. Kant, *Von dem ersten Grunde des Unterschiedes der Gegenden im Raume*, Gesammelte Werke, op. cit., vol. 2, pp. 375–383.

his critical philosophy, which found its earliest expression, as far as the notions of space and time are concerned, in his Inaugural Dissertation of 1770, *On the Form and Principles of the Sensible and the Intelligible World*,[56] Kant abandoned the Leibnizian and Newtonian points of view and regarded space and time as pure intuition (*intuitus purus*).

The Dissertation contains Kant's explanation of why he rejected Leibniz's theory of time; it shows that his reasons for doing so were intimately related to the notion of simultaneity. Kant begins his discussion on the concept of time with the declaration that "the idea of time does not emerge from the senses but is presupposed by them" and he substantiates this statement by arguing that "it is only by means of the representation of time that one recognizes whether events are simultaneous or successive" and that "succession does not generate time but rather appeals to it." On the basis of these considerations Kant now launches his first attack against Leibniz's theory by saying that "it is therefore very bad to conceive time—as if it were acquirable through experience—by defining it as the order of existents which follow one after the other. For I do not understand the meaning of *after* unless I have already the idea of time, because *one after another* is what occurs at *different times* just as *simultaneous* is what occurs *at the same time*."[57] According to Kant, Leibniz's definition, which, as Kant reads it, defines time as the series or order of consecutive existents, is a vicious circle because the notion of consecutiveness or succession presupposes the notion of time.

Kant continues his charge by directing it against Leibniz's conception of simultaneity. Referring to Leibniz's deduction of time from the sequence of states, Kant points out that "the falsity of this conception reveals itself already by the vicious circle in its definition of time and in addition by

[56] I. Kant. *De mundi sensibilis atque intelligibilis forma et principiis*, Gesammelte Werke, op. cit., vol. 2, pp. 385–419. English translation in W. J. Eckoff, *Kant's Inaugural Dissertation of 1770* (New York: Columbia College, 1894); J. Handyside, *Kant's Inaugural Dissertation and Early Writings on Space* (Chicago: Open Court, 1929).

[57] "*Idea temporis non oritur, sed supponitur a sensibus*. Quae enim in sensus incurrunt, utrum simul sint, an post se invicem, nonnisi per ideam temporis repraesentari potest; neque successio gignit conceptum temporis sed ad illum provocat. Ideoque temporis notio, veluti per experientiam acquisita, pessime definitur per seriem actualium *post* se invicem exsistentium. Nam, quid significet vocula *post*, non intelligo, nisi praevio iam temporis conceptu. Sunt enim *post* se invicem, quae exsistunt *temporibus diversis*, quemadmodum *simul* sunt, quae exsistunt *tempore eodem*." Reference 56, pp. 400–401.

its complete disregard of *simultaneity,* one of the most important consequences of time."[58] Kant here adds a footnote that deserves to be quoted in full.

> Things which are simultaneous are so not because they are not successive to each other. For although the removal of succession disjoins the connection which existed in virtue of the temporal sequence, it does not give thereby immediate rise to another true relation like that given by the conjunction of all that there is at the same instant of time. Simultaneous existents are connected by the same moment of time just like successive existents are by diverse moments of time. Even though time has only one dimension, the ubiquity of time (to use Newton's expression), in virtue of which all imaginable occurrences occur *at the same time,* endows reality with another dimension, so to say, in so far as they are connected by the same point of time. For if time is represented by an infinitely long straight line and all simultaneous occurrences at a given moment are represented by a tranversely drawn straight line through that point on the time line, the thus generated surface represents the phenomenal world with respect both to its substance and its accidents.[59]

This footnote reveals, perhaps more than the text itself, what Kant had been thinking when he wrote this passage. Kant's denial of the thesis, which he ascribes to Leibniz, that nonsuccessiveness (non succedunt) between states or events is a sufficient condition for their simultaneity (simultanea) suggests within the context of Kant's geometrization of temporal relations that he might have been thinking of two series of evolving states, say S and S′, represented by two parallel straight lines. Before simultaneity has been

[58]Posterioris autem sententiae falsitas, cum circulo vitioso in temporis definitione obvia luculenter semet ipsam prodat, et praeterea *simultaneitatem,* maximum temporis consectarium, plane negligat." Ibid., p. 400.

[59]"*Simultanea* non sunt ideo talia, quia sibi non succedunt. Nam remota successione tollitur quidem coniunctio aliqua, quae erat per seriem temporis, sed inde non statim oritur *alia* vera relatio, qualis est coniunctio omnium in momento eodem. Simultanea enim perinde iuguntur eodem temporis momento, quam successiva diversis. Ideo, quanquam tempus sit unius tantum dimensionis, tamen *ubiquitas* temporis (ut cum Newtono loquar), per quam *omnia* sensitive cogitabilia sunt *aliquando,* addit quanto actualium alteram dimensionem, quatenus veluti pendent ab eodem temporis puncto. Nam si tempus designes linea recta in infinitum producta, et simultanea in quolibet temporis puncto per lineas ordinatim applicatas: superficies, quae ita generatur, representabit *mundum phenomenon,* tam quoad substantiam, quam quoad accidentia." Ibid., p. 401.

defined no state of S′ is successive to any given state of S and vice versa. Hence Leibniz's definition would allow the decree of an arbitrarily chosen state of S′ to be simultaneous with a given state of S, and simultaneity would turn out to be merely a matter of convention. But because simultaneity is a "vera relatio" for Kant, it cannot be only the negation of the serial order of succession but also must be a positive relatedness of existents conjoined by belonging to the same instant of time and thereby a mode of time on a par with succession. In modern terminology we might say that this footnote tells us that by regarding simultaneity as a "true relation" Kant would have rejected the conventionality thesis concerning distant simultaneity, which will be discussed in detail later on. The footnote also has historical importance, because Kant's geometrical representation, as described above, may well be regarded as anticipating what in modern times is called a "space–time diagram."

In the Dissertation of 1770 Kant rejected Leibniz's definition of simultaneity as the antithesis of succession but did not offer a rigorous alternative definition. He only called it a "conjunction of all in the same instant of time" (coniunctio omnium in momento eodem). All that can be inferred from this description is that Kant, contrary to Leibniz, who ascribed to simultaneity logical priority over time, reversed this relation and subordinated simultaneity under the notion of time. As we will see, Kant conceived simultaneity just like succession and duration or permanence, as a mode (modus) of time. When using the term "conjunction" Kant did not specify in his Dissertation what precisely conjoins the participants of this conjunction and makes them simultaneous.

Kant dealt with this question about ten years later in *Critique of Pure Reason,*[60] published in 1781 and revised in 1787. In *Transcendental Aesthetics,* the part of the *Critique* that studies the principles of sensibility a priori, Kant declared: "Time is not an empirical concept deduced from any experience, for neither coexistence nor succession would enter into our perception, if the representation of time were not given a priori. Only when this representation a priori is given, can we imagine that certain things happen at the same

[60] I. Kant, *Kritik der reinen Vernunft* (Riga: Hartknoch, 1781, 1787); *Critique of Pure Reason* (N. Kemp Smith, transl.) (New York: Macmillan, 1929, 1970); (F. M. Müller, transl.) (New York: Macmillan, 1911; Doubleday, 1961).

time (simultaneously) or at different times (successively)."[61] This statement is clearly incompatible with Leibniz's theory of the logical priority of simultaneity over time.

The concept of simultaneity also plays an important role in what Kant calls "Analogies of Experience" in *Transcendental Analysis* of *Critique*. These are the principles or rules by which the mind organizes the temporal order of perceptions, without which objective experience would be impossible. Because, as Kant explains, there are three modi of time, namely permanence or duration, succession, and simultaneity, "there will be three rules of all relations of phenomena in time, by which the existence of every phenomenon with regard to the unity of time is determined, and these rules will precede all experience, nay, render experience possible."[62]

The argument by which Kant proves the *First Analogy*, that is, the assertion that "in all changes of phenomena the substance is permanent and its quantum is neither increased nor diminished,"[63] can be stated in brief as follows. Simultaneity and succession, and hence all changes of phenomena, can be represented only in time as the substratum. Time itself, therefore, cannot change, but time itself is not experienced and thus must be represented by an abiding substrate, which Kant identifies with substance (or matter).

The *Second Analogy* declares that all change or succession of phenomena occurs according to the law of connection between cause and effect.[64] Mere perception leaves the objective relation of successive phenomena undetermined. For their determination it is necessary to conceive the relation between the two states in such a way that it should be determined thereby with necessity, which of the two should be taken as coming first, and which as second, and not conversely. Kant explains the ordering function of causality by the examples of ob-

[61] "Die Zeit ist kein empirischer Begriff, der irgend von der Erfahrung abgezogen worden. Denn das Zugleichsein oder Aufeinanderfolgen würde selbst nicht in die Wahrnehmung kommen, wenn die Vorstellung der Zeit nicht *a priori* zugrunde läge. Nur unter deren Voraussetzung kann man sich vorstellen, dass einiges zu einer und derselben (zugleich) oder in verschiedenen Zeiten (nacheinander) sei." I. Kant, op. cit., B 46. Kant did not use the term "Gleichzeitigkeit" but used instead "Zugleichsein" (coexistence); for although the term "Gleichzeitigkeit" already existed in the German language (see note 14 in chapter 1), it had not yet gained common currency at the time Kant wrote the *Critique*. For details concerning Kant's use of these terms see D. Krallmann and H. A. Martin, *Wortindex zu Kants gesammelten Schriften* (Berlin: W. de Gruyter, 1967), vol. 1, p. 434 ("gleichzeitig") and vol. 2, p. 1077 ("zugleich", "Zugleichsein").

[62] See note 60, B 219.

[63] Ibid., B 224.

[64] Ibid., B 232.

serving a ship gliding down a stream, where "the perception of its place below follows the perception of its place higher up in the course of the stream." In contrast to David Hume, who tried to eliminate causality in favor of spatiotemporal contiguity or succession in time, Kant reinstated the causal connection as a necessary condition for determining the order of successive events.

Finally and most importantly, the *Third Analogy* or the *Principle of Coexistence, According to the Law of Reciprocity or Communion,*[65] as it is called in the second edition of the *Critique,* states that "all substances, so far as they can be perceived as coexisting in space, are always affecting each other reciprocally," or are "in thoroughgoing reciprocal interaction." To understand this principle of simultaneity, return to Kant's example of the ship driven downstream by the river. Here the order of perceptions is regulated by the law of causality, in particular, the order cannot be inverted, for it is impossible in the apprehension of this phenomenon to perceive the ship first below and afterward higher up the stream. If looking at a house, however, the parts of which are coexisting, our perception can begin in the apprehension of the roof and end with that of the basement, or it can begin below and end above. Thus, in the perception of simultaneous things the order is arbitrary. One may first observe the moon and afterward the Earth, or first the Earth and afterward the moon, to mention another of Kant's examples. This arbitrariness in the order of perception does not yet warrant the simultaneity of the perceived objects, however, that is, "if the one is there, the other also must be there in the same time and by necessity." Kant tries to avoid this objection by resorting to the idea of unity in which all appearances are connected because the determinations of one substance are grounded in the other substance. Such a relation between substances, however, is the relation of influence; "and if, conversely also, the first contains the ground of determinations in the latter, the relation is that of community or reciprocity. Hence the coexistence of substances in space cannot be known in experience otherwise but under the supposition of reciprocal action: and this is therefore the condition also of the possibility of things themselves as objects of experience."[66]

To conclude our lengthy exposition of Kant's varying conceptions of simultaneity we summarize his final doctrine of this notion that he presented in chapter 2 of book II of *Critique of Pure Reason.* Kant's starting point is the

[65] Ibid., B 256.
[66] Ibid., B 258.

recognition that the subjective temporal order of human sense appearances does not necessarily agree with the objective temporal order of physical phenomena, for example, when we see the lightning before we hear the thunder. To construct the objective temporal order use must be made mainly of what Kant called the *Second Analogy of Experience* or the *Principle of the Succession of Time, According to the Law of Causality*, and also of the *Third Analogy* or the *Principle of Coexistence, According to the Law of Reciprocity or Community*. According to the *Principle of the Succession of Time*, an event e_1 is judged to precede temporally another event e_2, if and only if, in accordance with a law of nature, e_1 is a cause of e_2 and e_2 is not a cause of e_1. According to the *Principle of Coexistence*, event e_1 is simultaneous with event e_2 if and only if a thoroughgoing mutual interaction exists between these two events. If we recall the footnote of Kant's Inaugural Dissertation we may also say that e_1 and e_2 are simultaneous if and only if neither precedes the other (i.e., in modern terminology, if both events lie in the same "simultaneity plane" or "line"). In accordance with the *Principle of Coexistence* this simultaneity can obtain either when (1) neither is e_1 a cause of e_2 nor is e_2 a cause of e_1, or when (2) e_1 is a cause of e_2 and e_2 is a cause of e_1. When considering the second possibility Kant probably thought of Newton's gravitational law of mutual and *instantaneous* attraction. As is well known, modern physics, which denies the existence of instantaneous interactions, also denies the existence of possibility (2).

Thus, to sum up, according to the former principle of temporal succession in accordance with the law of causality (Second Analogy) event e_1 is said to precede temporally event e_2 if and only if e_1 is a cause of e_2 and e_2 is not a cause of e_1. In accordance with the last-mentioned principle, the *Principle of Coexistence According to the Law of Reciprocity or Communion* (or "*thoroughgoing reciprocal interaction*") e_1 is simultaneous with e_2 if and only if both events are subject to a thoroughgoing mutual interaction. Hence, the former principle says that e_1 and e_2 are simultaneous if one of the two possibilities is realized: (1) neither is e_1 a cause of e_2 nor is e_2 a cause of e_1, or (b) e_1 is a cause of e_2 and e_2 is a cause of e_1. If t_1 denotes the time of the occurrence of event e_1 and t_2 denotes that of e_2 then if e_1 is the cause of e_2, $t_1 \leq t_2$, and if e_2 is the cause of e_1 then $t_2 \leq t_1$, hence $t_1 = t_2$.

This is the method that Kant applies in the Third Analogy of Experience (B 146). Because time itself is not the object of experience, "there must be something besides their mere existence by which e_1 determines its place in time for e_2, and e_2 for e_1, because thus only can these two substances [events] be represented empirically as coexistent."

A typical example of the second possibility, which Kant probably had in mind, is the mutual interaction described by Newton's law of gravitation according to which two distant masses attract each other in a "thoroughgoing interaction." Modern physics, which revokes the existence of instantaneous actions at a distance, also denies the second possibility.

To illustrate how greatly critics differ on Kant's treatment of simultaneity we review in brief only an extremely derogatory and an extremely laudatory assessment. Arthur Schopenhauer, who highly respected Kant but scoffed at him when disagreeing with him, rejected Kant's conception of simultaneity both in *The Fourfold Root of the Principle of Sufficient Reason* and in his opus magnum *The World as Will and Idea.* In the former, after having analyzed the notion of causality and shown that it implies succession, Schopenhauer concluded that the notion of reciprocity, as used by Kant, is unacceptable, for it assumes "that the effect is again the cause of its cause so that the subsequent is simultaneous with the antecedent."[67] In the latter he became more explicit: "The conception of reciprocity ought to be banished from metaphysics. For I now intend, quite seriously, to prove that there is no reciprocity in the strict sense, and that this conception, which people are so fond of using, just on account of the indefiniteness of the thought, is seen, if more closely considered, to be empty, false and invalid. . . . It implies that both the states *a* and *b* are cause and that both are effect of each other; but this really amounts to saying that each of the two is the earlier and also the later; thus it is an absurdity."[68]

Unlike Schopenhauer, George Lechalas, an engineer by profession and philosopher by inclination, praised Kant with having made "decisive progress" in the study of the nature of time: for he "has developed the essential points of the causal theory of time and we can but complete the considerations presented in the Second and Third Analogies." Referring to Kant's statement about reciprocity, Lechalas pointed out that these ideas can obviously be applied to the phenomenon of the mutual gravitational attraction between two bodies, for these stand, one with respect to the other, in the twofold relation of cause and effect, either of them accelerating the other and being accelerated by it

[67] A. Schopenhauer, *Ueber die vierfache Wurzel des Satzes vom zureichenden Grunde* (Rudolstadt: Hofbuchhandlung, 1813; Frankfurt a. M.: Hermann, 1813, 1847; Leipzig: Reclam, n.d.), p. 55.

[68] A. Schopenhauer, *Die Welt als Wille und Vorstellung* (Leipzig: Brockhaus, 1819, 1879), pp. 544–545.

in reverse.[69] Lechalas thus contended that, by applying Kant's approach to simultaneity, which he interpreted as invoking a reciprocal causal interaction, it is possible to define distant simultaneity on the basis of Newton's law of gravitation and the equality between action and reaction. Let P and P' be two bodies or particles interacting with each other in accordance with Newton's law of gravitation. Assuming that the dynamical state of each particle is fully determined by properties not explicitly containing the time variable, we denote by Z_m ($m = 1, 2, 3 \ldots$) the states of P and by Z'_n ($n = 1, 2, 3 \ldots$) the states of P'. Let F_m be the force exerted on P by P', when P is in state Z_m. Correspondingly, let F'_n be the force exerted on P' by P, when P' is in state Z'_n. For a given state $Z_{m'}$ with its associated $F_{m'}$ select that $F_{n'}$ which satisfies the equation $F_{m'} = F'_{n'}$ (action = reaction). Then state $Z_{m'}$ of P is simultaneous with state $Z_{n'}$ of P' and distant simultaneity has been established.

Lechalas's definition of simultaneity was fully approved by Henryk Mehlberg[70] but criticized by van Fraassen mainly because Lechalas did not prove "that only one possible ordering of the states is compatible with the laws of mechanics."[71] Lechalas's reasoning may also be charged with involving a vicious circle, for in classical mechanics the state of a particle is defined by its position and its velocity (or momentum), but the determination or measurement of velocity presupposes the concept of simultaneity, a fact that has been ignored, in general, but which plays an important role in modern discussions of the concept of simultaneity.

Nevertheless, Lechalas has to be credited, as van Fraassen also concedes, with having insisted that a theory of temporal order and simultaneity "ought to utilize the concepts of physics rather than the concepts of any philosophical system."[72] It is not accidental, perhaps, that the year 1895, in which the first edition of Lechalas's *Étude* was published, marks the turning point at which physical considerations began systematically to replace philosoph-

[69] " . . . la nature du temps, à laquelle Kant nous paraît avoir fait faire un progrès décisif . . . ". " . . . Kant lui-même a développé les points essentiels de la théorie causale du temps, et nous n'aurons qu'à compléter les considérations exposées dans sa seconde et sa troisième *analogies*." (It is probably in this statement that the term "théorie causale du temps" ["causal theory of time"] has been used for the first time). G. Lechalas, *Étude sur l'Espace et le Temps* (Paris: Alcan, 1895, 1909), pp. 278, 290.

[70] H. Mehlberg, "Essai sur la théorie causale du temps," *Studia Philosophica* **1,** 119–260 (1935); **2,** 111–231 (1937).

[71] B. C. van Fraassen, note 50, pp. 54–57.

[72] Ibid., p. 57.

ical argumentations in the study of the concept of simultaneity. It would be wrong, however, to regard Lechalas as the very first who proposed the definition of temporal order by the use of physical concepts. This idea had already been alluded to by Kant in his precritical period when he wrote: "It is proved that there would be no space and no extension, if substances had no force whereby they can act outside themselves. For without a force of this kind there is no connection, with this connection no order and without this order no space (or time for that matter).[73]

Although Kant's philosophy of time and, in particular, his thesis of its apriority were topics widely discussed both by his followers and his opponents in the nineteenth century, leading philosophers of that period, like Fichte, Hegel, Lotze, or Renouvier, made no comments on the concept of simultaneity. Still, there are some exceptions that are worth mentioning. One example is Shadworth Hollway Hodgson's treatise *Time and Space*. Following Kant, Hodgson regarded time as "a necessary concomitant of conscious experience" and said that time and space, as formal modes of consciousness, are "different but inseparable and simultaneous."[74] Of course, he did not mean that time and space are simultaneous, a statement that would assume the existence of a hypertime in which ordinary time and space are at the same hypertime. He only said that the consciousness of time and the consciousness of space are simultaneous in (ordinary) time, for he added: "in all time is involved space as its accompaniment, in all space there is involved time as its element." He may therefore be credited with having anticipated Hermann Minkowski's famous Cologne address of 1908 that contains the statement: "nobody ever noticed a place except at a time, or a time except at a place."[75] But Hodgson, in contrast to Minkowski, did not merge time and space into a four-dimensional manifold, although he came close to it when he declared: "our feelings in time are never presented or represented separate from the provisional accompaniment of space . . . owing to their constant association by the simultaneous exercise of the different senses, or to some laws of nature which are the objective aspects of that association."[76]

Adolf Trendelenburg in *Logische Untersuchungen* (Logical Inquiries) presented an approach similar to the idea of a Minkowski space–time manifold,

[73] I. Kant, note 56 (1929), p. 10.

[74] S. H. Hodgson, *Time and Space* (London: Longman, 1865), p. 118.

[75] H. Minkowski, "Raum und Zeit," *Physikalische Zeitschrift* **10**, 104–111 (1909).

[76] S. H. Hodgson, op. cit., p. 117.

but this time based on the concept of simultaneity and its graphical representation in note 59 in Kant's Dissertation. According to Trendelenburg the notion of time is not a priori but is derived from the concept of motion,[77] which is the basis of all our percepts and concepts. "If only one dimension, length, is ascribed to time, then it must appear as if there exists only a succession of moments and no simultaneity, as if only 'one-after-another' but no 'at-the-same-time' were possible. For the idea of simultaneity involves the concept of 'at the side of,' so that temporal simultaneity seems to contain implicitly a notion of two-dimensional breadth and the one-dimensionality of time to have attached to it another dimension. . . . The appearance of the second dimension, which represents simultaneity, arises from the connection of time with space through the action of motion."[78]

Max Eyfferth, in *Über die Zeit* (On Time), an expanded version of his doctoral dissertation at the University of Berlin, devoted a whole chapter, entitled "The division of time into temporal succession and simultaneity," to the discussion of whether simultaneity should be regarded as a second dimension of time. He distinguished between "subjective time" and "objective time" and contended that even the former could not exist if we would not possess in our consciousness both the perceptions of simultaneity and succession. In the subsection "On the two dimensions of time" he declared: "In the absence of simultaneity, that is, if all our apperceptions would occur only one after the other, we would be unable to conceive time. Remembrance would then be impossible. . . . Without simultaneity we could not compare anything. If, however, without comparison thinking is impossible, it is impossible also without simultaneity." He also tried to disprove the possible objection that an analogy between the two-dimensionality of time and the three-dimensionality of space could not be maintained, because only the dimensions of space can be interchanged but not those of time.[79]

[77] A. Trendelenburg, *Logische Untersuchungen* (Berlin: 1840; Leipzig: Hirzel, 1870; Hildesheim, Olms, 1964), vol. 1, p. 151.

[78] "Der Schein der zweiten Dimension, welcher die Gleichzeitigkeit begleitet, ensteht durch die Verknüpfung der Zeit mit dem Raume vermöge der Bewegung." Ibid., vol. 1, p. 230.

[79] "Ohne Gleichzeitigkeit würden wir nichts mit einander zu vergleichen im Stande sein. Wenn aber ohne alles Vergleichen kein Denken möglich ist, so ist es auch ohne Gleichzeitigkeit unmöglich." M. Eyfferth, *Über die Zeit* (Berlin: Henschel, 1871), pp. 46, 50. In support of his contention he mentions the Leipzig theologian Christian Hermann Weisse, who in an essay, published 1865 in the *Zeitschrift für Philosophie und philosophische Kritik* **46,** 201–208 (1865), also proposed to regard time as a product of two dimensions, but on theological reasons.

Even though these dimensionality speculations may be regarded as a faint anticipation of the multidimensional space–time diagrams of modern physics, the treatment of the notion of simultaneity was still carried out wholly within the framework of purely philosophical considerations. An important step in the transition from a *philosophical* to a *physical* treatment of the concept of simultaneity was also made by the Tübingen philosopher Christoph Sigwart in his monumental two-volume publication *Logic*.[80] Because Sigwart defined "logic" as the "technical science of thought, directing us how to arrive at certain and universally valid propositions,"[81] his *Logic* was not so much an investigation into the psychology of thought as an attempt to establish a methodology of finding the rules that confer objective validity on thought.

Sigwart agreed with Kant that time is a priori, insofar as it is a necessary ingredient of consciousness, and that it is a form, insofar as its mode of connection is independent of any specific content. But he claims that Kant's theory of time as an a priori and purely subjective condition of human perception "is not sufficient. We need also the determination in an objective time of a point of time which shall be *the same for all*; and we need a common measure of time according to which every particular fact of consciousness has its place assigned to it."[82] For "subjective statements about the facts of sensation as given in self-consciousness are not complete until the time-determination involved in them has been objectively fixed."[83] "It is necessary, in order that even our purely subjective statements about what is contained in our consciousness may be fully determined, that our own subjective time should be referred to a TIME-SYSTEM which *is common to all*, and so far objective."[84] But how can such an objective *time-system*, common to all, be established? Sigwart's answer to this question emphasizes the role that the notion of simultaneity plays for this purpose and also hints, by its reference to *external perceptions*, about the need for physical considerations. For Sigwart says: "Because the correspondence of one individual consciousness with another is only possible by means of external sensation, thus must depend upon external perceptions which are shared by all, and which occur *simultaneously* for all."[85]

[80] Ch. Sigwart, *Logik* (Tübingen: Mohr, vol. 1, 1873; vol. 2, 1878); *Logic* (London: Sonnenschein, 1895).

[81] Ch. Sigwart, op. cit. (1895), vol. 1, p. 1.

[82] Ibid., vol. 1, p. 305 (italics added).

[83] Ibid., vol. 2, p. 235.

[84] Ibid., vol. 2, p. 236.

[85] Op. cit. (1878), vol. 2, p. 334; (1895), vol. 2, pp. 236–237.

As these statements show, Sigwart assigned to the notion of simultaneity an indispensable role in the objectification of time for the establishment of a "time-system." Although he does not offer a formal definition of "simultaneity," he describes it as "reducing the Now of one man to comparison and coincidence with the Now of others."[86] To find such points of coincidence, he continues, we must be certain that "different people are simultaneously conscious of the same fact . . . and since the reference of the conscious content of one person to that of another is only possible through the external world, this must be where a phenomenon which is external for both is simultaneously perceived."[87]

In his 1905 seminal paper on relativity Einstein declared that all our judgments involving time "are always judgements of simultaneous events" and he constructed a "time-system," or as he called it "a common time," only after having defined an individual *A-time* and an individual *B-time*. It is therefore no exaggeration to say that Sigwart's "work is of such critical merit as to entitle him to be regarded as a conspicuous link between the work of Kant and the physical work of the Relativists."[88]

[86] Ibid. (1895), p. 238.
[87] Ibid. (1895), p. 238.
[88] J. A. Gunn, *The Problem of Time* (London: G. Allen and Unwin, 1929), p. 170.

CHAPTER SIX

The Transition to the Relativistic Conception of Simultaneity

Modern science and, in particular, modern physics—and with it the modern conceptions of simultaneity—are deeply indebted to several intellectual developments of the nineteenth century. Foremost among these was the discovery of non-Euclidean geometry. By abolishing the monopoly of Euclidean geometry this discovery stimulated an intense interest in critically re-examining not only the accepted ideas of space and time but also the general principles of scientific methodology. Two of these principles had a decisive impact on the development of the modern concept of simultaneity: (1) the positivistic tendency of demetaphysicizing the Newtonian concepts of absolute space and time, with Ernst Mach as its main representative, and (2) the doctrine, now generally called conventionalism, associated primarily with the name of Henri Poincaré.

Much has been written about Mach's rejection of Newton's concept of absolute or substantival space and absolute motion. In contrast, rather little[1]

[1] For example, M. Čapek, in *The Concepts of Space and Time* (Dordrecht: Reidel, 1976), quotes a lengthy excerpt from Mach's *Science of Mechanics* concerning his rejection of absolute space but not a single word about his rejection of absolute time.

has been written about Mach's rejection of Newton's absolute time, which he called a "superfluous *metaphysical* concept,"[2] and still less about how Mach conceived what are generally regarded as temporal concepts such as the concept of simultaneity.

As a matter of fact, as his writings testify,[3] the notion of time occupied Mach's attention much more than the notion of space. Although the notion of simultaneity was not dealt with explicitly in any of his books or essays, it implicitly underlay his very conception of time. According to Mach, time as an independent reality, like Newton's absolute time, does not exist. It is nevertheless used so frequently in physics or in ordinary experience because it serves as a coordinator among different processes. Psychological time, that is, time as sensation, is obtained "by the connection of that which is contained in the province of our memory with that which is contained in the province of our sense-perception. When we say that time flows on in a definite direction or sense, we mean that physical events generally (and therefore also physiological events) take place only in a definite sense. . . . In all this there is simply expressed a peculiar and profound connection of things."[4]

Mach's favorite example to illustrate the role of time as a linking mediator between different events was the thermodynamic cooling-down process of a hot body and the mechanical process of free fall. Because these two processes are described by equations that contain the time variable, "the time can be eliminated from them and the temperature can be determined by the distance of fall. The elements [processes] then reveal themselves simply as interdependent."[5] "If we once made clear to ourselves that we are concerned only with the ascertainment of the *interdependence* of phenomena . . . all metaphysical obscurities disappear."[6]

[2] "ein müssiger *metaphysischer* Begriff." E. Mach, *Die Mechanik in ihrer Entwicklung* (Leipzig: Brockhaus, 1883, 1921), p. 217; *The Science of Mechanics* (Chicago: Open Court, 1893, 1960), p. 273.

[3] See E. Mach, "Untersuchungen über den Zeitsinn des Ohres," *Sitzungsberichte der Kaiserlichen Akademie der Wissenschaften, Mathematisch-Naturwissenschaftliche Classe, Wien,* **51,** 133–150 (1865); *Die Mechanik in ihrer Entwicklung* (note 2) chapter 2, section 6; *Die Analyse der Empfindungen* (Jena: Fischer, 1885, 1922), chapter XII ("Die Zeitempfindung"); *The Analysis of Sensations* (Chicago: Open Court, 1914); *Die Principien der Wärmelehre* (Leipzig: Barth, 1896, 1919); *Erkenntnis und Irrtum* (Leipzig: Barth, 1905, 1920); *Knowledge and Error* (Dordrecht: Reidel, 1976), chapter XXIII, "Physiological Time in Contrast to Metrical Time," and chapter XXIV, "Space and Time Physically Considered."

[4] E. Mach, *Die Mechanik in ihrer Entwicklung,* note 2 (1921), pp. 218–219; (1960), pp. 274–275.

[5] E. Mach, *Die Analyse der Empfindungen,* note 3 (1922), p. 286.

[6] Note 2 (1921), p. 219; (1960), pp. 275–276.

It is not surprising that Mach never used the notion of distant simultaneity, for his rejection of time necessarily implied the rejection of simultaneity as a temporal concept. The nearest substitute can be found in his essay *Space and Time Physically Considered* in which he wrote: "The world remains a whole so long as no element is isolated, but all parts are connected, if not immediately then at least mediately through others. The concordant behaviour of members not immediately connected (the unity of space and time) then arises only apparently by failure to notice the mediating links."[7]

Although Mach never discussed the notion of simultaneity per se, it might be possible to extrapolate from his writings how he would have conceived this notion. According to Mach's philosophy of physical time, the intercorrelation of events is not determined by their relations to some specific moments of time, because these moments do not exist on their own. Hence, the coexistence or copresence of events, which are usually called simultaneous events, is not the consequence of their coincidence with a certain moment of time, for, again, such a moment does not exist in its own right. In other words, so-called simultaneous events simply happen to coexist without the intermediacy of any temporal relation to a common moment of time.[8] In short, the demetaphysicizing of the Newtonian concept of time led Mach to what may be called the detemporalization of simultaneity.

Mach's criticisms of the Newtonian concepts of space and time profoundly influenced the physicists and philosophers of the late nineteenth and the early twentieth centuries. His ideas about what we have called the detemporalization of simultaneity, however, were almost completely ignored although they agree with some modern theories according to which the notion of simultaneity has logical precedence to that of time.

In 1884, one year after the publication of Mach's *Die Mechanik in ihrer Entwicklung*, the Irish engineer James Thomson published a paper in which he drew attention, probably for the first time, to the problem of ascertaining

[7] Note 3 (1976), p. 347; (1920), p. 444.

[8] Unlike the English word "simultaneous," the German equivalent "gleichzeitig," denoting literally "at the same time" (gleich = equal; zeit = time), indicates explicitly a relation to a moment of time and would therefore have been inapplicable, or at least linguistically unaccommodating, to Mach in the present context. See the remarks on *simul* and "Gleichzeitigkeit" in chapter 1.

the simultaneity of locally separated events. In an essay published in the *Proceedings of the Royal Society of Edinburgh* he wrote:

> Men have very good means of knowing in some cases, and of imagining in other cases, the distance between the points of space simultaneously occupied by the centres of two balls; if, at least, we be content to waive the difficulty as to imperfection of our means of ascertaining or specifying, or clearly idealising, simultaneity at distant places. For this we do commonly use signals by sound, by light, by electricity, by connecting wires or bars, or by various other means. The time required in the transmission of the signal involves an imperfection in human powers of ascertaining simultaneity of occurrences in distant places. It seems, however, probably not to involve any difficulty of idealising or imagining the existence of simultaneity.[9]

Thomson obviously realized that the establishment of distant simultaneity poses a problem because of the transmission time of the signal employed. He even seems to have realized that the measurement of this transmission time requires knowledge of simultaneity. Had he further pursued this train of ideas, he would have easily anticipated the circularity involved that Poincaré dealt with fourteen years later.

Although Mach's critique of the Newtonian concepts of space and time was only indirectly influenced by the discovery of non-Euclidean geometries, namely only insofar as the existence of different geometries stimulated a revision of the foundations of science, Poincaré's epistemological study of these concepts, including the concept of simultaneity, was directly connected with this discovery. According to Poincaré the very existence of consistent alternative geometries shows that their axioms, which determine whether a geometry is Euclidean (parabolic), Lobachevskian or Bolyaian (hyperbolic), or Riemannian (elliptic), are neither synthetic judgments a priori nor experimental conclusions but merely more or less arbitrary conventions.[10] Our choice of conventions, of course, is limited by the requirement to avoid contradictions and the requirement of simplicity or

[9] J. Thomson, "On the law of inertia, the principle of chronometry and the principle of absolute clinural rest, and of absolute rotation," *Proceedings of the Royal Society of Edinburgh* **12,** 568–578 (1884). Quotation on p. 569.

[10] H. Poincaré, "On the foundations of geometry," *Monist* **9,** 1–43 (1898); *La Science et l'Hypothèse* (Paris: Flammarion, 1902), part 2, chapter 3; *Science and Hypothesis* (London: Scott, 1905; New York: Dover Publications, 1952).

economy. Although Euclidean geometry, according to Poincaré, will always be the most convenient geometry, it would be wrong to say that it is the *true* geometry, just as it would be wrong to say that the metric system (the decimal system of length and weight) is the *true* system of measures. It would be possible to construct a "dictionary" of the geometrical terms used in the different geometrical systems by which one could "translate" theorems of one geometry into theorems of another geometry. Such a "dictionary" would even guarantee the lack of contradiction of any of these systems if only one of them, for example, the Euclidean geometry, is free of contradictions.

Although Poincaré distinguished between geometry and mechanics insofar as the latter contains experimental laws that are not conventions, he argued that the *principles* of mechanics are conventions just like the axioms of geometry. We accept Newton's three laws as the foundations of mechanics, he contended, because they are the simplest laws but not because they are *true*.[11] By applying this argument to Newton's "First Law," the law of inertia, Poincaré would have been able to extend his conventionalism from the realm of geometrical or spatial conceptions to that of temporal conceptions such as the equality of two intervals of time. According to the law of inertia, as stated in Newton's *Principia*, "every body continues in its state of rest, or of uniform motion in a right line, unless compelled to change that state by forces impressed upon it" or, expressed in brief, a free particle moves always with constant velocity. But if such a particle covers equal distances in equal time intervals, the problem, discussed by Isaac Barrow,[12] of how to verify the equality of two temporally separated time intervals seems to easily find its solution. It would suffice to measure the equal distances covered by such a particle to ensure the equality of the time intervals corresponding to these distances. If as Poincaré contends, however, the law of inertia is merely a convention, it is not necessarily true that the time intervals under discussion are "really" equal in duration.

[11] That Newton's *Second Law*, in any of its formulations, is not a law but rather a definition of the concept of "force" or of "mass" has been argued long before Poincaré. See M. Jammer, *Concepts of Force* (note 52 in chapter 5), chapter 11. Modern textbooks of mechanics also emphasize this fact. Thus, for example, M. Alonso and E. J. Finn in *Fundamental University Physics* (Reading, Massachusetts: Addison-Wesley, 1967), vol. 1, p. 159, state that "*Newton's second law of motion* . . . is more a definition than a law."

[12] See chapter 5, notes 4 and 6.

In his essay, *La Mesure du Temps,*[13] published in 1898, Poincaré explained in detail why he regarded any statement about the equality of two intervals of time or about the simultaneity of spatially separated events as merely a convention, that is, as a matter of definition rather than of facts. Starting with the psychological remark that "we have not a direct intuition of the equality of two intervals of time" he based his contention of the conventionalism of the first statement on the argument that

> "if another way of measuring time would be adopted, the experiments on which Newton's law is founded would none the less have the same meaning. Only the enunciation of the law would be different, because it would be translated into another language. . . . Time should be so defined that the equations of mechanics may be as simple as possible. In other words, there is not one way of measuring time more true than another; that which is generally adopted is only more *convenient.* Of two watches, we have no right to say that the one goes true, the other wrong; we can only say that it is advantageous to conform to the indications of the first.[14]

The analogy between these temporal arguments and Poincaré's arguments for his geometrical conventionalism is obvious. Poincaré concluded his discussion on the epistemological status of statements concerning the equality of time intervals by pointing out—and rightly as we know—that this issue had already been dealt with by others before him.

Turning thereafter to the concept of simultaneity Poincaré remarked that this "second difficulty has up to the present attracted much less attention; yet it is altogether analogous to the preceding; and even, logically, I should have spoken of it first." Poincaré then offered what may be regarded as the first modern monograph on the concept of simultaneity, which therefore deserves to be discussed in detail. He began it by asking the following questions: "Two psychological phenomena happen in two different consciousnesses; when I say they are simultaneous, what do I mean? When I say that a physical phenomenon, which happens outside of every consciousness, is before or after a

[13] H. Poincaré, "La Mesure du Temps," *Revue de Métaphysique et de Morale* **6,** 1–13 (1898); reprinted in H. Poincaré, *La Valeur de la Science* (Paris: Flammarion, 1905, 1923), pp. 35–58; "The Measure of Time" in *The Value of Science* (New York: The Science Press, 1907; Dover Publications, 1958); pp. 26–36; also in H. Poincaré, *The Foundations of Science* (New York: The Science Press, 1913), pp. 223–234; and in M. Čapek (note 1), pp. 317–327.

[14] Ibid. (1923), pp. 44–45.

psychological phenomenon, what do I mean?" As an example he asked what it means to say that the explosion of the supernova creating a new star, observed by Tycho Brahe in 1572, happened before the discovery of America, that is, before the formation of the visual image of the isle of Española in the consciousness of Christopher Columbus. According to Poincaré "a little reflection suffices to understand that all these affirmations have by themselves no meaning. They can have one only as the outcome of a convention."[15] To prove this contention Poincaré analyzes several examples of using the notions of simultaneity, antecedence, or succession and shows that they involve explicitly or implicitly certain conventional assumptions. Concerning the often made statement that two "facts" should be regarded as simultaneous if the order of their succession could be interchanged, he pointed out that "this definition would not suite two physical facts which happen far from one another, and that, in what concerns them, we no longer even understand what this reversibility would be; besides, succession itself must first be defined."[16]

To show that statements about the succession or antecedence of facts or events depend ultimately on conventions Poincaré referred in the sequel to the vicious circle inherent in defining cause and effect in terms of a temporal sequence and a temporal sequence in terms of cause and effect: "We say now *post hoc, ergo propter hoc;* now *propter hoc, ergo post hoc*; shall we escape from this vicious circle?" Poincaré's answer was "only by convenience and simplicity."[17] He then studied the role of definitions or conventions in the work of scientists, especially of those who like Roemer measure the velocity of light.

> When an astronomer tells me that some stellar phenomenon, which his telescope reveals to him at this moment happened, nevertheless, fifty years ago, I seek his meaning, and to that end I shall ask him first how he knows it, that is, how he has measured the velocity of light. He has begun by *supposing* that light has a constant velocity, and in particular that its velocity is the same in all directions. That is a postulate without which no measurement of this velocity could be attempted. This postulate could never be verified directly by experiment; it might be contradicted by it if the results of different measurements were not concordant. We should think ourselves fortunate that this contradiction has

[15] "Il suffit d'un peu de réflexion pour comprendre que toutes ces affirmations n'ont par elles-mêmes aucun sens. Elles ne peuvent en avoir un que par suite d'une convention." Ibid., p. 46.

[16] Ibid., p. 48.

[17] Ibid., p. 53.

> not happened and that the slight discordances which may happen can be readily explained. . . . This postulate assumed, let us see how the velocity of light has been measured. You know that Roemer used eclipses of the satellites of Jupiter, and sought how much the event fell behind its prediction. But how is the prediction made? It is by the aid of astronomical laws: for instance Newton's law.[18]

Poincaré contended, however, that the observed facts could be accounted for just as well if we attributed to the velocity of light a slightly different value from that adopted and supposed a slightly different form of Newton's law, which might be more complicated. "So for the velocity of light a value is adopted, such that the astronomic laws compatible with this value may be as simple as possible." Concerning the notion of simultaneity he added that "it is difficult to separate the qualitative problem of simultaneity from the quantitative problem of the measurement of time; no matter whether a chronometer is used, or whether account must be taken of a velocity of transmission, because such a velocity could not be measured without *measuring* time." Poincaré concluded the essay by emphasizing that our notions of the equality of durations and of distant simultaneity are based on rules that are not imposed on us and could be replaced by other rules that might complicate the enunciation of the laws of physics, mechanics, and astronomy. We therefore choose these rules, not because they are true but because they are the most convenient, and we may recapitulate them as follows: "The simultaneity of two events, or the order of their succession, the equality of two durations, are to be so defined that the enunciation of the natural laws may be as simple as possible. In other words, all these rules, all these definitions are only the fruit of an unconscious opportunism."[19] As this statement shows, identifying a synchronization procedure for defining or testing the simultaneity of events is, according to Poincaré, from the purely logical point of view, a matter of free choice or convention, but from the scientific point of view it is a choice constrained by the requirement of producing a physical system which is as simple as possible.

Poincaré's article *La Mesure du Temps* obviously was primarily a philosophical or, more precisely, an epistemological study of the nature of time and, in particular, of the concept of simultaneity whose objective meaning it called into question. It did not deal with the more physical problem concerning the possibility of synchronizing clocks that could be used to test

[18] Ibid., pp. 53–55.
[19] Ibid., pp. 57–58.

whether spatially separated events are simultaneous. Poincaré ignored this clock synchronization problem, though it was intimately connected with the notion of simultaneity, because it would involve an issue he did not discuss in this 1898 essay. As a firm believer in the ether theory he thought that, to resolve such a problem, one must consider clocks not only at rest but also in motion relative to the ether: a satisfactory solution could be obtained, therefore, only within the context of a study of the relativity of motion.

One of the most perplexing problems of those days was the question of how to explain the experimental undetectability of "the absolute motion of matter, or rather the relative motion of ponderable matter with respect to the ether."[20] For Poincaré this undetectability was the manifestation of an empirical principle that he called at first "the principle of relative motion"[21] and later, in his 1904 Saint Louis address before the International Congress of Arts and Sciences, "the principle of relativity." The principle of relativity states, as he put it, that "the laws of physical phenomena should be the same, whether for an observer fixed, or for an observer carried along in a uniform movement of translation; so that we have not and could not have any means of discerning whether or not we are carried along in such a motion."[22] To understand how this issue is connected with the problem of distant simultaneity, recall that in 1895 Hendrik Antoon Lorentz[23] simplified the mathematical treatment of electromagnetic phenomena in a reference system, moving with the velocity v in the positive direction of the x axis relative to the

[20] "L'expérience a révélé une foule de faits qui peuvent se résumer dans la formule suivante: il est impossible de rendre manifeste le mouvement absolu de la matière, ou mieux le mouvement relatif de la matière pondérable par rapport à l'éther," H. Poincaré, "A propos de la théorie de M. Larmor," *L'Éclairage Électrique* **5,** 5–14 (1895); reprinted in *Oeuvres de Henri Poincaré* (Paris: Gauthier-Villars, 1934–1956), (1954), vol. 9, pp. 395–423. Quotation on p. 412.

[21] "Principe du mouvement relatif" H. Poincaré, "La théorie de Lorentz et le principe de la réaction," *Archives néerlandaises des Sciences exactes et naturelles* **5,** 252–278 (1900); *Oeuvres* (op. cit.), pp. 464–488.

[22] "Le principe de la relativité, d'après lequel les lois des phènomènes physiques doivent être les mêmes, soit pour un observateur fixe, soit pour un observateur entrainé dans un mouvement de translation uniforme; de sorte que nous n'avons et ne pouvons avoir aucun moyen de discerner si nous sommes, oui ou non, emportés dans un pareil mouvement." H. Poincaré, "L'état actuel et l'avenir de la Physique mathèmatique," *Bulletin des Sciences Mathématiques* **28,** 302–324 (1904); reprinted in H. Poincaré, *La Valeur de la Science* (Paris: Flammarion, 1905, 1923), pp. 170–211. Quotation on pp. 176–177. *The Foundations of Science* (1913), p. 300; also in *The Monist* **15,** 1–24 (1905).

[23] H. A. Lorentz, *Versuch einer Theorie der electrischen und optischen Erscheinungen in bewegten Körpern* (Leiden: Brill, 1895); reprinted in H. A. Lorentz, *Collected Papers* (Hague: Nijhoff, 1935–1939), vol. 5, pp. 1–137.

ether by introducing what he called the "local time" t'; it differs at x' from the "true time" by the amount $v\, x'/c^2$. Poincaré, following Lorentz, realized that the laws of physics satisfy the principle of relativity if they are formulated in terms of "local time." He therefore proposed the following synchronization procedure for spatially separated clocks.

> Imagine two observers who wish to adjust their timepieces by optical signals; they exchange signals, but as they know that the transmission of light is not instantaneous, they are careful to cross them. When station B perceives the signal from station A, its clock should not mark the same hour as that of station A at the moment of sending the signal, but this hour augmented by a constant representing the duration of the transmission. Suppose, for example, that station A sends its signal when its clock marks the hour 0, and that station B perceives it when its clock marks the hour *t*. The clocks are adjusted if the slowness equal to *t* represents the duration of the transmission, and to verify it, station B sends in its turn a signal when its clock marks 0; then station A should perceive it when its clock marks *t*. The timepieces are then adjusted. And in fact they mark the same hour at the same physical instant, but on the one condition, that the two stations are fixed. Otherwise the duration of the transmission will not be the same in the two senses, since the station A, for example, moves forward to meet the optical perturbation emanating from B, whereas the station B flees before the perturbation emanating from A. The watches adjusted in that way will not mark, therefore, the true time; they will mark what may be called the *local time*, so that one of them will be slow of the other. It matters little, since we have no means of perceiving it. All the phenomena which happen at A, for example, will be late, but all will be equally so, and the observer will not perceive it, since his watch is slow; so, as the principle of relativity requires, he will have no means of knowing whether he is at rest or in absolute motion.[24]

Because the synchronization of spatially separated clocks is an operational procedure to establish distant simultaneity, the conventionality of simultaneity implies that the synchronization procedure must also be a matter of convention. Considering therefore alternative synchronization procedures Poincaré raised the question: "What would happen if one could communicate by non-luminous signals whose velocity of propagation differed from that of light? If, after having adjusted the watches by the optical procedure,

[24] H. Poincaré, *Oeuvres* (note 20), vol. 9, p. 486. *The Foundations of Science*, pp. 306–307.

we wished to verify the adjustment by the aid of these new signals, we should observe discrepancies which would render evident the common translation of the two stations. And are such signals inconceivable, if we admit with Laplace that universal gravitation is transmitted a million times more rapidly than light?"[25] Poincaré's treatment of clock synchronizations agreed, as we see, with his 1898 epistemological comments on the concept of simultaneity insofar as he admitted the possibility of several physical procedures to establish distant simultaneity, provided they agreed with each other and contributed to the simplification of the physical theory.[26] One may wonder why Poincaré did not dispense with the distinction between "true" and "local" time and did not continue in his physical studies the approach taken in his philosophical essay of 1898. As a student of Poincaré's work aptly remarked: "While Poincaré may well have used a conventionalistic position while talking *about* the nature of physics, in his own work in theoretical physics he was anything but a conventionalist."[27]

In *La dynamique de l'électron*,[28] published in 1908, three years after the birth of Einstein's special theory of relativity, Poincaré discussed once more the concept of simultaneity but without adding anything new in substance. The only remarkable feature it displayed in this context was Poincaré's total disregard of Einstein's 1905 paper on the theory of relativity to which of all Einstein's precursors he has come closest to anticipating.[29]

[25] H. Poincaré, *Oeuvres*, vol. 9, p. 487. *The Foundations of Science*, p. 308.

[26] Note that, as the quotations show, Poincaré never used the French term "synchronizer" or any of its derivations, although such terms were current in French at that time, as P. Larousse's *Grand Dictionnaire Universel du XIX Siècle* (Paris, 1866–1876) testifies. Instead he usually applied the term "régler" (to regulate, to adjust), which in its standard combination "régler sa montre" means "to set one's clock right" and only by implication "to synchronize one's clock."

[27] S. Goldberg, "Henri Poincaré and Einstein's theory of relativity," *American Journal of Physics* **35,** 934–944 (1967).

[28] H. Poincaré, "La dynamique de l'électron," Revue générale des Sciences pures et appliquées **19,** 386–402 (1908).

[29] Much has been written on the role of Poincaré in the early development of relativity and, in this context, on his use of the notions of local time and the ether. The reader interested in further details may find valuable information in the following articles: G. Holton, "On the origin of the special theory of relativity," *American Journal of Physics* **28,** 627–638 (1960); Ch. Scribner, Jr., "Henri Poincaré and the principle of relativity," *American Journal of Physics* **32,** 672–678 (1964); S. Goldberg, "Henri Poincaré and Einstein's theory of relativity," *American Journal of Physics* **35,** 934–944 (1967); A. I. Miller, *Albert Einstein's Special Theory of Relativity* (Reading, Massachusetts: Addison-Wesley, 1981); A. Borel, "Henri Poincaré and special relativity," *L'Enseignement Mathématique* **45,** 281–300 (1999); J. Reignier, "Éther et mouvement absolu au 19e siècle," *Revue des Questions Scientifiques* **170,** 261–282 (1999).

CHAPTER SEVEN

Simultaneity in the Special Theory of Relativity

From Albert Einstein's letters, written between 1898 and 1902 to his fiancée Mileva Marić, we know that as a student Einstein had already been deeply interested in the ether theories of electrodynamics and in the problem of the detectability of the Earth's motion through the supposedly immobile ether. We also know from his correspondence with his lifelong friend Michele Besso[1] and from a remark made by Maurice Solovine about what he had read and discussed when he met with Einstein in Bern that Einstein had been "profoundly impressed"[2] by Poincaré's *Science and Hypothesis* in which the 1898 article *La Mesure de Temps* is briefly mentioned. Whether Einstein ever read this article is not known.

After having studied Heinrich Hertz's reformulation of Maxwell's electrodynamics Einstein wrote in August 1899 to Mileva that he was becoming

[1] P. Speziali, *Albert Einstein—Michele Besso: Correspondence 1902–1955* (Paris: Hermann, 1972), p. 464.

[2] "La Science et l'Hypothèse de Poincaré, un livre qui nous a profondément impressionés et tenus en haleine pendant de longues semaines . . . " in M. Solovine (ed.), *Albert Einstein: Lettres à Maurice Solovine* (Paris: Gauthier-Villars, 1956), p. VIII (introduction).

more and more convinced that the electrodynamics of moving bodies, as currently presented, is not correct, and that it should be possible to present it in a simpler way. All his attempts to construct such a theory on the basis of the relativity principle and the principle of the invariance of the velocity of light were thwarted by the apparently irreconcilable conflict between the light principle and the rule of the addition of velocities as used in mechanics. In an impromptu talk on the creation of the theory of relativity, delivered at Kyoto University on 14 December 1922, Einstein reportedly gave the following account:

> Why do these two concepts contradict each other? I realized that this difficulty was really hard to resolve. I spent almost a year in vain trying to modify the idea of Lorentz in the hope of resolving this problem. By chance a friend of mine [Michelo Besso] in Bern helped me out. It was a beautiful day when I visited him with this problem. I started the conversation with him in the following way: "Recently I have been working on a difficult problem. Today I come here to battle against that problem with you." We discussed every aspect of this problem. Then suddenly I understood where the key to this problem lay. Next day I came back to him and said to him, without even saying hello, "Thank you. I've completely solved the problem. An analysis of the concept of time was my solution." Time cannot be absolutely defined, and there is an inseparable relation between time and signal velocity. With this new concept I could resolve all the difficulties completely for the first time. Within five weeks the special theory of relativity was completed.[3]

That it was indeed a new conception of time that played such a crucial role had been emphasized by Einstein in 1907 when he wrote in an essay summarizing his new theory: "It turned out, surprisingly, that it was only necessary to provide a sufficiently precise formulation of the notion of time in order to overcome the difficulty encountered."[4]

[3] A. Einstein, "How I created the theory of relativity," *Physics Today* **35,** 45–47 (August 1982). This is a translation into English by Yoshimasa A. Ono of Jun Ishiwara's Japanese translation of Einstein's talk delivered in German. Concerning the authenticity of the translation, see H. J. Haubold and E. Yasui, "Jun Ishiwaras Text über Albert Einsteins Gastvortrag an der Universität zu Kyoto am 14. Dezember 1922," *Archive for History of Exact Sciences* **36,** 271–279 (1986).

[4] "Es zeigte sich aber uberraschenderweise, dass es nur nötig war, den Begriff der Zeit genügend scharf zu fassen, um über die soeben dargelegte Schwierigkeit hinweg zu kommen." A. Einstein, "Über das Relativitätsprinzip und die aus demselben gezogenen Folgerungen," *Jahrbuch der Radioaktivität und Elektronik* **4,** 411–462 (1907). Quotation on p. 413. *The Collected Papers of Albert Einstein* (Princeton, New Jersey: Princeton University Press, 1989), vol. 2, pp. 432–484; *English Translations* (Princeton, 1989), vol. 2, pp. 252–311. Quotation on p. 253.

Because the acknowledgment at the end of Einstein's famous 1905 paper, which presents his new conception of time and of simultaneity, in particular, mentions only Michele Angelo Besso to whom he is "indebted for several valuable suggestions," as stated in his Kyoto lecture, it would be interesting to know in what respect precisely Besso should be credited with having assisted Einstein in resolving "all the difficulties completely for the first time." Unfortunately no documentary evidence apparently exists that could answer this question.

Albrecht Fölsing recently suggested a possible answer in his biography of Einstein. He speculated that the two friends "had before them one of Poincaré's papers in which he presented his method for the synchronization of clocks as being equivalent to Lorentz's 'local time'—either his 1904 lecture in St. Louis or his contribution to the Lorentz *Festschrift.*" That the latter must have been available to Einstein in Bern Fölsing deduced from the fact that Einstein quoted it a year later, though in a different context. Fölsing also considered the possibility that Einstein could have read the contents of Poincaré's St. Louis lecture "in a copy, hot off the press, of a collection of essays called *The Value of Science.*" In addition, Fölsing deemed it likely

> that in their conversation Einstein and Besso discovered some aspects of Poincaré's synchronization procedure that may have escaped Poincaré himself. How would it be—the two friends, by then skeptical about "true time," might have asked—if the time defined by Poincaré's experiment was not just a mathematical device for Lorentz's "local time" but in fact everything that a physicist could expect of a meaningful concept? Admittedly this would give a different "time" for every inertial system, but the constancy of the velocity of light for any observer would in that case be inherent in Poincaré's definition of simultaneity and would not, as with Lorentz, have to be forcibly brought about by a laborious adjustment to theory.[5]

The most striking evidence, that it was the notion of time and specifically that of simultaneity which started the final development of this theory, was provided by Einstein's first paper on it, his seminal 1905 essay, "On the dy-

[5] A. Fölsing, *Albert Einstein* (New York: Viking, 1997), pp. 176–177. Originally published in German (Frankfurt am Main: Suhrkamp, 1993), pp. 201–292.

namics of moving bodies."[6] After a short introduction in which Einstein says that the insufficient consideration of the relationship between rigid bodies (systems of coordinates), clocks, and electromagnetic processes lies at the root of the difficulties that the electrodynamics of moving bodies at present encounters, the first section of the main text, carrying the subtitle "§ 1. Definition of Simultaneity," began with an analysis of the notions of simultaneity and time: "If we want to describe the motion of a material point, we give the values of its coordinates as a function of time. However, we should keep in mind that for such a mathematical description to have physical meaning, we first have to clarify what is to be understood here by 'time.' We have to bear in mind that all our propositions involving time are always propositions about *simultaneous events*." Einstein explains this by an example. "If, for instance, I say 'the train arrives here at 7 o'clock,' that means more or less, 'the pointing of the small hand of my clock to 7 and the arrival of the train are simultaneous events,' "—a statement of which Leopold Infeld once commented that it was "the simplest sentence I have ever encountered in a scientific paper."[7]

This example, Einstein declared, suggested that it might be possible to overcome all difficulties involved in the definition of "time" simply by substituting "position of the small hand of my clock" for "time." Such a definition would indeed be sufficient, Einstein continued, if time had to be defined only for the place where the clock is located, but this definition ceases to be satisfactory as soon as we have to connect temporally events that occur at different places, or what amounts to the same, events occurring at locations remote from the clock.

The identification of the time of an event with the reading of a clock may suggest, Einstein pointed out, "that we could content ourselves with the time values determined by an observer stationed together with the clock at the origin of the coordinate system, and coordinating the corresponding positions of the hands with light signals, given out by every event to be timed, and reaching him through empty space" or, in brief, to define the time of an event as the time at which an observer *sees* the event.

[6] Einstein, "Zur Elektrodynamik bewegter Körper," *Annalen der Physik* **17,** 891–921 (1905). Reprinted in *The Collected Papers of Albert Einstein* (Princeton, New Jersey: Princeton University Press, 1987), vol. 2, pp. 276–306; "On the electrodynamics of moving bodies," *The Collected Papers of Albert Einstein—English translations* (by A. Beck), (Princeton, New Jersey: Princeton University Press, 1989), vol. 2, pp. 140–171. Also in H. A. Lorentz, A. Einstein et al., *The Principle of Relativity* (London: Methuen, 1923; New York: Dover Publications, 1952), pp. 37–65.

[7] L. Infeld, *Albert Einstein—His Work and Influence on our World* (New York: Scribner's Sons, 1950), p. 27.

Einstein, however, rejected this method of associating time variables with events on the grounds that "such a coordination has the disadvantage that it is not independent of the standpoint of the observer with the clock, as we know from experience. We arrive at a much more practical determination along the following line of thought."

Having thus argued for the priority of the notion of simultaneity over that of time Einstein assumed that the concept of the simultaneity of spatially contiguous events or, as we have called it "local simultaneity," posed no physical problem. Einstein also assumed tacitly that instead of defining distant simultaneity it suffices to define the synchronism of spatially separated clocks. For, obviously, distant events are simultaneous if and only if the readings of synchronized clocks at their positions are the same. Because of its historical importance Einstein's 1905 clock-synchronization procedure or, equivalently, his definition of distant simultaneity is quoted *in extenso*.

> If there is a clock at point *A* of space, then an observer located at *A* can evaluate the time of the events in the immediate vicinity of *A* by finding the clock-hand positions that are simultaneous[8] with these events. If there is also a clock at point *B*—we should add, "a clock of exactly the same constitution as that at *A*"—then the time of the events in the immediate vicinity of *B* can likewise be evaluated by an observer located at *B*. But it is not possible to compare the time of an event at *A* with one at *B* without a further stipulation; thus far we have only defined an "*A*-time" and a "*B*-time" but not a "time" common to *A* and *B*. The latter can now be determined by establishing *by definition* that the "time" needed for the light to travel from *A* to *B* is equal to the "time" it needs to travel from *B* to *A*. For, suppose a ray of light leaves from *A* toward *B* at "*A*-time" t_A, is reflected from *B* toward *A* at "*B*-time" t_B, and arrives back at *A* at "*A*-time" t_A'. The two clocks are synchronous by definition if $t_B - t_A = t_A' - t_B$.[9]

For the sake of later references the German original of the third last sentence, beginning with "The latter can now be determined . . . ", should be quoted. It reads: "Die letztere Zeit kann nun definiert werden, indem man *durch Definition* festsetzt, dass die 'Zeit', welche das Licht braucht, um von *A*

[8] The term "simultaneous" used here refers, of course, to "local simultaneity" and not to "distant simultaneity." Otherwise a *circulus in definiendo* would be involved.

[9] See note 6 (1905), pp. 893–894; (1952), pp. 39–40; (1989), p. 142. Italics in original.

nach *B* zu gelangen, gleich ist der 'Zeit', welche es braucht, um von *B* nach *A* zu gelangen."[10]

Einstein concluded this first paragraph of his 1905 paper with a statement of the light postulate which in the present context he formulated as follows: "In agreement with experience we further assume the quantity $2\ AB\ /\ (t'_A - t_A) = c$ to be a universal constant—the velocity of light in empty space."

Note that in Einstein's clock-synchronization or simultaneity definition the term "time" was used with three different meanings. In the combination "*A*-time" or "*B*-time" it denotes the *reading* or *date* indicated by the clock at *A* or *B*, respectively; in the expression "the 'time' needed for the light to travel . . . " it denotes a *time interval*; and in the phrase "a 'time' common to *A* and *B*" it denotes what is often called *coordinate time*, that is, the set of all time coordinates associated with a coordinate system.

Keeping the exact wording of Einstein's simultaneity definition in mind, we find it appropriate now to digress into some of the historical aspects of the acceptance of this definition. This digression, which will also contribute to a deeper understanding of the definition, deals with the strange story of how two almost identical, and yet mutually independent, misrepresentations of Einstein's simultaneity definition gained considerable importance in the physical and philosophical literature on relativity.

By about 1913 the theory of relativity had attracted widespread attention and easily accessible collections of the original papers on this issue were in demand. Hermann Minkowski's much discussed popular Cologne lecture *Space and Time*, published 1909 as a separate reprint with an introduction by the Halle mathematician August Gutzmer (and sold for 0.80 Mark) had already been out of print for a long time. When a second edition, to be published in the B. G. Teubner series *Fortschritte der mathematischen Wissenschaften in Monographien*, was being considered, Arnold Sommerfeld suggested to Otto Blumenthal, a professor of mathematics at the Technical University of Aachen and editor of this series, that he also include articles by Lorentz and Einstein. The first edition of *Das Relativitätsprinzip*, a collection of original papers by Lorentz, Einstein, and Minkowski, was thus published in 1913. It became a best-seller, and every three years or so, a new and enlarged edition appeared on the market, the fourth in 1923.

In the fall of 1920 a young English mathematician, George Barker Jeffery, wrote a letter to Einstein in which he complained about the short-

[10] Note 6 (1905), p. 895.

age of English-written mathematical treatises on relativity. He suggested that

> the publication of an English translation of a carefully selected group of your papers would probably provide a better exposition of the subject from the point of view of the serious mathematical student than any now existing in English and would certainly prove a record of great historical worth showing the way in which the theory grew up in the mind of the creator. I should like to hear how a proposition of this kind appeals to you. I have consulted in a preliminary way with a colleague of mine, Dr. W. Perrett who is a distinguished German scholar and he would be prepared to collaborate with me in the work of translation on the basis that he should overlook the literary side of the work while I looked after the mathematics. Dr. Perrett is a doctor of philosophy of Heidelberg and I am a doctor of science of London. We are respectively lecturers in German and in Applied Mathematics in the University of London.

In a postscript Jeffery added: "If you approve of this suggestion I hope that the work may be an official translation with possibly an introductory chapter from yourself and published under some such title as *Relativity and the Theory of Gravitation* by Prof. Albert Einstein, being an authorised translation of the original papers by Dr. G. B. Jeffery and Dr. W. Perrett."[11]

Jeffery, a gifted mathematician who was to become in 1926 a Fellow and in 1938 Vice President of the Royal Society, was interested primarily in the general theory of relativity. When writing this letter he had just completed his first paper on this subject; it dealt with some mathematical issues related to the deflection of light in a gravitational field.[12] Recall that only a few months earlier Einstein's prediction of such a deflection had been publicly confirmed when on 6 November 1919, the Royal Society announced in London the official results of the two famous solar eclipse expeditions, an announcement that made Einstein an overnight international celebrity. These circumstances explain Jeffery's eagerness to enlist Wilfrid Perrett, an erudite author of many literary works,[13] to collaborate with him on the project of translating Einstein's papers on relativity.

[11] Letter from G. B. Jeffery to A. Einstein, dated London, 14 October 1920; Einstein Archive (National and University Library, Jerusalem), reel 13-432.

[12] G. B. Jeffery, "On the path of a ray of light in the gravitational field of the sun," *Philosophical Magazine* **40,** 327–329 (1920).

[13] W. Perrett, *The Story of King Lear from Geoffrey of Monmouth to Shakespeare* (Berlin: Mayer und Müller, 1904); *Some Questions of Phonetic Theory* (London: University Press, 1916); *Poetickay: an Essay towards the Abolition of Spelling* (Cambridge: Heffers, 1920).

Einstein replied that he had no objection to such a translation and added that Teubner had published his more important essays in a collection of papers that he would mail to Jeffery; but also asked Jeffery to include the papers by Lorentz and Minkowski.[14] This then is the history of how the Perrett and Jeffery translation of the German original *Das Relativitätsprinxip* was published in 1923, first by Methuen and Company in London and by Dodd, Mead and Company in New York, and subsequently in numerous reprintings by Dover Publications in New York.[15] It was not the first English translation[16] of Einstein's 1905 relativity paper nor its last,[17] but because it is undoubtedly the most widely circulated English translation in the world, an error committed in it could have serious consequences.

Such an error was committed by Jeffery and Perrett, and specifically in their translation of Einstein's definition of distant simultaneity. For his sentence beginning with the words "The latter can now be determined . . . ", quoted above also in its original German version,[18] was translated by them as follows: "For the latter cannot be defined at all unless we establish *by definition* that the 'time' required by light to travel from *A* to *B* equals the 'time' to travel from *B* to *A*." Comparison of this version with the original shows that Perrett and Jeffery presented the equal-time stipulation as a *necessary* condition for the establishment of clock-synchrony or simultaneity, whereas Einstein propounded it only as a *sufficient* condition. In other words, Einstein did not commit himself on whether alternative possibilities of defining distant simultaneity exist, whereas according to Perrett and Jeffery he denied such a possibility.

That the Perrett and Jeffery translation differs from the original has been noted previously by Charles Scribner Jr., who stated[19] that the translators "by an unnecessary elaboration of the text" have "somewhat altered its exact

[14] Letter from Einstein to Jeffery, dated Berlin, 14 December 1920; Einstein Archive, reel 13-436.

[15] A. Einstein, H. A. Lorentz, H. Minkowski, and H. Weyl, *The Principle of Relativity—with Notes by A. Sommerfeld, translated by W. Perrett and G. B. Jeffery* (New York: Dover Publications).

[16] The earliest English translation of Einstein's 1905 relativity paper is found in *The Principle of Relativity: Original Papers by A. Einstein and H. Minkowski*, translated by M. N. Saha and S. N. Bose (Calcutta: University of Calcutta, 1920).

[17] It has been translated into English also by M. H. Shamos (1929), C. Kittel et al. (1965), C. Kacser (1967), L. Pearce Williams (1968), C. W. Kilmister (1977), H. M. Schwartz (1977), A. I. Miller (1981), and A. Beck (1989), see note 6.

[18] See note 6.

[19] Ch. Scribner Jr., "Mistranslation of a passage in Einstein's original paper on relativity," *American Journal of Physics* **31,** 398 (1963).

meaning." For Scribner the translation was defective because it seemed to him to suggest that distant simultaneity—and by implication "time"—are definable only by means of the propagation of light. To prove his point Scribner referred to Einstein's Stafford Little Lecture of 1921 in which Einstein defended himself against the charge that his theory ascribes an excessive role to the propagation of light by founding upon it the definition of such a fundamental notion like "time."

Before showing that this mistranslation involves a much more profound issue let us try to identify the cause of this error. The explanation of his error seems to hinge on Einstein's use of the word "nun" in his statement, "Die letztere Zeit kann nun definiert werden, indem man *durch Definition* festsetzt . . . " The German adverb "nun," etymologically related to the Latin "nunc" and the English "now," of course literally means "now" or "at the present time," but is frequently employed in a nontemporal sense as a paratactical connective between two sentences and can therefore be omitted in such cases without any change in the logical contents of the sentences it connects. Shortly before he used this adverb "nun" Einstein emphasized that "it is not possible to compare the time of an event at *A* with one at *B* without a further stipulation," expressing thereby a necessary condition. It is very likely, therefore, that whoever translated this passage, having in mind the just-quoted necessary condition, misread "nun" as "nur," which in German means "only" or "solely" and denotes a necessary condition.[20]

That such a misrepresentation of only a single letter can have serious consequences, even if a translation is not involved at all, is shown by Hugo Dingler's misquotation of Einstein's 1905 definition of distant simultaneity. Dingler, who because of his insistence on a constructive-axiomatic foundation of physics is often regarded today as the progenitor of "protophysics," published in 1921 *Physik und Hypothese*[21] in which he criticized Einstein's theory of relativity. One of his main critical arguments is based on Einstein's definition of distant simultaneity, or rather on his quotation of it which reads as follows: "Herr Einstein says: so far we have not defined a time common for *A* and *B*. He continues: This latter can only be defined through decreeing by definition that the 'time' required by light to travel from *A* to *B* equals the 'time' to travel from *B* to *A*. The expression 'can only' proves that the

[20] The Saha and Bose translation, mentioned in note 16, does not contain this error.

[21] H. Dingler, *Physik und Hypothese—Versuch einer induktiven Wissenschaftslehre nebst einer kritischen Analyse der Fundamente der Relativitätstheorie* (Berlin: W. de Gruyter, 1921).

natural definition of simultaneity was, in fact, unknown to Herr Einstein. This expression is, however, incorrect not only with respect to the natural definition of simultaneity but also in so far as there exist infinitely many possibilities of other determinations."[22] Again, Einstein's "nun" has been misquoted by "nur." We can safely rule out the possibility that Dingler's misquotation was the cause of the Perrett and Jeffery mistranslation, for Dingler's book was virtually unknown outside Germany and Perrett and Jeffery never deviated from the text of the fourth edition of *Das Relativitätsprincip*, with the exception of Einstein's simultaneity definition. That their mistranslation was an unintentional result of sheer inadvertence can hardly be doubted. The same is probably true for Dingler's misrepresentation.[23]

After this lengthy historical digression let us return to Einstein's own words that follow his definition of distant simultaneity. He wrote:

> We assume "that it is possible for this definition of synchronism to be free of contradictions, and to be so for arbitrarily many points, and that the following relations are therefore generally valid: (1) If the clock in *B* is synchronous with the clock in *A*, then the clock in *A* is synchronous with the clock in *B*. (2) If the clock in *A* is synchronous with the clock in *B* a well as with the clock in *C*, then the clocks in *B* and *C* are also synchronous relative to each other." In other words, Einstein *assumes* that synchrony, and therefore also simultaneity, is a symmetric and transitive relation. Einstein then summarizes his clock-synchronization or simultaneity procedure as follows: "With the help of some physical (thought) experiments, we have thus laid down what is to be understood by synchronous clocks at rest that are situated at different places, and have obviously obtained thereby a definition of 'synchronous' and of 'time.' The 'time' of an event is the reading obtained simultaneously with the event from a clock at rest that is

[22] "Herr Einstein sagt: wir haben bisher keine für A und B gemeinsame Zeit definiert. Er fährt fort: 'Die letztere kann nur definiert werden, indem man durch Definition festsetzt, dass die 'Zeit', welche das Licht braucht, um von A nach B zu gelangen, gleich ist der 'Zeit', welche es braucht, um von B nach A zu gelangen." Das Wörtchen 'kann nur' beweist, dass Herrn Einstein die natürliche Definition der Gleichzeitigkeit tatsächlich unbekannt war. Diese Wörtchen sind aber nicht nur in Hinblick auf die natürliche Definition der Gleichzeitigkeit unrichtig, sondern auch in der, dass es unbegrenzt viele Möglichkeiten anderer Festsetzungen gibt." Ibid., p. 162.

[23] Dingler's critique of Einstein in 1921 was not yet biased, as it was later, by antisemitic tendencies. In his 144-page-long book *Die Kultur der Juden—Eine Versöhnung zwischen Religion und Wissenschaft* (Leipzig: Neuer Geist Verlag, 1919), Dingler expressed a high opinion of Jewish thought and tradition. See on this issue G. Wolters, "Hugo Dingler," *Science in Context* **2,** 359–367 (1988) and Wolters' book *Mach I, Mach II, Einstein und die Relativitätstheorie* (Berlin: W. de Gruyter, 1987), pp. 260, 264, and 272.

located at the place of the event and that for all determinations is in synchrony with a specified clock at rest."

Needless to say, the term "simultaneously" in this sentence has again only the meaning of "local simultaneity.[24] This statement also shows that according to Einstein the notion of synchrony or simultaneity logically precedes that of time. Einstein concluded the first paragraph of this essay with the statement of the light postulate, which he formulates in these words: "Based on experience, we also postulate that the quantity $2\,AB\,/\,(t'_A - t_A) = c$ is a universal constant (the velocity of light in empty space)."

In the second paragraph of his 1905 relativity paper Einstein proved that simultaneity, as defined in the first paragraph, is a *relative* concept. This means, as Einstein explained, that "we must not ascribe *absolute* meaning to the concept of simultaneity; instead, two events that are simultaneous when observed from some particular coordinate system can no longer be considered simultaneous when observed from a system that is moving relative to that system."[25] In these words Einstein for the first time offered an explicit definition of what he means when he speaks of a *relative* concept. As we see, it is a concept whose validity depends on the coordinate system chosen or, in brief, it is a frame-dependent concept. Simultaneity is no longer a binary relation between two events, as in Newtonian or classical physics, but a ternary relation depending also on the coordinate-frame involved. Furthermore, Einstein's proof of the relativity of simultaneity in this paragraph is the first ever published rigorous proof of the relativity of a physical concept. The relativity of simultaneity became also the first major subject of dispute between proponents and opponents of the theory of relativity.[26]

Einstein's proof of the relativity of simultaneity is based on the relativity principle and on the light principle and consists of thought experiments involving length measurements in two inertial reference systems $S(x, y, z, t)$ and $S'(x', y', z', t')$ in standard configuration.[27]

[24] See note 8.

[25] "Wir sehen also, dass wir dem Begriffe der Gleichzeitigkeit keine *absolute* Bedeutung beimessen dürfen, sondern dass zwei Ereignisse, welche, von einem Koordinatensystem aus betrachtet, gleichzeitig sind, von einem relativ zu diesem bewegten System aus betrachet, nicht mehr als gleichzeitige Ereignisse aufzufassen sind." Note 5 (1905), p. 897; (1989), p. 145.

[26] See chapter 8.

[27] This includes alignment of the x, y, and z axes of S with the x', y', and z' axes of S' and movement of the x' axis along the x axis with constant velocity v.

Einstein began the second paragraph, "On the relativity of lengths and times," by stating the two postulates on which his theory was based: (1) *the principle of relativity*, according to which "the laws by which the states of physical systems undergo changes are not affected, whether these changes of state be referred to the one or the other of two systems of coordinates in uniform translatory motion" and (2) *the light principle*, according to which "any ray of light moves in the 'stationary' system of coordinates with the determined velocity c, whether the ray be emitted by a stationary or by a moving body."[28] Although these two principles had been stated briefly in the introduction of the paper, preceding § 1, they are repeated in § 2, because they contain the notions of "uniform translatory motion" and "velocity of light," concepts that involve, if even only implicitly, the notion of *time* that was defined only at the end of § 1.

In the sequel Einstein demonstrated the relativity of simultaneity, as defined in § 1, namely that events that are simultaneous in a coordinate system $S(x, y, z, t)$ are not simultaneous in a coordinate system $S'(x', y', z', t')$ that is in motion relative to S. Because Einstein's proof of the relativity of distant simultaneity evoked strong objections, especially by philosophers, his first demonstration[29] of it deserves to be quoted in some detail.

Using his favorite technique, thought experiments, Einstein assumed a rigid rod, with end points A and B, lying at rest along the x' axis of a coordinate system S', which is moving in standard configuration with velocity v relative to a system S. System S is equipped with clocks synchronized by the method described in § 1. He also assumed that the ends of the rod, A and B, carry clocks

> that are synchronous with the clocks of the system at rest, i.e., whose readings always correspond to the "time of the system at rest" at the locations they happen to occupy: hence, these clocks are "synchronous in the system at rest." We further imagine that each clock has an observer co-moving with it, and that

[28] "1. Die Gesetze, nach denen sich die Zustände der physikalischen Systeme ändern, sind unabhängig davon, auf welches von zwei relativ zueinander in gleichförmiger Translationsbewegung befindlichen Koordinatensystemen diese Zustandsänderungen bezogen werden.
2. Jeder Lichstrahl bewegt sich im 'ruhenden' Koordinatensystem mit der bestimmten Geschwindigkeit c, unabhängig davon, ob dieser Lichstrahl von einem ruhenden oder bewegten Körper emittiert ist. Hierbei ist Geschwindigkeit = Lichtweg / Zeitdauer, wobei 'Zeitdauer' im Sinne der Definition des § 1 aufzufassen ist." Note 6 (1905), p. 895.

[29] Ibid., § 2.

these observers apply to the two clocks the criterion for synchronism formulated in § 1. Suppose a ray of light starts out from A at time t_A, is reflected from B at time t_B, and arrives back at time t'_A. Taking into account the principle of the constancy of the velocity of light, we find that

$$[c(t_B - t_A) = r_{AB} + (t_B - t_A)\,v \quad \text{and} \quad c(t'_A - t_B) = r_{AB} - (t'_A - t_B)\,v \text{ or } \quad]$$
$$t_B - t_A = r_{AB} \,/\, (c - v) \quad \text{and} \quad t'_A - t_B = r_{AB} \,/\, (c + v),$$

where r_{AB} denotes the length of the moving rod, measured in the system at rest. The observers co-moving with the moving rod would thus find that the two clocks do not run synchronously while the observers in the system at rest would declare them synchronous. Thus we see that we must not ascribe *absolute* meaning to the concept of simultaneity; instead, two events that are simultaneous when observed from some particular coordinate system can no longer be considered simultaneous when observed from a system that is moving relative to that system.[30]

Einstein thus arrived at the important conclusion that distant simultaneity is a relativistic concept and that, almost paradoxically, its *relativity* is a consequence of the *invariance* of the velocity of light (i.e., of the postulate of the constancy of the velocity of light [*Einstein's light postulate*]).

As we see, Einstein's proof of the relativity of simultaneity is based on the mathematical fact that the equations just derived imply that for nonzero velocities v the synchrony equation

$$t_B - t_A = t'_A - t_B \tag{7.1}$$

cannot be satisfied. Einstein's synchrony equation (7.1) can be given many equivalent formulations. It can be written in the form

$$t_B = {}^1\!/\!_2(t_A + t'_A) \tag{7.2}$$

which shows that an event occurring at B at time t_B is simultaneous with an event occurring at A at the time ${}^1\!/\!_2(t_A + t'_A)$. For reasons to be explained later, this equation may also be written in the form

$$t_B = t_A + {}^1\!/\!_2(t'_A - t_A). \tag{7.3}$$

Finally, with the clock at B replaced by a mirror, figure 7.1 illustrates schematically how the "velocity of light" or more precisely the "two-way velocity" or

[30] Note 6 (1905), pp. 896–897; (1989), pp. 144–145; (1952), p. 42.

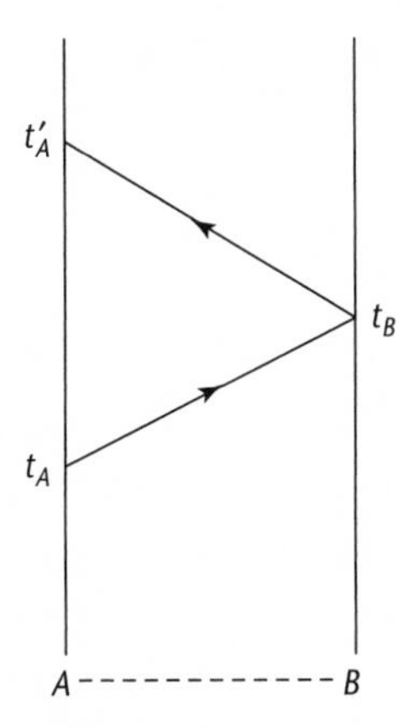

Figure 7.1

"round-trip velocity" of light had been measured by Fizeau, Foucault, Michelson and others as the ratio between the length of the total path traversed by the light, that is 2 *d*, and the total time $t'_A - t_A$ required for the traversal:

$$c = 2\, d \,/\, (t'_A - t_A). \tag{7.4}$$

From the last two equations we obtain by substitution

$$t_B = t_A + d \,/\, c, \tag{7.5}$$

a most convenient formulation of the criterion of what is generally called "standard signal synchrony." The broken line between *A* and *B* in figure 7.1 represents the constant distance between the two clocks located at the points *A* and *B* of an inertial system and the vertical lines represent the readings of these clocks. The lines with arrows indicate the path of the light rays emitted from *A* toward *B* and reflected to *A*. Once the concept of time, more precisely of the time of an inertial reference system, has been defined, the figure may be regarded as a Hermann Minkowski diagram[31] in which the vertical lines are the "world lines" of the clocks.[32]

Before proceeding to Einstein's later discussions of the concept of simultaneity let us make a few comments on his definition of this concept and the

[31] H. Minkowski, "Raum und Zeit" (Vortrag, gehalten auf der 80. Naturforscherversammlung zu Köln am 21. 9. 1908) *Physikalische Zeitschrift* **10,** 104–111 (1909); "Space and time," note 6 (1952), pp. 73–91.

[32] The direction from bottom to top along the vertical lines corresponds to the temporal direction from past to future.

clock-synchronization equation (7.1). The notions of simultaneity and of clock synchronization are intimately related, because spatially separated events are defined as simultaneous if and only if synchronized clocks at the locations of these events indicate the same readings when the events occur. Hence, every definition of clock synchronization is a definition of simultaneity and vice versa. Because these definitions require observing each clock at its place at the moment when the event occurs there, it obviously involves thereby local simultaneity. It follows, therefore, that Einstein's definition of simultaneity is ultimately a reduction of the concept of distant simultaneity to the concept of local simultaneity, which is regarded as a physically unproblematic notion. Einstein himself seems to have been aware of this, for he says in a footnote: "We shall not discuss here the imprecision that is inherent in the concept of simultaneity of two events taking place at (approximately) the same location and that also must be surmounted by an abstraction."[33]

In the sequel Einstein declared that we *assume*[34] that this definition of synchronism "is free from contradictions, and possible for any number of points: and that the following relations are universally valid: (1) If the clock at B synchronizes with the clock at A, the clock at A synchronizes with the clock at B. (2) If the clock at A synchronizes with the clock at B and also with the clock at C, the clocks at B and C also synchronize with each other." In other words, Einstein *assumed* that simultaneity, as defined by him, is a symmetric and transitive relation. Whether Einstein was right to call (1) and (2) "assumptions" will be discussed in chapter 11, which deals extensively with such questions.

The operational procedure of synchronizing spatially separated clocks or, equivalently, the operational definition of distant simultaneity, which Einstein proposed in his seminal 1905 relativity paper, may be called a one-ray simultaneity definition because it deals with the emission (at point A) and reflection (at point B) of only one and the same ray of light. Although undoubtedly the first simultaneity definition *published* by Einstein, it was probably *preceded* in Einstein's mind by another (two-ray) definition of simultaneity, which he published twelve years later in his popular exposition *Über*

[33] "Die Ungenauigkeit, welche in dem Begriffe der Gleichzeitigkeit zweier Ereignisse an (annähernd) demselben Orte steckt und gleichfalls durch eine Abstraktion überbrückt werden muss, soll hier nicht erörtert werden." Note 6 (1905), p. 893; (1989), p. 278; (1989, English translation), p. 141.

[34] "Wir nehmen an . . . " Note 6 (1905), p. 894.

die spezielle und die allgemeine Relativitätstheorie.[35] Two arguments seem to support this contention. First in the introduction to this book Einstein explicitly declared that he presents in it his ideas "in the sequence and connection in which they actually originated." Second, this contention also agreed with the conversations that Einstein had with the psychologist Max Wertheimer, a colleague of his in Berlin, who claimed that Einstein's way of thinking confirms the theses of *Gestalt* psychology. In fact, Wertheimer in his report on these conversations, *Einstein: the Thinking that led to the Theory of Relativity,*[36] commented on Einstein's just-quoted 1917 exposition that "what Einstein here says . . . is similar to the way his thinking proceeded." Because we intend to present Einstein's writings on the concept of simultaneity chronologically in the order in which they were published we will deal with his 1917 definition of distant simultaneity in a later chapter.

Neither Einstein's conversations with Wertheimer nor his autobiographical notes inform us whether Einstein's operational definition of distant simultaneity was the result of a sudden inspiration, as his 1922 Kyoto Lecture seems to suggest, or had been motivated by other factors. Such a factor may have been his study of Poincaré's *La Science et l'Hypothèse* (1902). Its German translation, *Wissenschaft und Hypothese,* published in 1904, which he probably read, contains an excerpt from Poincaré's essay *La Mesure du Temps,* which, as mentioned in chapter 6, anticipated to some extent Einstein's conceptions of simultaneity and time. As Maurice Solovine recalls, Einstein and his friends at the "Olympia Academy" have been discussing this essay "for weeks."

Another possible stimulus, recently proposed by Peter Galison,[37] may have been Einstein's daily work at the patent office in Bern where he had the opportunity to examine technical patents designed for the synchronization of clocks, which were used at that time at railway stations or public buildings. Typ-

[35] A. Einstein, *Über die spezielle und die allgemeine Relativitätstheorie—Gemeinverständlich* (Braunschweig: Vieweg, 1917, 14th edition 1922); *Relativity: The Special and the General Theory—A Popular Exposition* (London: Methuen, 1920; New York: Holt, 1921).

[36] M. Wertheimer, *Productive Thinking* (New York: Harper and Brothers, 1945; enlarged edition, 1959), pp. 213–233. The reservations expressed by A. I. Miller in his essay "Albert Einstein and Max Wertheimer: A Gestalt Psychologist's View of the Genesis of Special Relativity Theory," published in *History of Science* **13,** 75–103 (1975), do not concern the issue presently under discussion.

[37] P. Galison, *Einstein's Clocks and Poincaré's Maps. The Empire of Time* (New York: W.W. Norton, 2002); *Einsteins Uhren, Poincarés Karten—Die Arbeit an der Ordnung der Zeit* (Frankfurt am Main: S. Fischer, 2003), chapter 5. The book contains photographs of the two clocks mentioned in the text (figs. 5.3 and 5.8).

ical examples were David Perret's patent for "Eine elektrische Installation für Zeitübertragung" (An electrical installation for the transmission of time), patent 27555 (1903), or L. Agostinelli's patent for an "Installation mit Zentraluhr, um die Zeit synchron an mehreren Orten anzuzeigen" (Installation with central clock for the synchronization of clocks at different places), patent 29073 (1904).

In fact, Einstein may have already encountered the problem of time synchronization on his daily walk to the patent office when passing near the famous tower clock on the Kramgasse, where he lived, and seeing the distant big clock on the tower of the church in Muri, a nearby suburb of Bern.

That two clocks like these played a decisive role in Einstein's thinking about synchrony is apparent from remarks made by Joseph Sauter, one of Einstein's colleagues at the patent office. Sauter recalled that Einstein repeatedly emphasized "the need of a new definition of the synchronism of two identical clocks which are spatially separated one from the other" and explained his new definition in terms of the following example which refers specifically to those two clocks.

"Avant toute autre considération théorique," Sauter declared,

> Einstein fait remarquer la nécessité d'une nouvelle définition du "synchronisme" de deux horloges identiques distantes l'une de l'autre; pour fixer les idées, me dit-el, supposons l'une des horloges sur une tour de Berne et l'autre sur une tour de Muri (l'ancienne annexe aristocratique de Berne). A l'instant où l'horloge de Berne marque midi juste, faisons parti de Berne un signal lumineux dans la direction de Muri; il arrivera à Muri quand l'horloge de Muri marquera un temps midi + t; à ce moment, réfléchisson le signal dans la direction de Berne; si au moment où il attaint Berne l'horloge de Berne marque midi + 2 t, nous dirons que les deux horloges son en synchronisme.[38]

Even if Einstein's conception of distant simultaneity should be regarded as the result not of abstract thought but of some mundane practical experience, as Galison contends, the idea of using this concept as the cornerstone for the construction of a new revolutionary theory was undoubtedly the work of a genius.

In his 1905 paper Einstein formulated the light principle for the two-way velocity of light, for he postulated that "the quantity $2\ AB\ /\ (t'_A - t_A) = c$ to be a universal constant—the velocity of light in empty space."[39] By refor-

[38] J. Sauter, "Comment j'ai appris à connaître Einstein," quoted in M. Flückinger, *Albert Einstein in Bern* (Bern: P. Haupt, 1974), p. 156.

[39] See note 10.

mulating it in his 1907 synoptic presentation of the theory for the one-way velocity of light he applied it also as a definition of synchrony or simultaneity as follows. As in his 1905 paper, to define the "time" of a coordinate system *S*, he imagined clocks at rest in *S* that are isochronous or, as he phrased it, "equivalent, i.e., the difference between the readings of two clocks shall remain unchanged if they are arranged next to each other."[40] For the totality of the clock readings to give us "time," Einstein continues,

> we need a rule according to which these clocks will be set relative to each other. We now assume *that the clocks can be adjusted in such a way that the propagation velocity of every light ray in vaccum—measured by means of these clocks—becomes everywhere equal to a universal constant c*, provided that the coordinate system is not accelerated. If *A* and *B* are two points at rest relative to the coordinate system, which are equipped with clocks and are separated by a distance *r*, while t_A is the reading of the clock in *A* at the moment when a ray of light propagating through a vaccum in the direction *AB* reaches point *A*, and t_B is the reading of the clock at *B* at the moment the ray reaches *B*, then we should always have $r / (t_B - t_A) = c$, whatever the motion of the light source emitting the light ray or the motion of other bodies may be.[41]

Although Einstein did not call the assumption that "the clocks can be adjusted in such a manner" a definition of clock synchronization or of simultaneity, he regarded it as such, for in the sequel he declared that "the aggregate of the readings of all clocks synchronized according to the above . . . we call . . . the time of the system." If indeed this assumption was meant to serve as a definition it may be criticized as involving a circularity because of its use of the concept of the one-way velocity of light, the determination of which requires two spatially separated synchronized clocks. Einstein's 1905 definition of simultaneity, based as it was on a two-way or round-trip propagation of light, emitted from *A* to *B*, where it is reflected to *A*, required only one clock, a clock at *A*, and did therefore not face this problem.

[40] Einstein, "Über das Relativitätsprinzip und die aus demselben gezogenen Folgerungen." *Jahrbuch der* Radioaktivität und Elektronik **4,** 411–462 (1907); *The Collected Papers* (1989), vol. 2, pp. 433–484; English translation, "On the relativity principle and the conclusions drawn from it," Ibid., (1989), vol. 2, pp. 252–311; H. M. Schwartz, "Einstein's comprehensive 1907 essay on relativity, part I," *American Journal of Physics* **45,** 512–517 (1977).

[41] Ibid., p. 256 (emphasis in German original).

It may well be that for this reason Einstein eliminated any reference to the velocity of light c in his 1910 essay, *The Principle of Relativity and its Consequences in Modern Physics*, which like his 1905 and 1907 essays begins with a definition of "time." To this end Einstein suggests the following procedure:

> First, we furnish ourselves with a means of sending signals, be it from *A* to *B*, or from *B* to *A*. This means should be such that we have no reason whatsoever to believe that the phenomena of signal transmission in the direction *AB* will differ in any way whatsoever from the phenomena of signal transmission in the direction *BA*. In that case there is, obviously, only one way of regulating the clock at *B* against the clock at *A* in such a manner that the signal traveling from *A* to *B* would take the same amount of time—measured with the clocks described above—as the signal traveling from *B* to *A*. If we denote by
>
> | | | | | |
> |---|---|---|---|---|
> | t_A | the reading of the clock at | *A* | at the moment signal | *AB* leaves *A* |
> | t_B | " | *B* | " | *AB* arr. at *B* |
> | t'_B | " | *B* | " | *BA* leaves *B* |
> | t'_A | " | *A* | " | *BA* arr. at *A* |
>
> then we have to set the clock at *B* against that at *A* in such a way that $t_B - t_A = t'_A - t'_B$.[42]

In short, in this essay Einstein declared that only this procedure of synchronizing spatially separated clocks is compatible with the light postulate. However, he admitted—and did so for the first time—the possibility of using, instead of light rays, "for example, sound waves that propagate between *A* and *B* through a medium that is at rest with respect to these points. . . . It does not make any difference whether we choose this or that kind of signals. If two kinds of signals were to produce discrepant results, we would have to conclude that, for at least one of the two kinds of signals, the condition of equivalence of the paths *AB* and *BA* was not satisfied"[43] (see fig. 7.2).

In contrast to his 1905 and 1907 definitions of simultaneity, Einstein's 1910 definition used two separate light rays but had the advantage of being

[42] Einstein, "Le Principe de Relativité et ses Conséquences dans la Physique Moderne," *Archives des Sciences Physiques et Naturelles* **29,** 5–28, 125–144 (1910); "The principle of relativity and its Consequences in Modern Physics," *The Collected Papers* (1993), vol. 3, pp. 117–142; English translation (1993), vol. 3, pp. 131–174.

[43] Op. cit., pp. 126–127. For the possibility of defining simultaneity by the use of acoustical signals see K. C. Kar, "Relativity in an acoustical world," *Indian Journal of Theoretical Physics* **18,** 1–11 (1970).

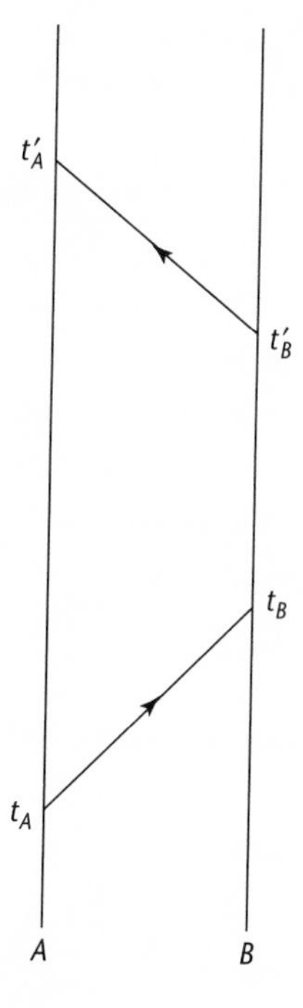

Figure 7.2

completely symmetric in the sense that all operations involved are the same at *A* as at *B*. Einstein's 1905 definition may be regarded as a special case of this third 1910 definition, namely, if t_B coincides with t'_B.

This diagram has historical importance insofar as it also illustrates the method of measuring the velocity of light proposed by Galileo and described in his *Discorsi*.[44] A person located at *A* uncovers at time t_A a lantern containing a light, which is seen by another person at *B* at time t_B. As soon as this other person sees it he uncovers his lantern at t'_B, the light of which is then observed by the person at *A* at the time t'_A. Galileo assumed that the time difference between t_B and t'_B is negligible for he said that "after a few trials the response will be so prompt that without sensible error the uncovering of one light is immediately followed by the uncovering of the other." Still, the question may be asked why a mirror, known since antiquity, was not used to re-

[44] G. Galilei, *Discorsi e dimostrazioni matematiche intorno à due nuove scienze* (Leiden: Elsevier, 1638); *Dialogues concerning Two New Sciences* (London: Macmillan, 1914; New York: Dover Publications, n.d.).

flect the light at *B*, as Einstein proposed in 1905. The following historical remark may perhaps serve as an explanation. In the early seventeenth century the "lapis bononiensis," a phosphorent mineral, was discovered near Bologna and its luminescent phenomena, especially its "afterglow," attracted much attention. In this context the idea, that even ordinary reflection in a mirror may not be instantaneous, gained wide popularity, as intimated, for example, by Fortunio Liceti.[45] This may well be the reason that Galileo did not apply the "one-ray" procedure involving a reflecting mirror as later Foucault, Fizeau, and others did in their measurements of the velocity of light.

Exactly three months after his appointment as Associate Professor at the University of Zürich, Einstein delivered on 16 January 1911 a lecture on the theory of relativity to the local Association of Natural Scientists. Pointing out that the classical conception of time as an independent variable of events did not suffice to determine the time of an event, he defined time, as he did before, as the set of the readings of synchronized identical clocks distributed and at rest throughout a coordinate system *S*. The problem of time was therefore again reduced to the problem of how to synchronize distant clocks or how to define distant simultaneity. Einstein explained in his lecture how to resolve this problem as follows.

> Let us imagine a clock (a balance wheel clock, for example) at the origin of a coordinate system *S*. Using this clock we can evaluate the time of events occurring directly at this point or in its immediate vicinity. However, events occurring at another point of *S* cannot be evaluated with this clock. If an observer standing next to the clock at the origin of *S* notes the time at which he received notice of the event in question by means of a ray of light, this time will not be the time of the event itself, but a time greater than the latter by the time of propagation of the light ray from the event to the clock. If we knew the velocity of propagation of light relative to the system *S* in the direction under consideration, it would be possible to determine the time of the event using the above clock; but the velocity of light can be measured only if the problem of the determination of time, which we are now discussing, has been solved. To measure the velocity of light in a given direction, we would have to measure the distance between points *A* and *B*, between which the light ray propagates, and further, the time of the emission of the light at *A* and the time of the arrival of

[45] Fortunio Liceti, *Litheosphorus sive de lapide bononiensi* (Utini: Schiratti, 1640).

> the light at B. Thus, time would have to be measured at different locations; however, this can be done only if the definition of time we are seeking has already been given. But if it is impossible in principle to measure a velocity, in particular the velocity of light, without recourse to arbitrary stipulations, then we are justified in making arbitrary stipulations regarding the velocity of light. We shall now stipulate that the velocity of the propagation of light in vacuum from some point A to some point B is the same as that from B to A. By virtue of this stipulation we are indeed in a position to regulate identically constructed clocks that we have arranged at various points at rest relative to the system S. For example, we will set the clocks at the points A and B in such a manner that the following will obtain: If a ray of light sent from A toward B at time t (measured by the clock at A) arrives at B at the time $t + a$ (measured by the clock at B), then conversely, a ray sent from B toward A at time t (measured by the clock at B) must arrive at A at time $t + a$ (measured by the clock at A). This is the rule according to which all clocks arranged in the system S must be regulated. If we follow this rule, we achieve a determination of time from the standpoint of the measuring physicist. That is to say, the time of an event is equal to the readings of the clocks located at the place of the event that are regulated according to the rule we just described.[46]

In contrast to Einstein's 1905 definition of distant simultaneity or clock synchronization criterion,[47] which was based on the equation $t_B - t_A = t'_A - t_B$, his 1911 criterion was based on the stipulation that the velocity of the light propagation from A to B equals that from B to A. But because Einstein explicitly declared that "the velocity of light can be measured only if the problem of the determination of time . . . has been solved" the question arises whether this 1911 synchronization criterion does not involve a logical circularity. Such a circularity is not avoided by the fact that the 1911 stipulation required only the equality of two velocities without the need of measuring their numerical values. As we shall see later on,[48] Einstein's argument that the measurement

[46] A. Einstein, "Die Relativitätstheorie," *Vierteljahresschrift der Naturforschenden Gesellschaft, Zürich* **56,** 1–14 (1911); *The Collected Papers* (1993), vol. 3, pp. 425–439; English translation (1993), vol. 3, pp. 340–350. It is here where Einstein used the term "Relativitätstheorie" for the first time in the title of an article. This article was reviewed by Liese Meitner in the *Naturwissenschaftliche Rundschau* **27,** 285–288 (1912).

[47] See equation (7.1).

[48] See, for example, the statement that "in order to measure a velocity . . . the simultaneity of distant events must already be known" in H. Reichenbach, *Philosophie der Raum-Zeit-Lehre* (Berlin: De Gruyter, 1928); *The Philosophy of Space and Time* (New York: Dover Publications, 1958), p. 126.

of a velocity presupposes the concept of distant simultaneity plays an important role in the debate about the conventionality of this concept.

That Einstein defined distant simultaneity, though only unwittingly, in 1911 without using any light rays at all was pointed out by Wolfgang Pauli, when he wrote that a thought experiment proposed by Einstein "shows that the determination of the simultaneity of spatially separated events . . . can be carried out with the help of measuring rods, without the use of clocks."[49]

Einstein's thought experiment, referred to by Pauli, was prompted by an article in which Vladimir Varičak argued that in Einstein's theory the Lorentz contraction was not a physically real effect but "only an apparent subjective phenomenon produced by the manner our clocks are regulated and lengths are measured."[50] In particular, Varičak contended, the relativistic length contraction is the result of Einstein's nonclassical method of defining distant simultaneity; for the length of a rod in motion relative to an inertial system S is defined as the distance between the simultaneous projections of the rod's end points in S. To refute Varičak's argument Einstein proposed the following "twin-rod" experiment.[51]

"Consider two equally long rods (when compared at rest) $A'B'$ and $A''B''$, which can slide along the x-axis of an unaccelerated coordinate system in the same direction as and parallel to the x-axis. Let $A'B'$ and $A''B''$ glide past each other with an arbitrarily large, constant velocity, with $A'B'$ moving in the positive, and $A''B''$ in the negative direction of the x-axis. Let the endpoints A' and A'' meet at a point A^* on the x-axis, while the endpoints B' and B'' meet at a point B^*. According to the theory of relativity, the distance A^*B^* will then be smaller than the length of either of the two rods $A'B'$ and $A''B''$, which fact can be established with the aid of one of the rods, by laying it along the stretch A^*B^* while it is in the state of rest." The text[52] does not state unambiguously whether the two velocities are supposed to be equal, but

[49] W. Pauli, *Theory of Relativity* (New York: Pergamon Press, 1958), p. 12. Originally published in vol. 5 of the *Encyklopädie der mathematischen Wissenschaften* (Leipzig: Teubner, 1921).

[50] V. Varičak, "Zum Ehrenfestschen Paradoxon," *Physikalische Zeitschrift* **12,** 169 (1911).

[51] Einstein, "Zum Ehrenfestschen Paradoxon," *Physikaliche Zeitschrift* **12,** 509–510 (1911); *The Collected Papers of Albert Einstein* (note 6), (1993), vol. 3, pp. 482–483; "On the Ehrenfest paradox" (Princeton, English translation) (1992), vol. 3, p. 378.

[52] "$A'B'$ und $A''B''$ sollen aneinander vorbeigleiten, wobei $A'B'$ im Sinne der positiven, $A''B''$ im Sinne der negativen x-Achse mit beliebig grosser konstanter Geschwindigkeit bewegt sei." Op. cit., p. 510.

most commentators, among them Christian Møller,[53] Francis W. Sears,[54] and Herman M. Schwartz,[55] assume that the velocities are equal; they conclude therefore that "by symmetry" or "by the principle of sufficient reason" these events, that is, the coincidences of A' with A'' and of B' with B'', must be simultaneous. But since the very requirement of the equality of the two velocities cannot be satisfied without a definition of simultaneity, they argue, the contention that such an experiment establishes simultaneity is begging the question. In 1972, however, John A. Winnie showed by a profound analysis that "contrary to a widespread view . . . the thought experiment does not require that the two rods be travelling at equal speeds . . . in order that the experiment serve its intended purpose."[56]

The intent of the experiment, as envisaged by Einstein, was merely to show that according to the theory of relativity the distance, in an inertial reference system S, between the event A^*, the meeting of A' and A'', and the event B^*, the meeting of B' and B'', is smaller than the rest length of either of the two rods. Winnie proved that this is indeed the case independently of whether the velocities of the rods are equal. He did not discuss the question whether these events are simultaneous, which they are only if these velocities are equal. Because the establishment of equal velocities requires synchronized clocks and thus by implication the notion of distant simultaneity, however, Einstein's twin-rod experiment does not establish distant simultaneity contrary to Pauli's assertion.

In a short article, written in the spring of 1914 for the daily *Vossische Zeitung,* Einstein presented a nontechnical review of the theory of relativity in which he stated that the notion of simultaneity, as conceived in classical physics, can no longer be maintained; but he gave no details of how it should be defined. He only pointed out that "the simultaneity of two events is not absolute, but instead can only be defined relative to one observer of a given state of motion." This relativity of simultaneity, he added, "is the most important, and also the most controversial theorem of the new theory of rela-

[53] C. Møller, *The Theory of Relativity* (Oxford: Clarendon Press, 1952), p. 46.

[54] F. W. Sears, "Simultaneity without synchronized clocks," *American Journal of Physics* **37,** 668 (1969).

[55] H. M. Schwartz, "A new method of clock synchronization without light signals," *American Journal of Physics* **39,** 1269–1270 (1971).

[56] J. A. Winnie, "The twin-rod thought experiment," *American Journal of Physics* **40,** 1091–1094 (1972).

tivity. It is impossible to enter here into an in-depth discussion of the epistemological and 'naturphilosophischen' assumptions and consequences which evolve from this basic principle."[57]

In a brief survey of the theory, published 1915 in *Die Kultur der Gegenwart*,[58] Einstein presented an exposition of the definition and relativity of distant simultaneity that is essentially identical with his 1910 treatment of these notions.

All of Einstein's definitions of simultaneity discussed so far were based on, or identical with, clock-synchronization procedures. By definition, events are simultaneous if and only if synchronized clocks at their locations indicate the same time when these events occur.

In his popular 1917 exposition of relativity, which appeared in many editions and numerous translations, Einstein presented for the first time a different definition of distant simultaneity which, as stated above, was a two-ray definition and did not make use of clocks. In the beginning of chapter 8, entitled "Über den Zeitbegriff in der Physik" (On the Idea of Time in Physics) of this book Einstein, who—as we know—liked to make use of thought experiments, assumed that "lightning has struck the rails on a railway embankment at two places *A* and *B* far distant from each other" and that "these two lightning flashes occurred simultaneously." He then asked how one can verify that these flashes occurred simultaneously. Stating that the concept of simultaneity "does not exist for the physicist until he has the possibility of discovering whether or not it is fulfilled in an actual case," Einstein showed the need of a definition of simultaneity such that "this definition supplies us with the method by means of which, in the present case, one can decide by experiment whether or not both lightning strokes occurred simultaneously."[59]

In compliance with these methodological precepts Einstein proposed the following definition of distant simultaneity: "By measuring along the rails, the connecting line *AB* should be measured up and an observer placed at the midpoint *M* of the distance *AB*. This observer should be supplied with an arrangement (e.g. two mirrors inclined at 90°) which allows him visually to

[57] A. Einstein, "Vom Relativitäts-Prinzip," *Vossische Zeitung*, 26 April 1914 (no. 209), pp. 1–2; *The Collected Papers* (1996), vol. 6, p. 4; English translation (1997), vol. 6, pp. 3–5.

[58] A. Einstein, "Die Relativitätstheorie," *Die Kulture der Gegenwart—Ihre Entwicklung und Ziele* (Braunschweig: Vieweg, 1915), vol. 1, pp. 708–710.

[59] Note 35, § 8.

observe both places A and B at the same time. If the observer perceives the two flashes of lightning at the same time, then they are simultaneous." Such a criterion, Einstein continued, would certainly be right if one would know "that the light by means of which the observer at M perceives the lightning flashes travels along the length $A \rightarrow M$ with the same velocity as along the length $B \rightarrow M$. But an examination of this supposition would only be possible if we already had at our disposal the means of measuring time. It would thus appear as though we are moving here in a logical circle." Einstein continued, nevertheless,

> this definition can be maintained, because in reality it assumes absolutely nothing about light. There is only *one* demand to be made of the definition of simultaneity, namely, that in every real case it must supply us with an empirical decision as to whether or not the conception that has to be defined is fulfilled. That this definition satisfies this demand is indisputable. That light requires the same time to traverse the path $A \rightarrow B$ as for the path $B \rightarrow A$ is in reality neither a *supposition nor a hypothesis* about the physical nature of light, but a *stipulation* which I can make of my own freewill in order to arrive at a definition of simultaneity.

As we will see in chapter 9, the "logical circle" referred to by Einstein in this passage played an important role in Hans Reichenbach's argumentation for the conventionality thesis of distant simultaneity. In fact, some advocates of this thesis credited Reichenbach with having discovered this "logical circle." As mentioned in chapter 3, Einstein's two-ray simultaneity definition was in principle identical with St. Augustine's criterion of simultaneity in his confutation of astrology; the two messengers in St. Augustine's argument, assumed to be running with the same velocity and meeting midway from their points of departure, are now replaced by the two rays of light meeting at the midpoint M of the line AB.

Compare Einstein's 1917 two-ray simultaneity definition with his 1905 one-ray definition. Although the 1917 procedure was, strictly speaking, a reduction of distant simultaneity to local simultaneity (the observer's inspection of the two mirrors inclined at 90°) and did not involve any clock, the 1905 definition was primarily a definition of clock synchronization and made use of the additional definition according to which two events are simultaneous if synchronized clocks located at their positions indicate the same time. Although both definitions are based on the assumption that the velocity of light c is the same in all directions (Einstein's light postulate that c is a universal constant), the 1917 definition had the following advantage. The isotropy of c does not exclude the theoretical possibility that the magnitude of c changes in the course of time. Clearly, the 1917 definition would be immune against such a change, but the 1905 definition would lose its applicability.

In his popular 1917 exposition of relativity Einstein also presented a simple demonstration of the relativity of distant simultaneity by the following "train/embankment" thought experiment (see fig. 7.3). A very long train S' travels along the rails on an embankment S with constant velocity v. The question is whether two events, for example, a lightning strike (or an explosion) e_A occurring at A and another e_B at B "which are simultaneous *with reference to the railway embankment*, are also simultaneous *relatively to the train*." To say that these lightning strikes are simultaneous relative to S means that the light rays emitted from them

> meet each other at the mid-point M of the length $A \rightarrow B$ of the embankment. But the events [at] A and B also correspond to positions A and B on the train. Let M' be the mid-point of the distance $A \rightarrow B$ on the travelling train. Just when the flashes of lightning occur (as judged from the embankment), this point M' naturally coincides with the point M, but it moves towards the right in the diagram with the velocity v of the train. If an observer sitting in the position M' in the train did not possess this velocity, then he would remain permanently at M, and the light rays emitted by the flashes of lighting A and B would reach him simultaneously, *i.e.*, they would meet just where he is situated. Now in reality (considered with reference to the railway embankment) he is hastening towards the beam of light coming from B, whilst he is riding on ahead of the beam of light coming from A. Hence the observer will see the beam of light emitted from B earlier than he will see that emitted from A. Observers who take the railway-train as their reference-body must therefore come to the conclusion that the lightning flash B took place earlier than the lightning flash A. We thus arrive at the important result: Events which are simultaneous with reference to the em-

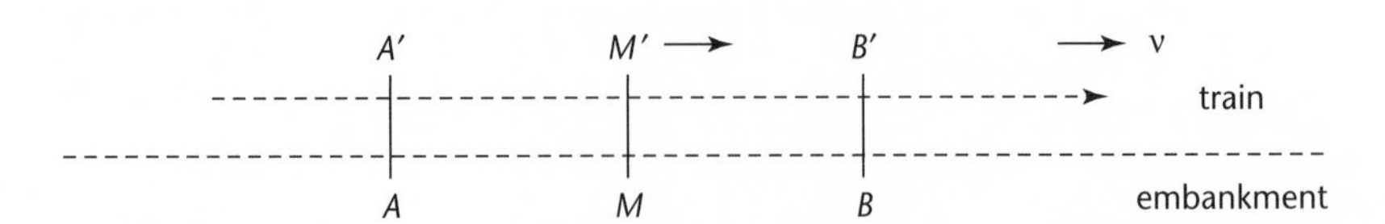

Figure 7.3

> bankment are not simultaneous with respect to the train, and *vice versa* (relativity of simultaneity). Every reference-body (coordinate system) has its own particular time; unless we are told the reference-body to which the statement of time refers, there is no meaning in a statement of the time of an event.[60]

It is clear that the simultaneity of the two lightning strikes, hitting *A* and *B*, is assured, in accordance with the definition of distant simultaneity presented in the same popular exposition, because the light rays emitted by the two strokes arrive simultaneously, in the sense of local simultaneity, at the midpoint *M* of the segment *AB* of the embankment.

As we will see at the end of chapter 8 this simple train/embankment thought experiment, contrived to prove the relativity of distant simultaneity, became, and remains, a subject of lively controversy, especially among philosophers of physics.

In "A brief outline of the development of the theory of relativity," written in 1921 for a special issue of the periodical *Nature* devoted to relativity, Einstein defined distant simultaneity as follows: "Let *A* and *B* be two points in a coordinate system *S*, for instance the endpoints of a rod, stationary in *S*, and let *M* be its midpoint. A light signal is emitted from *M* in all directions. The principle of the constancy of the velocity of light dictates us to conclude that the arrival of the signal in *A* and its arrival in *B* are simultaneous events." The proof of the relativity of simultaneity is similar to that given in Figure 7.3.[61]

In his Stafford Little Lectures, delivered in May 1921 at Princeton University and published 1922 as *The Meaning of Relativity*, Einstein gave the following definition of simultaneity: "Let us suppose that we place similar clocks at points of the system *S*, at rest relatively to it, and regulated according to the following scheme. A ray of light is sent out from one of the clocks, U_m, at the instant when it indicates the time t_m, and travels through a vacuum a distance r_{mn}, to the clock U_n; at the instant when this ray meets the clock U_n the latter is set to indicate the time $t_n = t_m + (r_{mn}/c)$. The principle of the constancy of the velocity of light then states that this adjustment of the clocks will not lead to contradictions." But in a footnote to this passage Einstein added: "Strictly speaking, it would be more correct to define simultaneity first,

[60] Note 35 (1917), pp. 16–18; (1920), pp. 25–26.

[61] For details see R. W. Lawson's translation of the German original, kept at the Morgan Library in New York, *Nature* **106,** 782–784 (1921).

somewhat as follows: two events taking place at the points *A* and *B* of the system *S* are simultaneous if they appear at the same instant when observed from the middle point, *M*, of the interval *AB*. Time is then defined as the ensemble of the indications of similar clocks, at rest relatively to *S*, which register the same simultaneously."[62]

The two sentences in this footnote are the most concise, yet comprehensive definitions of the concepts of distant simultaneity and time ever given by Einstein. They clearly show that he assigned logical priority to the notion of simultaneity over that of time. They contain the last definition of distant simultaneity that Einstein ever wrote in his life. It is identical with his 1917 simultaneity definition and agrees therefore again with St. Augustine's criterion of simultaneity.

True, in his *Autobiographical Notes*, which he wrote in 1945 at the age of 67, Einstein refers to the notion of simultaneity on two occasions. First, when he explained why it had not been recognized earlier that to different inertial systems belong different "times." "One would have noticed this long ago," he wrote, "if, for the practical experience of everyday life, light did not appear (because of the high value of *c*) as the means for the statement of absolute simultaneity." Later on, when he referred to the major new insights, which physics owes to the special theory of relativity, he mentioned as the first new insight that "there is no such thing as simultaneity of distant events"—meaning of course the classical notion of simultaneity, and to emphasize the importance of this new insight he added that "consequently there is also no such thing as immediate action at a distance in the sense of Newtonian mechanics."[63] Nowhere in these *Notes* does Einstein explicitly define the notion of distant simultaneity.

I have purposely quoted Einstein's various definitions of distant simultaneity or, equivalently, his various formulations of the synchronization of distant clocks in detail not only because of their intrinsic historical importance but also to enable the reader to verify for himself the correctness of the following statements.

All of Einstein's definitions of distant simultaneity were ultimately based on the assumption that a light signal propagates with the same velocity from

[62] Einstein, *The Meaning of Relativity* (Princeton, New Jersey: Princeton University Press, 1921; London: Methuen, 1922, 1950), p. 26.

[63] A. Einstein, "Autobiographical Notes," in P. A. Schilpp (ed.), *Albert Einstein: Philosopher-Scientist*, (Evanston, Illinois: The Library of Living Philosophers, 1949), pp. 55 and 61.

A to B as from B to A, although the term "velocity" was not yet being used but only indirectly indicated by the readings of the clocks in *A* and *B*. In fact, if Einstein had already explicitly used the concept of "velocity," his reasoning would have been vitiated by a vicious circularity, because the notion of velocity presupposes the concept of "the time of a stationary system" or briefly "system time," in contrast to what he had called "*A-time*" or "*B-time*." But this concept is defined only afterward, at the end of § 1.

In his earlier formulations, as in 1905, 1910, 1912, and 1915, it was one and the same space interval *AB* that was traversed by the light signals; in his later formulations, in 1917 and 1922, two separate but equally long space intervals, *AM* and *MA*, were applied. Another, but related, issue is the fact that, in 1905, for example, only one light ray and its reflection had been used, whereas later, as in 1910, 1917, and 1922, two separate light flashes were used. This variety of Einstein's formulations of clock synchronization or of distant simultaneity and the fact that according to Einstein the "time" of a coordinate system *S* is the ensemble of the indications of similar synchronized clocks, at rest in *S*, raised the question of whether Einstein did not offer at least "*two* different definitions of time and time relations."[64] True, Einstein never discussed the problem of whether his various definitions are logically or physically equivalent, probably because he thought that this equivalence is self-evident. The same holds, of course, for his different demonstrations of the relativity of simultaneity.

In anticipation of chapter 9, in which the problem of the conventionality of the concept of simultaneity is dealt with extensively, we discuss part of this problem in the present context for the following reason. The preceding detailed review of Einstein's various, but equivalent, definitions of the concept of simultaneity enables us to examine whether he regarded this concept as referring to something factual and to an empirically determinable datum or whether he regarded it as merely a more or less arbitrary convention. Thus, it enables us to examine whether conventionalists, like Hans Reichenbach or Adolf Grünbaum,[65] were justified in claiming Einstein as one of their own.

Conventionalists base their claim mainly on the facts that Einstein entitled § 1 of his 1905 paper, in which he introduced the concept of simul-

[64] As claimed, for example, by Harald Nordenson in *Relativity—Time and Relativity* (London: George Allen and Unwin, 1969).

[65] See, for example, A. Grünbaum, *Philosophical Problems of Space and Time* (New York: A. A. Knopf, 1963; Dordrecht: Reidel, 1973), p. 343.

taneity, "*Definition of Simultaneity*," and he used the term "stipulation" (Festsetzung), when he wrote that a time common to A and B cannot be obtained "unless we stipulate *by definition* (indem man *durch Definition* festsetzt) that the 'time' required by light to travel from A to B equals the 'time' it requires to travel from B to A."[66] In his Zürich[67] and his Princeton lectures[68] he referred to it as an "arbitrary stipulation," and in a little known letter in 1924 to André Metz he wrote that the theory of relativity involves conventions, and among them is the concept of simultaneity.[69]

The opponents of the conventionality thesis, eager to show that Einstein was not a conventionalist, quoted § 2 of the 1905 paper, which states that the theory is based on the principles of relativity and the constancy of the velocity of light. But if every ray of light moves with the same velocity c, they claimed, then clearly space is isotropic with reference to the velocity of light, $c_{AB} = c_{BA}$ and equation (7.1), that is, $t_B - t_A = t_A' - t_B$ is not a convention. This argument ignores the fact, however, that the formulation of the light principle in § 2 of that paper is given after the definition of time, as stated at the end of §, and based implicitly therefore on a convention.

Furthermore, they claimed that, in paragraph 7 of his 1917 book, Einstein obviously referred to the one-way velocity of light when he declared that the "velocity of light is the same in all colors, because if it were not the case, the minimum of emission would not be observed simultaneously for different colors during the eclipse of a fixed star by its dark neighbour. By similar considerations, based on observations of double stars, the Dutch astronomer De Sitter was able to show that the velocity of propagation of light cannot depend on the velocity of motion of the body emitting the light. The assumption that this velocity is dependent on the direction 'in space' is in itself improbable." Had Einstein been a conventionalist, they claimed, he would not have called the anisotropy of space "improbable."

This conclusion is contradicted, however, because in the 1924 letter to André Metz, the author of the popular exposition *La Rélativité*, Einstein explicitly stated that the theory of relativity involved conventions and that one of these conventions is "simultaneity."[70]

[66] Note 6 (1905), p. 894 (italics in original).
[67] Note 46.
[68] Note 62.
[69] Letter from Einstein to A. Metz, 27 November 1924; Einstein Archive, reel 18-255.
[70] Note Letter from Einstein to A. Metz of 27 November 1924, Einstein Archive, reel 18-255.

Summing up, it seems that, as far as the concept of distant simultaneity is concerned, Einstein can be classified as a conventionalist, who however sometimes made statements not wholly consistent with the position.

What have not yet been discussed, despite their decisive importance for this chapter, are the following two questions. Einstein, it will be recalled, told his friend Besso[71] in 1905 that by an analysis of the concept of simultaneity he "completely solved the problem." Similarly, in his 1922 Kyoto lecture he declared that "an analysis of the concept of time was my solution."[72] And in his 1907 survey article he emphasized that "it was only necessary to provide a sufficiently precise formulation of the notion of time in order to overcome the difficulty encountered."[73] The two questions to be dealt with are, therefore: (1) What, precisely, was the difficulty that Einstein faced when constructing his relativity theory? (2) How, precisely, did Einstein resolve this difficulty?

Although Einstein hinted at the answers to these questions in several of his published papers, a fully detailed reply to these questions can be found only in what is generally called *Einstein's 1912 Manuscript on the Special Theory of Relativity*.[74] The first part of it, including § 7, entitled "Apparent Incompatibility of the Principle of the Constancy of the Speed of Light with the Relativity Principle," was probably written, as the color of the ink indicates, in the winter of 1911 and spring of 1912 in Prague, where Einstein lectured at the Karl-Ferdinand University, and the rest of it in 1912 in Zürich after his appointment at the Swiss Federal Institute of Technology (ETH).[75]

In sections preceding § 7 Einstein reviewed Lorentz's reformulation of Maxwell's electromagnetic theory and showed that it leads to the experimentally well confirmed conclusion, called in § 5 "the principle of the constancy of the velocity of light." According to it the velocity of light (*in vacuo*) is the same for all observers, independently of the velocity of the source of light relative to the observer. In § 6, entitled, "The Principle of Relativity,"

[71] Chapter 7, note 3.

[72] Ibid.

[73] Chapter 3, note 4.

[74] A. Einstein, "Spezielle Relativitätstheorie" in *The Collected Papers of Albert Einstein* (1995), vol. 4, pp. 9–108; *Einstein's 1912 Manuscript on the Special Theory of Relativity* (New York: George Braziller, 1996). The manuscript was purchased by the Jacob E. Safra Philanthropic Foundation and is on display at the Israel Museum in Jerusalem.

[75] For further historical details concerning the date and genesis of this unique manuscript, see the introduction to the just-quoted facsimile edition by Braziller.

Einstein formulates this principle, using von Laue's terminology, as follows: "Every coordinate system that is in uniform translational motion relative to a justified system ["ein berechtigtes System"] is again a justified system. The equations of motion of any system are the same with respect to all such justified systems."[76] Einstein then reminds us that the coordinates of two such systems, $S(x, y, z, t)$ and $S'(x', y', z', t')$, when in standard configuration, that is, the x' axis gliding along the x-axis with a constant velocity v and their origins O and O' coinciding at time $t = t' = 0$, satisfy the transformation equations (usually called "the Galileian equations"[77]):

$$x' = x - v\,t \quad y' = y \quad z' = z \quad t' = t \tag{c}$$

In § 7, finally—and this is an important stage in his exposition—Einstein points out that the "following three things are incompatible with one another (a) the relativity principle, (b) the principle of the constancy of the velocity of light, (c) the transformation equations."

More precisely, the difficulty Einstein faced was the following incompatibility: (1) on the one hand, experiments verify that all electromagnetic or optical phenomena proceed in the same way in all inertial systems and thus confirm the relativity principle (a); (2) on the other hand, the equally well confirmed basic equations of electrodynamics or optics are *not* invariant in a transition from one inertial system to another under the transformations (c) and thus conflict with (a); and (3) the relativity principle (a) conflicts with the transformation equations (c). Thus, for example, a light ray, traveling along the x axis in the inertial system S (x, y, z, t) with the velocity $c = dx/dt$ should according to (b) travel in the inertial system $S'(x', y', z', t')$ with the same velocity c. But according to the transformation equations (c) its velocity in S' is $dx'/dt' = c - v$ contrary to (a).

Clearly, (a) and (b) are incompatible as long as (c) remains valid. Einstein's decision to reject (c) to save (a) and (b), a choice that required a revision of the accepted concept of time, was indeed the decisive step in the construction of

[76] Einstein apparently had just read Max von Laue's *Das Relativitätsprinzip* (Braunschweig: Vieweg, 1911), the first edition of his well-known text *Die Relativitätstheorie* (also published by Vieweg, 5th edition, 1952). In this 1911 text von Laue used the term "ein berechtigtes System" (a justified system) to denote what Ludwig Lange in his 1885 essay "Über das Beharrungsgesetz," *Leipziger Berichte* **37,** 333–351, especially p. 337, had already called an "inertial system" (Inertialsystem).

[77] The term "Galileian equations" was coined by Philipp Frank in his paper "Die Stellung des Relativitätsprinzips im System der Mechanik," *Wiener Sitzungsberichte* **118,** 373–446 (1909).

his theory. As he wrote in the 1912 manuscript: "One arrives at the theory that is now called 'the theory of relativity' by keeping (a) and (b) but rejecting (c). In what follows it will become evident that it is *possible* to proceed in this way."[78] To reject (c), however, implies a revision of the concepts of simultaneity and time. Einstein was fully aware of this, for he wrote: "An analysis of the physical concepts of time and space revealed that *in reality an incompatibility of the principle of relativity with the law of the propagation of light does not exist.*"[79]

The rejection of (c), that is, of the equations of the Galileian transformation, requires their replacement by other equations that relate the space–time coordinates of one inertial system $S(x, y, z, t)$ to those of another such system $S'(x', y', z', t')$, the so-called Lorentz transformations. Einstein accomplished this task in § 8 and § 9 of his manuscript.[80] In § 8, "The physical meaning of spatial and temporal determinations," Einstein defined the concept of "time" in an inertial system S in essentially the same way as he had done two years earlier in "*Le Principe de Relativité et ses Conséquences dans la Physique Moderne.*"[81] To show that it is possible to synchronize any number of clocks with a master clock and that each clock can be synchronized with each other to establish synchrony and, therefore, simultaneity throughout a given reference system, Einstein appealed tacitly to the symmetry and (almost tacitly) to the transitivity of these relations and concluded that "the result of synchronizing the clocks can be characterized as follows: each clock is synchronized with each of the others."[82]

Chapter 11, which deals extensively with the symmetry and transitivity of the simultaneity relation, presents a more detailed analysis of the establishment of synchrony throughout a firm reference system, and explains the precise meaning of the parenthetical expression "almost tacitly."

The result that each clock at rest in a "system of rest" can be synchronized with each of the other clocks at rest in this system is of utmost importance

[78] "Dass diese Art des Vorgehens *möglich* ist wird sich im Folgenden zeigen." Note 74, § 7, italics in original.

[79] "Durch eine Analyse der physikalischen Begriffe von Zeit und Raum zeigt sich, *dass in Wahrheit eine Unvereinbarkeit des Relativitätsprinzips mit dem Ausbreitungsgesetz des Lichtes gar nicht vorhanden sei,* dass man vielmehr durch systematisches Festhalten an diesen beiden Gesetzen zu einer logisch einwandfreien Theorie gelange." A. Einstein, *Über die spezielle und die allgemeine Relativitätstheorie* (Braunschweig: Vieweg, 1920), p. 13; 21st edition, (1969), p. 20. Italics in original.

[80] Note 74, p. 23.

[81] Note 42.

[82] Note 74, p. 23.

for Einstein's construction of his theory because it enabled him to define the concepts of simultaneity and of time in "a system of rest." He had done so already in 1905 when he wrote in his pioneering paper: "It is essential that we have defined time by means of clocks at rest in a system at rest; because it belongs to the system at rest, we designate the time just defined as 'the time of the system at rest' ['die Zeit des ruhenden Systems']."

But what is this "system at rest"? Einstein defined it at the very beginning of § 1 in his 1905 paper as follows: "Consider a coordinate system in which the equations of Newtonian mechanics hold good. In order to render our presentation more precise and to distinguish this system of coordinates verbally from others which will be introduced later on, we will designate this system as the 'system at rest.' "[83] What Einstein defined was, of course, what is now generally called an "inertial system," because it is a coordinate system in which Newton's First Law, the law of inertia, holds true.

The logical structure of Einstein's construction of his theory of relativity is therefore based on the following three definitions: (a′) an *inertial system S* is defined as a coordinate system in which "the equations of Newtonian mechanics hold good"; (b′) a clock at rest in S at point A and another at point B are *synchronized* if the condition $t_B - t_A = t'_A - t_B$ (see fig. 7.1) is satisfied; and two events, one at A and the other at B, are *simultaneous* if synchronized clocks at rest at their places agree when the events occur; (c′) the *time* of the system S is defined as the set of all simultaneities, or more precisely, as is explained in chapter 11, as the quotient set of all events modulo the simultaneity relation, that is, (E/σ), where E denotes the set of all events and σ denotes the equivalence relation "simultaneity."

The three definitions or conditions (a′), (b′), and (c′) are obviously logically interdependent in the sense that (c′) presupposes (b′) and (b′) presupposes (a′). Together with postulates (a) and (b), as stated in the 1912 manuscript, they enabled Einstein to prove the relativity of simultaneity and to derive the Lorentz transformations. In particular, they led him to the relativistic velocity composition theorem that resolved all the difficulties he had encountered previously. In fact, these three definitions, culminating in the definition of the *time* of an inertial system, enabled Einstein to complete his theory and to present it in his

[83] "Es liege ein Koordinatensystem vor, in welchem die Newtonschen mechanischen Gleichungen gelten. Wir nennen dies Koordinatensystem zur sprachlichen Unterscheidung von später einzuführenden Koordinatensystemen und zur Präzisierung der Vorstellung das 'ruhende System.' " Note 6 (1905), p. 892.

1905 essay which, as stated earlier, has been called "possibly the most important scientific paper that has been written in the twentieth century."[84]

But, alas, the three definitions (a′), (b′), and (c′), presented in this paper, and in similar formulations in later papers, as the foundations of the theory, contain, if critically examined, an apparently unnoticed logical flaw. Unlike the three postulates (a), (b), and (c), as denoted in Einstein's 1912 manuscript, the three definitions (a′), (b′), and (c′) are not incompatible but rather contain, as we shall presently see, a vicious logical circle.

As will be recalled, the 1905 paper begins with the definition of an inertial system[85] as "a system of coordinates in which the equations of Newtonian mechanics hold good." Because this paper, as is well known, disproves the validity of Newtonian mechanics, it contains a logical paralogism, for it starts with the premise of the validity of Newtonian mechanics and concludes with a refutation of it.

Einstein himself seems to have recognized that his qualification of "the system of coordinates" as one for which "the equations of Newtonian mechanics hold good" poses a problem in the present context. In fact, when his 1905 paper was reprinted in *The Principle of Relativity—A Collection of Essays*,[86] a footnote was added to the words "hold good," namely, "i.e. to the first approximation." It apparently refers to the fact that for $c \rightarrow \infty$ the Galileian transformations of Newtonian mechanics are an approximation of the Lorentz equations. It has usually been assumed that Arnold Sommerfeld, who initiated the publication of this collection, wrote the footnote. But as Ian McCausland[87] has shown, it was probably Einstein who wrote it or was at least

[84] Introduction, note 11.

[85] As Hugo Seeliger in his article "Über die sogenannte absolute Bewegung," (Münchner Berichte **36,** 1906), p. 89, reports, the term "inertial system" (Inertialsystem), was introduced by L. Lange in 1885 and was used quite frequently in 1887; see, for example, *Mind* **12,** 151 (1887). Einstein seems to have still avoided its use in 1916 when he called a coordinate system, in which the inertial law is valid, a "Galileisches Koordinatensystem." See *Über die spezielle und die allgemeine Relativitätstheorie* (Braunschweig: Vieweg, 1917, 1920 [8th edition]), p. 8. This seems to indicate that Einstein had never read Lange's publications. In any case, Einstein's library contained none of Lange's publications.

[86] H. A. Lorentz, *Das Relativitätsprinzip, eine Sammlung von Abhandlungen* (edited by O. Blumenthal) (Leipzig: Teubner, 1913). The footnote was also published in the English editions, see note 6 (1923, 1952), p. 38.

[87] "It seems reasonable to assume, unless there is clear evidence to the contrary, that Einstein's cooperation included the addition of the footnote in question." I. McCausland, "Einstein and Special Relativity: Who Wrote the Added Footnotes?," *The British Journal for the Philosophy of Science* **35,** 60–61 (1984).

consulted whether it should be added. In any case, whoever added the footnote seems to have realized that such a definition of an inertial system in the present context poses a problem.

True, the Newtonian equations of motion are a first approximation of the relativistic equations. The added footnote, however, does not resolve the following logical error. Any reference to the Newtonian equations and, in particular, to the First Law, even if stated only as a first approximation, necessarily implies the use of the concept of velocity which, in turn, presupposes the notion of simultaneity. The added footnote, therefore, does not remove the logical circularity, namely that the definition of simultaneity presupposes conditions which themselves involve the notion of simultaneity.

Einstein, as Abraham Pais once recalled from conversations with him, had been well aware of the fact that the concept of the velocity of an object with respect to an inertial system S presupposes the notion of distant simultaneity or synchronization; it does so because the determination of the velocity of the motion of an object through a given distance Δx requires the measurement of the duration of motion, or flight time Δt, of the object through that distance, an operation which requires the use of synchronized clocks stationed at the endpoints of Δx and at rest relative to S.[88] It follows therefore logically that any reference to an inertial system and hence implicitly, according to Einstein's definition of it, to "the equations of Newtonian mechanics," including the First Law, involves the notion of simultaneity or synchrony. But, as mentioned earlier, according to Einstein, the notions of synchrony and of time [i.e., (b′) and (c′)] presuppose (a′), that is, the concept of an inertial system. We thus face again a logically vicious circle.

This raises the question of whether the logical impasse can be avoided by simply omitting (a′), that is, by not restricting (b′) and (c′) to inertial systems. Unfortunately such an attempted solution of the difficulty is of no avail simply because, as we shall see in due course, in noninertial systems, as encountered in general relativity and cosmology, Einstein's method of defining standard simultaneity does not work in general.

This conclusion raises, in turn, the question of whether inertial reference systems can be defined without any reference to a coordinate system "in which the equations of Newtonian mechanics hold good."

[88] See, for example, M. Podlaha et al., "Zur Problematik der Geschwindigkeitmessung," *Philosophia Naturalis* **16,** 315–317 (1977).

The best, and in principle perhaps the only, known alternative definition of an inertial system, which has been claimed not to involve any metrical concept of velocity or of temporal duration, is the definition that Ludwig Lange proposed in 1885. According to Lange a coordinate system in which "three particles, projected from the same origin in different non-coplanar directions and left to themselves ("sich selbst überlassen"), move in straight lines is an inertial system."[89] Two arguments refute such an attempt to vindicate Einstein's approach. (1) The apparently innocuous condition of "being left to themselves" means of course, if expressed in physical terms, "not acted upon by forces." But the notion of "force" is introduced only in the electrodynamical part of the 1905 paper and, moreover, is commented upon in the Blumenthal edition (1913) "as not advantageous, as was first shown by M. Planck. It is more to the point to define force in such a way that the laws of momentum and energy assume the simplest form."

In any case, because the notion of force necessarily implies either the concept of momentum or that of acceleration, and because the concepts of acceleration, momentum, and force also imply, in their turn, at least implicitly, the concept of velocity, which, as has been shown, implies the notion of simultaneity, Lange's definition of an inertial system does not eliminate the vicious circularity under discussion. (2) Moreover, as F. L. Gottlob Frege in his critique[90] of Lange's book pointed out in 1891, expressions like "a body moves" (ein Körper bewegt sich) or "is at rest" (ist in Ruhe) have a meaning only with respect to a "reference system" (Bezugssystem) relative to which the body is in motion or at rest. Similarly, in 1923 Einstein in his address acknowledging the Nobel Prize, a lecture in which the concept of an "inertial frame" plays a major role, declared that "motion can only be conceived as the relative motion of bodies,"[91] but Lange did not specify such a reference system in his definition of an inertial system.

[89] L. Lange, "Ueber das Beharrungsgesetz," *Leipziger Berichte* **37,** 333–351 (1885). *Die geschichtliche Entwicklung des Bewegungsbegriffs* (Leipzig: Hirzel, 1886). For details see M. Jammer, *Concepts of Space* (Cambridge, Massachusetts: Harvard University Press, 1954, pp. 138–139; 1969, pp. 140–141); H. P. Robertson and T. W. Noonan, *Relativity and Cosmology* (Philadelphia: Saunders, 1968), p. 13; R. Torretti, *The Philosophy of Physics* (Cambridge: Cambridge University Press, 1999), p. 53.

[90] F. L. G. Frege, "Über das Trägheitsgesetz," *Zeitschrift für Philosophie und philosophische Kritik* **98,** 145–161 (1891).

[91] A. Einstein, "Fundamental ideas and problems of the theory of relativity," Lecture delivered to the Nordic Assembly of Naturalists at Gothenburg, 11 July 1923; published in *Nobel Lectures (Physics), 1901–1921* (New York: Elsevier, 1967), pp. 479–490.

Furthermore, in the first edition of his well-known and frequently reprinted text *Die Relativitätstheorie*[92] Max von Laue praised Lange's definition of an inertial system as "a major achievement" (eine Grosstat), because it eliminates the "somewhat ghostlike" (etwas gespenstisch) Newtonian notions of absolute space and absolute time. However, in an essay published two years later he wrote: "Truly speaking, an inertial system cannot be defined by the observation of the motions of bodies left to themselves simply because such bodies are not available for us."[93] Surely, the idea of using empirically unavailable bodies would certainly contradict the Machian empiricism in which the first paragraph of Einstein's 1905 paper was written.

In a lecture, delivered 1966 at the Ernst-Mach-Institute in Freiburg, Martin Strauss defined "inertial frames" as "precisely those frames in which the 4-dimensional 'lightgeometry' coincides with the chronogeometry of rigid rods and mechanical clocks. . . . This formulation can also be used as an 'operational definition' [viz., an operational criterion] of 'inertial frame.' "[94] Clearly, the terms *4-dimensional lightgeometry*, and even more explicitly *chronogeometry*, indicate that the notion of an inertial frame includes that of "*time.*" In a similar vein, but in a different context, Peter Janich, one of the leading protophysicists and, as such, deeply interested in an operational definition of the concept of an inertial system, wrote in 1978 that "it is generally not controversial that *inertial systems* cannot be defined *only* by kinematical statements."[95] But any dynamical statement involves the notion of force and hence, as explained previously, also the notions of time and simultaneity.

Needless to say, the preceding critical remarks concerning a vicious circle in Einstein's famous 1905 relativity paper were not intended to disprove the validity of this theory nor to belittle Einstein's merit for having created it and having profoundly changed thereby the fundamental concepts and laws of classical physics. In fact, these new laws are empirically verified daily in every high-energy laboratory.

[92] M. von Laue, *Das Relativitätsprinzip* [later named *Die Relativitätstheorie*] (Braunschweig: Vieweg, 1911; 5th edition, 1952), § 1.

[93] M. von Laue, "Das Relativitätsprinzip," *Jahresberichte der Philosophie* **1,** 243–273 (1913).

[94] "On the logic of 'inertial frame' and 'mass,' " in M. Strauss, "*Modern Physics and its Philosophy* (Dordrecht-Holland: Reidel, 1972), pp. 119–129. Quotation on p. 121.

[95] P. Janich, "Die Protophysik der Zeit und das Relativitätsprinzip," *Zeitschrift für allgemeine Wissenschaftstheorie* **9,** 343–347 (1978); reprinted in J. Pfarr (ed.), *Protophysik und Relativitätstheorie* (Mannheim: BI Wissenschaftsverlag, 1981), pp. 179–183 (italics in original).

The above-mentioned critique was presented merely because of its intimate relation to the notion of distant simultaneity, which, as has been shown, played a critical role in Einstein's construction of the special theory. This does not mean, however, that the special theory of relativity could not have been arrived at in a way that does not involve the notion of simultaneity as a fundamental cornerstone. It could be formulated, for instance, at least theoretically, as a special case of the general theory of relativity, namely, as the case of a universe completely lacking any gravitational sources. Obviously, in such a derivation of the special theory the concept of distant simultaneity would not, and could not, play a constitutive role in the construction of the theory.

The theoretical possibility of alternative constructions of the special theory raises the question of what precisely prompted Einstein to begin his construction of the theory with a definition of distant simultaneity. Many historians and philosophers of physics, among them John Earman, Clark Glymour, Stanley Goldberg, Gerald Holton, Arthur Miller, Robert Rynasiewicz, Roberto Torretti, and most recently Bernd Lukoschik,[96] studied the problem of how much the writings of Hume, Kant, Mach, Lorentz, Poincaré, Föppl, and possibly others[97] had influenced Einstein's construction of his theory.

Einstein himself acknowledged in this context that "the reading of Hume, along with Poincaré and Mach, had some influence" on this development.[98] But the specific problem of why this development began just with an analysis of the notion of simultaneity, as Einstein acknowledged in his Kyoto lecture,[99] was actually dealt with only by Einstein himself when he wrote in his *Autobiographical Notes*: "The type of critical reasoning which was required for the discovery of this central point was decisively furthered, in my case, especially by the reading of David Hume's and Ernst Mach's writings."[100] However, as far as a critical analysis of the notion of distant simultaneity as the first step in the construction of a new theory is concerned, Einstein declared: "It is not improbable that Mach would have hit on relativity theory when in his time—when he was in fresh and youthful spirit—physicists would have

[96] B. Lukoschik, "Realismus und Instrumentalismus im Weltbild des frühen Einstein," *Philosophia Naturalis* **39,** 111–140 (2002).

[97] Important, but so far not sufficiently studied, sources were probably the writings of Emil Cohn such as his essay "Zur Elektrodynamik bewegter Systeme," *Berliner Berichte* **40,** 1294–1303, 1404–1416 (1904).

[98] See C. Seelig (ed.), *Helle Zeit, Dunkle Zeit* (Zürich: Europa Verlag, 1956), note 1, p. 464.

[99] Note 3.

[100] Note 63, p. 53.

been stirred by the question of the meaning of the constancy of the speed of light. In the absence of this stimulation, which flows from Maxwell-Lorentzian electrodynamics, even Mach's critical urge did not suffice to raise a feeling for the need of a definition of simultaneity for spatially distant events."[101] As this statement shows, in 1916 Einstein firmly believed that an analysis of the notion of distant simultaneity was an inevitable step in the construction of the theory of relativity.

Recall, however, that in 1910 Waldemar von Ignatowski[102] derived the structure of the Lorentz transformations and hence the special theory of relativity without invoking the light postulate by assuming only the principle of relativity, the isotropy and homogeneity of space, and the reciprocity of the transformations. One year later Philipp Frank and Hermann Rothe[103] accomplished this task by confining themselves to only two even more restricted postulates. Of course, neither Ignatowski nor Frank and Rothe were able to identify the invariant velocity contained in the resulting transformations with the velocity of light. From then[104] until today[105] numerous other derivations of the Lorentz transformations or their equivalents have been proposed without invoking the light postulate or any reference to the concept of simultaneity. These facts are mentioned to stress again that the above-mentioned criticism of the logical structure of Einstein's 1905 relativity paper is in no way intended to criticize the special theory of relativity itself.

Note that some prominent philosophers of physics deprecate the importance of Einstein's 1905 operational definition of simultaneity for the construction of the special theory of relativity. Thus, for example, John D. Norton, in a recently published profound study declared: "What I would like to

[101] A. Einstein, "Ernst Mach," [Eulogy] *Physikalische Zeitschrift* **17,** 101–104 (1916); *Collected Papers* (1996), vol. 6, 278–281; English translation (1997), vol. 6, 141–145.

[102] W. von Ignatowski, "Einige allgemeine Bemerkungen zum Relativitätsprinzip," *Physikalische Zeitschrift* **12,** 972–976 (1910).

[103] P. Frank and H. Rothe, "Über die Transformation der Raum-Zeitkoordinaten von ruhenden auf bewegte Systeme," *Annalen der Physik* **34,** 825–855 (1911).

[104] For the literature on derivations of the Lorentz transformations, see H. Arzeliès, *Relativistic Kinematics* (Oxford: Pergamon Press, 1966), pp. 80–82.

[105] Recent examples are N. David Mermin, "Relativity without light," *American Journal of Physics* **52,** 119–124 (1984); L. H. Kauffman, "Transformations in Special Relativity," *International Journal of Theoretical Physics* **24,** 223–236 (1985); J. H. Field, "Space-time invariance: Special relativity as a symmetry principle," *American Journal of Physics* **69,** 569–575 (2001); Y. Friedman and Y. Gofman, "Relativistic linear spacetime transformations based on symmetry," *Foundations of Physics* **32,** 1717–1736 (2002).

suggest is that it is entirely possible that thoughts about clocks and their synchronization by light signals played no essential role in Einstein's discovery of the relativity of simultaneity . . . thoughts of light signals and clock synchronization most likely played a role only a brief moment, some five or six weeks prior to the completion of the paper. . . . We should not allow the excitement of this moment to obscure the fact that its place in Einstein's pathway is momentary in comparison to the years of arduous exploration that preceded it."[106]

In reply to Norton's comments, it should be pointed out that, of course, Einstein did not start the conceptual development of his theory by studying the meaning of the concept of simultaneity or of time but that it was rather, as stated earlier, his dissatisfaction with the conceptual status of Maxwell's theory prevailing at that time that prompted him to revise its foundations. But to do so he was struck by the apparent irreconcilability between the two postulates, relativity and the invariance of the velocity of light; it was indeed a sudden and unpremeditated intuition that helped him to resolve this difficulty. Because it served him, so to speak, as a catalyst without which his theory, at least at that time, could not have been generated, its critical importance cannot be overestimated.

[106] J. D. Norton, "Einstein's Investigations of Galilean Covariant Electrodynamics Prior to 1905," *Archive for History of Exact Sciences* **59,** 45–105 (2004).

CHAPTER EIGHT

The Reception of the Relativistic Conception of Simultaneity

Einstein's definition of distant simultaneity had far-reaching consequences. It led not only to a new conception of time, which in itself would have been a major innovation, but also to a radical break with classical physics and philosophy. First, the definition implies the renunciation of the previously generally accepted idea that there is no upper limit to physically attainable velocities, for if arbitrarily high velocities were admitted, the classical notion of distant simultaneity, obtained by an instantaneous transmission of information, would not have to be discarded. Second, the acknowledgment of a finite upper bound of such velocities is incompatible with classical mechanics, for according to Newton's second law of motion a constant force, acting for a sufficiently long time, could accelerate a given mass to an arbitrarily high velocity. The new conception of distant simultaneity entails, therefore, the abandonment of classical mechanics and of all philosophical systems based on it.

These consequences were not recognized immediately with the publication of Einstein's 1905 paper. His 1905 method of deriving the Lorentz transformations, which explain all relativistic effects, from his definition of distant simultaneity was soon superseded by mathematically simpler methods that did not explicitly involve the notion of distant simultaneity. This notion consequently lost much of its prominence in the thinking of the physicists, who were more interested in the experimental consequences and empirical confirmations of the theory than in its conceptual foundations.

Philosophers, before the early 1920s, did not engage themselves with the foundations of relativity either. The mathematically minded philosopher Hugo Bergmann, when he reviewed a text on relativity written by Alexander Brill[1] in 1913, complained that despite the fact that philosophical conceptions, like space and time, play an important role in the foundations of the theory, philosophers had not yet acquainted themselves with the new theory.[2] This is not to say, however, that the concept of time played no role in the philosophical literature of those days. Metaphysicians, like F. H. Bradley, B. Bosanquet, J. Royce, or G. Gentile, dedicated a large part of their writings to the problem of time and especially to the question of its reality. J. M. E. McTaggart's attempt at refuting the reality of time, published at first in the 1908 issue of *Mind,* is still today a subject of lively debates. Samuel Alexander began his Gifford Lectures with the statement that "all the vital problems of philosophy depend for their solution on the solution of the problem what Space and Time are"; and in his monumental work *Space, Time and Deity* he wrote that "to realize the importance of Time as such is the gate of wisdom."[3] The writings of M. Guyau and J. Ward stressed more the psychological aspects of the idea of time. In all these writings, however, the notion of simultaneity was hardly given serious attention, if at all. The outstanding exceptions were an essay by the philosopher Louis Dominique Joseph Armand Dunoyer, who called the notion of simultaneity "la porté philosophique de la théorie [relativiste]"[4] and, of course, the treatise *Durée et Simultanéité*[5] in which Henri Bergson challenged the relativistic conception of simultaneity for reasons to be discussed.

[1] A. Brill, *Das Relativitätsprinzip* (Leipzig: Teubner, 1912). This is the second textbook on relativity ever published. The first was Max von Laue's *Das Relativitätsprinzip* (Braunschweig: Vieweg, 1911).

[2] H. Bergmann, "Rezension," *Zeitschrift für Philosophie und philosophische Kritik* **150,** 211–212 (1912).

[3] A. Alexander, *Space, Time and Deity* (London: Macmillan, 1920), vol. 1, p. 36.

[4] L. Dunoyer, "Einstein et la relativité," *La Revue Universelle* **9,** 179–188, 814–835 (1922).

[5] H. Bergson, *Durée et Simultanéité* (Paris: Alcan, 1922); *Duration and Simultaneity* (New York: Bobbs-Merrill, 1965).

The following discussion of the acceptance of Einstein's early papers on relativity is confined, of course, only to responses that refer to the notion of distant simultaneity or to the related notion of the synchronization of spatially separated clocks. Because this synchronization involves the question of whether the "one-way" velocity of light or of any other signal can be determined experimentally, our discussion also involves issues related to the problem of the conventionality thesis of the concept of distant simultaneity. Remember that in the first five years or so after the publication of Einstein's 1905 paper "On the electrodynamics of moving bodies," Einstein's new theory was regarded, in general, as a modification of the prevailing theory of electrodynamics, that is, as an elaboration of the theory of Hendrik Antoon Lorentz. Some reviewers even spoke of "the Einstein-Lorentz theory," for they thought that the main purpose of Einstein's paper was a new derivation of the Lorentz transformation. It was only natural, therefore, that the German periodical for abstracts in physics, *Fortschritte der Physik* (*Advances in Physics*), listed during those years all abstracts of papers on relativity under the heading "electrodynamics."

The earliest review of the 1905 paper was an abstract in the *Fortschritte der Physik*, written by Siegfried Valentiner, who worked at the *Physikalisch-Technische Reichsanstalt* in Berlin. He states that the theory is based on two assumptions, the principle of relativity (which says that the laws of physics are the same in all inertial systems) and the principle of the constancy of the velocity of light (which says that the velocity of light does not depend on whether its source is at rest or in uniform motion relative to an inertial system). Valentiner also declares that in the new theory "time is determined by clocks which, placed at different locations in the reference system, are synchronized in accordance with the equal-time stipulation" (which stipulates that the one-way velocity of light from A to B equals its one-way velocity from B to A). The abstract concludes with referring the reader for further details to the original paper the exposition of which is called "logical and concise."[6]

At about the same time the *Fortschritte der Mathematik* also published an abstract of Einstein's 1905 relativity paper. It was written by Curt Grimm, an assistant of Paul Drude, the author of an influential textbook on optics and the editor of the *Annalen der Physik*, which published Einstein's paper. In contrast to Valentiner, Grimm listed three basic assumptions on which Einstein

[6] *Fortschritte der Physik* **61,** 280–281 (1906).

based his theory: the relativity principle, the independence of the light velocity of its source, and "the definition of common time for two synchronized clocks, according to which the 'time,' required by the light to propagate from one clock to the other equals the 'time' required for the return trip."[7] That the light postulate, by invoking the concept of velocity, a ratio between distance and time, already involves the notion of coordinate time was not realized, so that this order of presentation violated the methodological rule not to define a scientific concept in terms of a theory that contains the definiendum.

As mentioned previously, Einstein's early work on relativity was often regarded as merely an elaboration of the Lorentz theory of the electromagnetic field. Lorentz, in contrast to Einstein, however, retained the concept of an all-pervasive stationary ether. It is not surprising, therefore, that some physicists tried to reconcile Einstein's theory with the theory of an ether. Foremost among these physicists was Emil Wiechert, the founder of the Göttingen school of geophysicists. In a lengthy article on the principle of relativity and the ether Wiechert devotes a whole paragraph (§ 21) to what he calls "physical simultaneity." Wiechert rejects the approach according to which the finite velocity of light leaves some liberty in the definition of distant simultaneity. According to his own theory "the velocity of light presents no natural limitation for connecting space and time. . . . The indeterminacy in the determination of simultaneity can be recognized from the practical point of view but has no fundamental significance for events of the world in general. . . . Only the assumption of a homogeneous and isotropic ether offers the possibility of an unambiguous definition of physical simultaneity, which corresponds to the determination of time in a Lorentz-system which is at rest relative to the ether."[8]

In 1912 the logician Paul Bernays, who with David Hilbert wrote the classical study on the foundations of mathematics, criticized the operational approach of the theory of relativity according to which the notion of simultaneity "becomes meaningful only by means of a physical procedure which decides when this notion can be applied in experience." That this cannot be true, argued Bernays, "can be seen by the very fact that the problem to find

[7] *Jahrbuch über die Fortschritte der Mathematik* **36,** 920–921 (1906).

[8] E. Wiechert, "Relativitätsprinzip und Äther," *Physikalische Zeitschrift* **12,** 689–707, 737–758 (1911).

a physical method for the determination of simultaneity could not arise if one would not have prior to the discovery of this method a concept of simultaneity." Another argument, said Bernays, which is independent of the former, runs as follows: because space does not offer an absolute determination of location, it is meaningless to say of two events occurring at different times that they occurred at the same place or at different places; interchanging space and time shows that it is meaningless to say that two spatially separated events occurred at the same time or at different times. This argumentation, declared Bernays, is inadmissible (*unzulässig*) because the analogy between space and time is not complete; time, for example, has a distinguished direction, from past to future, whereas such a distinction does not exist for spatial directions. These considerations, Bernays concluded, "suffice to show that the claim, the theory of relativity offers new insights into the relation between space and time, is unfounded."[9]

Like Emil Wiechert in Germany, Ebenezer Cunningham in England, a student of Joseph Larmor and author of the first book-length account of the special theory of relativity in English,[10] tried to make the ether compatible with the principle of relativity. Like Wiechert he claimed that the undetectability of the ether does not disprove its existence. In 1914, shortly before his monograph appeared, he published a paper with the same title, *The Principle of Relativity*, in the prestigious journal *Nature*, in which he showed how the acceptance of the "hypothesis of relativity" makes it impossible for us "to determine uniquely whether two events are or are not simultaneous." Assuming the situation depicted in figure 7.1 of chapter 7, but imagining that the two points A and B, separated by the distance d, are moving relative to the ether with the same velocity v, Cunningham argues as follows:

> Since the relative velocity of the light on the outward journey [from A to B] is $c - v$, we have $t_2 - t_1 = d/(c - v)$, and similarly since the relative velocity on the return is $c + v$, $t_3 - t_2 = d/(c + v)$. From these equations we obtain $t_2 = \frac{1}{2}(t_1 + t_3) + d\,v/(c^2 - v^2)$. Now if the velocity v were zero, we should have the result that the moment of reflection at B is simultaneous with the moment $\frac{1}{2}(t_1 + t_3)$, that is, with the moment at A midway between those of emission and return of the ray. But if the velocity v is unknown, which is the hypothesis with which we are

[9] P. Bernays, "Über die Bedenklichkeiten der neueren Relativitätstheorie," *Abhandlungen der Fries'schen Schule* (Neue Folge) **4,** 459–482 (1912).

[10] E. Cunningham, *The Principle of Relativity* (Cambridge Cambridge University Press, 1914).

dealing, then we cannot say from this experiment what instant at A is simultaneous with the instant t_2 at B.[11]

These considerations led Cunningham to conclude that the concept of simultaneity "does not become definite until we have assigned a definite velocity to a certain point, which may conveniently be our point of observation." This indefiniteness also applies to the notion of the "length of a body" because it is defined as "the distance between two points of our universal frame of reference, with which the ends of the body 'simultaneously coincide.' "

Cunningham's article was soon criticized severely by Alfred A. Robb, known as the first theoretician who tried to construct the relativistic space–time structure on causality relations. He pointed out that the notion of "definite velocity," which has to be used to define simultaneity, implies the terms "length and time," which according to Cunningham presupposed the concept of simultaneity. Robb therefore concluded his critique with the question: "What are Mr. Cunningham's fundamental concepts?"[12]

Note that in his mathematical argumentation Cunningham applied the classical and not the relativistic addition law of velocities. This leads us to our next remarks.

Einstein's 1905 proof of the relativity of distant simultaneity, involving the above-quoted equation[13] $t_B - t_A = r_{AB}/(c - v)$ and $t'_A - t_B = r_{AB}/(c + v)$, was criticized by Hans Strasser, Halmar J. Mellin, G. H. Keswani, Karl Stiegler, Deuk Son Kim, Ulrich Hoyer, and others,[14] on the basis that the division of r_{AB} by $c - v$ or $c + v$, respectively, presupposes the classical nonrelativistic theorem of the addition of velocities or, equivalently, the Galilean transfor-

[11] E. Cunningham, "The principle of relativity," *Nature* **93,** 378–379, 408–410 (1914). Quotation on p. 409. See also paragraph 24, "On the idea of simultaneity," in E. Cunningham, *Relativity, The Electron Theory and Gravitation* (London: Longmans, Green and Co., 1921), pp. 28–31.

[12] A. A. Robb, "The principle of relativity," *Nature* **93,** 454 (1914).

[13] See chapter 7, note 30.

[14] H. Strasser, *Die Grundlagen der Einsteinschen Relativitätstheorie* (Bern: Haupt, 1922); *Einsteins spezielle Relativitätstheorie—eine Komödie der Irrungen* (Bern: Bircher, 1923). H. J. Mellin, "Das Lichtproblem," *Annales Academiae Scientiarum Fennicae* **24,** 9–18 (1925). G. H. Keswani, "Origin and concept of relativity II," *British Journal for the Philosophy of Science* **16,** 19–32 (1965). K. Stiegler, "On errors and inconsistencies in Einstein's 1905 paper 'Zur Elektrodynamik bewegter Körper', *Proceedings of XIIIth International Congress of the History of Science, Moscow, 1971* (Moscow: Naouca, 1974), section 6, pp. 53–63. D. S. Kim, "Kritische Überlegungen zur Relativitätstheorie," Lecture delivered at the Congress of the Deutsche Physikalische Gesellschaft, Münster, 16 March 1984. U. Hoyer, "Die Grundlagen der Relativitätstheorie," *Zeitschrift für allgemeine Wissenschaftstheorie* **17,** 1–18 (1986).

mation in blatant violation to the second postulate of the invariance of the velocity of light. Thus Mellin of the Technical University of Helsinki wrote in 1925: "If Einstein divides r_{AB} by $c - v$ or $c + v$, respectively, he makes use of the addition rule of velocities. According to the principle of the constancy of light he should have divided r_{AB} in both cases by c."[15]

All these authors apparently interpreted these equations as having been obtained by Einstein as an application of the relation "time = distance/velocity," in which the velocity $c - v$ (respectively, $c + v$) is the velocity of light relative to the moving rod and is obtained by the classical velocity-addition law. Had this been Einstein's approach, these criticisms would have been fully justified. Einstein did not add or subtract velocities, however, he added or subtracted distances. He explicitly emphasized that all the metric quantities under discussion were taken relative to the inertial system S and not relative to that of the rod. To remove any misunderstanding let us retrace Einstein's reasoning in obtaining the first of the two equations. The physical process begins with the emission of a light signal at the S-time t_A and ends with its reception at the S-time t_B. During the S-time interval $t_B - t_A$ the signal moves with velocity c so that it traverses in S the distance $c(t_B - t_A)$, which is composed of the distance r_{AB}, as measured in S, and in addition of the distance $v(t_B - t_A)$ by which the B end of the rod recedes in front of the approaching signal. Hence, by addition of these two segments, $c(t_B - t_A) = r_{AB} + v(t_B - t_A)$ which yields the first of the two above-quoted equations. A similar argument establishes the second equation. No use has been made of the classical law of the addition of velocities, and Einstein's 1905 proof of the relativity of distant simultaneity is therefore free of the logical inconsistency imputed to it by those critics.

The story repeated itself with Einstein's 1916 relativity proof of simultaneity.[16] Let us quote again only Mellin, who claimed:

> According to the principle of the constancy of light velocity the same result will be obtained also if the observer passes through the midpoint of *AB* when the signals are emitted. To see this, one has only to regard the observer as "the system of rest," the system of the light sources as "the moving system," and to apply the said principle to the case. It follows that Einstein's criterion of simultaneity is satisfied just as it is in the case the observer is at rest in the midpoint

[15] H. J. Mellin, op. cit., p. 14.
[16] See figure 7.3 in chapter 7.

> of *AB*, so that the simultaneity of the arrival of the signals in the one case is inseparably connected with the simultaneity of their arrival in the other case. According to the principle under discussion simultaneity is therefore *not relative*.[17]

Mellin seems to have been unaware that his reasoning contains a *petitio principii*, for he takes it for granted that the emission of the light signals also occurs simultaneously ("beim Abgang der Signale") for the moving observer.

A similar elementary logical error was committed by Oskar Kraus, a professor of law at the University of Prague and sympathizer with Hans Vaihinger's "As-If" philosophy. Vaihinger's at-that-time influential *The Philosophy of 'As-If'*[18] advocated an extreme form of pragmatism according to which fundamental principles of science, for example, are merely pure fictions or rather statements that, although lacking objective truth, prove useful when used for some definite purpose. It suffices to show that such statements are contradictory with either internal or external data. This is exactly what Kraus tried to show for the statements of the theory of relativity and, in particular, for its assertion of the relativity of distant simultaneity. If event e_1, he argued, is simultaneous with event e_2 and event e_2 is simultaneous with event e_3, then e_1 must be simultaneous with e_3; because relativity admits the possibility that this is not the case, the theory, says Kraus, asserts a proposition and its negation and thus violates the principle of contradiction.[19]

The most famous and perhaps also the most serious opponent of Einstein's relativistic conception of distant simultaneity and its implied multiplicity of time series was the French philosopher Henri Bergson. Although Bergson greatly admired Einstein's theory, which he called "not only a new physics, but also, in certain respects, a new way of thinking,"[20] he severely criticized the relativity of distant simultaneity and regarded it as a fiction or illusion.

The notion of time was always at the center of Bergson's philosophical thinking. In his juvenile *Time and Free Will*,[21] in 1910, he had distinguished

[17] H. J. Mellin, op. cit., p. 10.

[18] H. Vaihinger, *Die Philosophie des 'Als-Ob'* (Leipzig: Meiner, 1922); *The Philosophy of 'As If'* (London: Kegan Paul, 1924).

[19] O. Kraus, "Fiktion und Hypothese in der Einsteinschen Relativitätstheorie," *Annalen der Philosophie* **2,** 335–396 (1921).

[20] H. Bergson, "Address," *Bulletin de la Société Française de Philosophie* **22,** 102 (1922). Reprinted in J. Langevin and M. Paty, "Le sejour d'Einstein en France en 1922," *Cahiers Fundamenta Scientiae, No. 93* (University of Strasbourg), 1979.

[21] H. Bergson, *Essai sur les Données immédiates de la Conscience* (Paris: Alcan, 1889); *Time and Free Will* (London: George Allen and Unwin, 1910, 1950).

between the time that we think about from the time that we experience and ascribed reality only to the latter, which "living and conscious" beings share in common.

In *Duration and Simultaneity*[22] Bergson confronted his philosophy of time with the philosophically significant conclusions of the theory of relativity. Confining ourselves to Bergson's critique of Einstein's notion of distant simultaneity we analyze Bergson's discussion of Einstein's thought experiment of the train moving along the embankment. After quoting Einstein's description of this experiment Bergson declared that the observer at M (see fig. 7.3 in chapter 7) on the embankment, who perceives the two flashes simultaneously, can only infer that the observer at M' in the train does not see them simultaneously, for he moves toward the flash coming from B and flees in front of the flash coming from A, but this proves only that from the viewpoint of the observer at M events simultaneous at M are not so for the observer at M'. More precisely, Einstein's reasoning merely leads to the conclusion that the observer at M judges that events simultaneous for him are not so for his colleague, but it does not prove that they really are not simultaneous for him. To prove the latter assertion it would have been necessary for the observer at M to know the perceptions of the observer at M'. Einstein, according to Bergson, committed the error of confounding the perceptions of the observer at M' with the judgment of the observer at M about them, a fallacy that Bergson referred to as "a mirage effect" (un effet de mirage). Moreover, Bergson even claimed that Einstein's thesis of the relativity of distant simultaneity is antirelativistic and hence self-defeating. By taking Einstein's proof literally, we are driven, according to Bergson, to the conclusion that if simultaneity is really relative it is so because the observer at M infers from his own observations without consulting his colleague at M' that events simultaneous for him are not so for his colleague. Einstein's proof thus singles out the embankment as that reference system that alone recognizes absolute reality and designates it as the system relative to which phenomena are seen as they really are. Bergson concludes therefore: "Einstein's theses not only do not contradict, they even confirm mankind's natural belief in a unique and universal time" and "rather than disproving the thesis of a unique time (and hence of absolute simultaneity) Einstein's theory leads to it and makes it more

[22] H. Bergson, *Durée et Simultanéité* (Paris: Alcan, 1922, 1923, 1926, 1931); *Duration and Simultaneity* (New York: Bobbs-Merrill, 1966).

intelligible." Or in other words, Einstein's theory, if fully pursued to its logical conclusions, only confirms Bergson's conviction that "there is only one real Time, and all the others are fictions."[23]

Bergson added that even if we admit the "simultaneity" propounded in the theory of relativity we must realize that it is not based on any immediate experience but on some regulation of clocks performed by optical signals. It is based, therefore, on our immediate experience of local simultaneity without which we would never be able to make use of clocks. True simultaneity is but an extension of local simultaneity.

In the discussion, which took place on 6 April 1922, at the Collège de France in Paris, Einstein replied to Bergson's presentation of these ideas that one has to distinguish between the time studied by the psychologist and the time used by the physicist. Einstein agreed that the notion of a universal time indeed has its roots in the psychological experience of simultaneity and is the first step toward objectivity. But our experiments in physics, which involve the high propagational velocity of light, show that the psychologically acquired concept of simultaneity leads to contradictions. The theory of relativity therefore ignores psychological time and works with the concept of an objective time that is different from that of psychological time. Einstein concludes his comments by stating that "there is no 'philosopher's time' which is both psychological and physical; there exists only a psychological time which is different from the time of the physicist."[24]

For André Metz's 1924 essay[25] on the relation between Bergson's philosophy of time and relativity, which also touches on the notion of distant simultaneity, we refer the interested reader to *Bergson and the Evolution of Physics*[26] for translations into English of this and other articles on Bergson's philosophy of time.

The French physicist Charles Nordmann, the author of several books on the philosophy of science, also discussed the debate between Einstein and Bergson. In the sixth chapter of his book *The Tyranny of Time* Nordmann reviewed Bergson's criticism of Einstein's thought experiment of the moving

[23] "Il y a un seul Temps réel, et les autres sont fictifs." Note 22 (1931), p. 107.

[24] "Il n'y a donc pas un temps des philosophes; il n'y qu'un temps psychologique différent du temps du physicien." A. Einstein, note 20 (1922), p. 107.

[25] A. Metz, "Le Temps d'Einstein et la philosophie: à propos de l'ouvrage de M. Bergson, Durée et Simultanéité," *Revue de Philosophie* **31,** 56–88 (1924).

[26] P. A. Y. Gunter (ed.), *Bergson and the Evolution of Physics* (Knoxville: The University of Tennessee Press, 1969).

train. After stating that "the theory of relativity is entirely founded on the relativity of simultaneity" Nordmann declared that Einstein's demonstration of it "does not suffice to prove that simultaneity is really relative." Nordmann proposed a slight modification of the experiment in which the issue concerning the movement of either one of the two observers with respect to the light rays and the question as to the judgment of either observer concerning the perception of the other can no longer be raised and which is therefore immune to Bergson's objections.[27]

The distinction between psychological time and physical time in the context of the conception of simultaneity, which Einstein mentioned in his 1922 debate with Bergson, had been emphasized earlier in 1915 by Moritz Schlick. Having started his career as a student of physics and having written his dissertation on a problem in optics, Schlick turned to philosophy and founded the so-called *Vienna Circle* of logical positivists in 1924. It had been argued, primarily by some Neo-Positivists, that the theory of relativity can have no validity with respect to the objective world because it contradicts our a priori intuition, the laws of which determine the objective world. In a paper entitled "The philosophical importance of the principle of relativity,"[28] Schlick rebutted this contention by pointing out that the theory does not contradict our intuition simply because our time intuition tells us nothing about the properties of time dealt with in relativity. The time of our intuition is psychological time, something purely qualitative and unmeasurable, whereas the time dealt with in the theory is something purely quantitative and measurable. In particular, Schlick remarks, our intuition cannot teach us whether the notion of simultaneity is absolute or relative, for it is impossible to experience intuitively the simultaneity of distant events, because at least one of two spatially separated events can become part of our knowledge only by the mediation of spatiophysical processes.[29]

In an introduction to Einstein's theory, written in 1917, Schlick explained the relativity of distant simultaneity by using the mathematics of Einstein's

[27] Ch. Nordmann, *Notre Maitre le Temps* (Paris: Hachette, 1923); *The Tyranny of Time* (London: Unwin, 1925).

[28] M. Schlick, "Die philosophische Bedeutung des Relativitätsprinzips," *Zeitschrift für Philosophie und philosophische Kritik* **159,** 129–175 (1915).

[29] "Es lehrt uns nichts darüber, ob der Begriff der Gleichzeitigkeit absolut oder relativ ist, denn Gleichzeitigkeit von Ereignissen an verschiedenen Orten wird niemals unmittelbar anschaulich erfahren, weil mindestens der eine von zwei räumlich getrennten Vorgängen nur durch Vermittlung räumlich-physischer Prozesse zu unserer Kenntnis gelangt." Ibid., p. 143.

thought experiment of the train and the embankment and concluded his explanation with the following remark: "All this, as one sees, is a necessary consequence of the synchronization of clocks in accordance with the principle, that the velocity of light is a constant, could not be accomplished without arbitrariness in a different manner."[30] Schlick did not deny, as we see, the possibility that the introduction of some arbitrary different assumptions could lead to alternative conceptions of simultaneity.

Apart from Schlick there were only few philosophers in those years who had sufficient knowledge of the mathematical physics that one needs to fully understand Einstein's work. An outstanding exception was Alfred North Whitehead, who, like Schlick, began his career as a mathematical physicist, became a logician, when writing with Bertrand Russell the momentous *Principia Mathematica*, and also a philosopher, who was deeply influenced by Bergson's insistence on the importance of immediate intuition. Consequently, Whitehead strongly rejected the modern scientist's bifurcation of nature, which distinguishes between nature as sensed and perceived and nature as mathematically and formally conceived by scientific theory, which leads to the fact that "scientific theory is shot through and through with notions which are frankly inconsistent with its explicit fundamental data."[31]

It was from this point of view that Whitehead criticized Einstein's definition of distant simultaneity on various counts. His first objection, made in 1919, concerned the use of light signals. Although he admitted their importance in our lives, he objected that "the very meaning of simultaneity is made to depend on them. There are blind people and dark cloudy nights, and neither blind people nor people in the dark are deficient in a sense of simultaneity. They know quite well what it means to bark both their shins at the same instant."[32]

Einstein, of course, was aware of the possibility of such a criticism. In his Stafford Little Lectures of 1921 he responded by pointing out that "in order

[30] M. Schlick, *Raum und Zeit in der gegenwärtigen Physik* (Berlin: Springer, 1817, 1919), p. 16.

[31] A. N. Whitehead, *An Inquiry concerning the Principles of Natural Knowledge* (Cambridge: Cambridge University, 1919), p. 15.

[32] Ibid., p. 53. Whitehead was not the first to charge Einstein with having exaggerated the role of light in the construction of the theory of relativity. Previously, in 1910 Woldemar von Ignatowsky criticized the excessive role of optics in the theory and proposed a derivation of the Lorentz transformation without any reference to optical phenomena to confer thereby upon the principle of relativity a much greater generality. Cf. his articles "Einige allgemeine Bemerkungen zum Relativitätsprinzip," *Physikalische Zeitschrift* **11,** 972–975 (1910), and "Das Relativitätsprinzip," *Archive der Mathematik und Physik* **17,** 1–24 (1911); **18,** 17–40 (1912).

to give physical significance to the concept of time, processes of some kind are required which enable relations to be established between different places. It is immaterial what kind of processes one chooses for such definition of time. It is advantageous, however, for the theory, to choose only those processes concerning we know something certain. This holds for the propagation of light *in vacuo* in a higher degree than for any other processes which could be considered, thanks to the investigations of Maxwell and H. A. Lorentz."[33]

It is easy to show, by means of a Minkowski diagram, for example, that if we used a signal faster than the fastest known signal, that is, light, in Einstein's definition of distant simultaneity, causal anomalies would arise. For instance, a light signal could arrive at its destination before it has left its point of departure. Although Einstein never used this argument in his rebuttal of the criticism, he must have known about it because he once said that if superluminal signals existed one would be able "to telegraph into the past."

For Whitehead such considerations were of minor importance, because he categorically denied that any measurement provides the true meaning of simultaneity. According to Whitehead the meaning of simultaneity cannot be obtained by any operational procedure but only by abstraction from immediate sense awareness. In fact, local simultaneity, which occurs when reading the clock in the vicinity of an event and which is also a constitutive element in Einstein's approach, is obtained precisely from immediate sense awareness. In contrast, Einstein's definition of *distant* simultaneity, Whitehead contended, is a classical example of the bifurcation of nature.

Whitehead's second criticism referred to the practical applicability of Einstein's definition. For, as Whitehead claimed, "the determination of simultaneity in this way is never made, and if it could be made would not be accurate; for we live in air and not *in vacuo*."[34] On the whole, however, Whitehead's critique resulted from his philosophical presuppositions, which were incompatible with the positivist presumptions adopted by Einstein in his construction of the special theory of relativity.

It may sound self-contradictory that Whitehead rejected Einstein's conception of distant simultaneity but affirmed the relativity of simultaneity in *The Principle of Relativity* and even called the relativization of simultaneity "a

[33] A. Einstein, chapter 7, note 62, p. 27.
[34] Note 31, p. 54.

heavy blow at the scientific materialism, which presupposes a definite present instant at which all matter is simultaneously real. In the modern theory there is no such a unique present instant."[35] A closer study of his writings resolves this apparent contradiction, however, for he explicitly stated that although he maintained what he called the "old fashioned belief in the fundamental character of simultaneity," he adapted it to "the novel outlook by the qualification that the meaning of simultaneity may be different in different individual experiences. Furthermore," he continued, "since I start from the principle that what is apparent in individual experience is a fact of nature, it follows that there are in nature alternative systems of stratification involving different meanings for time and different meanings for space. Accordingly two events which may be simultaneous in one instantaneous space for one mode of stratification may not be simultaneous in an alternative mode."[36] In a talk delivered in 1926 he declared that the term "simultaneity" is ambiguous, because it may mean causally unrelated occasions and also occasions that are apprehended with "presentational immediacy. If we identify these two meanings we are reduced to the classical view of time as strictly serial. If we hold that the presentationally immediate occasions are only some among the occasions which are not causally related, we can include modern relativity-theory in this doctrine of time."[37]

The profound divergence between Einstein and Whitehead is well illustrated by a report that Filmer S. C. Northrop, an ardent student of Whiteheadian philosophy, presented regarding a conversation he once had with Einstein. When Einstein confessed to him that he simply did not understand Whitehead, Northrop replied that

> there is no difficulty in understanding him. When Whitehead affirms an intuitively given meaning for the simultaneity of spatially separated events, he means immediately sensed phenomenological events, not postulated public physically defined events. On this point he is clearly right. We certainly see a flash in the distant visual sky now, while we hear an explosion beside us. His reason for maintaining that this is the only kind of simultaneity which is given arises from

[35] A. N. Whitehead, *Science and the Modern World* (New York: Macmillan, 1925, 1953), p. 172.

[36] A. N. Whitehead, *The Principle of Relativity with Applications to Physical Science* (Cambridge: Cambridge University Press, 1922), p. 67.

[37] A. N. Whitehead "Time," *Proceedings of the Sixth International Congress of Philosophy* (Cambridge, Massachusetts, 1926), p. 63.

> his desire, in order to meet epistemological difficulties, to have only one continuum of intuitively given events, and to avoid the bifurcation between these phenomenal events and the postulated physically defined public events.

Whereupon Einstein replied: "Oh! Is that what he means? That would be wonderful! So many problems would be solved were it true! Unfortunately, it is a fairy tale. Our world is not as simple as that." And after a moment's silent reflection he added: "On that theory there would be no meaning to two observers speaking about the same event."[38]

In his contribution to *The Encyclopedia of Philosophy*,[39] Adolf Grünbaum severely criticized Whitehead's attempt to base distant simultaneity on sensed coincidence. First, Grünbaum pointed out, "sentient observers stationed at different space points of the same inertial system will render inconsistent verdicts on the sensed simultaneity of a given pair of separated events," which makes the simultaneity relation dependent on the observer's position in one and the same inertial system. Simultaneity based on sense coincidence could be ascribed therefore only to events "whose simultaneity would be compatible with equal physical one-way velocities of influence chains issuing from these events and meeting at a sentient observer." This, however, would involve the employment of one-way velocities the measurement of which required synchronized clocks and thus would lead to a vicious circle. "Consequently," concluded Grünbaum, "sensed coincidence must be rejected as a basis for physical simultaneity at a distance."

The relation of Bergson's and Whitehead's philosophies of time to Einstein's conception of simultaneity has become the subject of numerous essays,[40] a discussion of which would lead us too far into technical details.

[38] F. S. C. Northrop, "Whitehead's philosophy of science," in P. A. Schilpp (ed.), *The Philosophy of Alfred North Whitehead* (Evanston, Illinois: North Western University Press, 1941), p. 204.

[39] A. Grünbaum, "Relativity Theory, Philosophical Significance of," in P. Edwards (ed.), *The Encyclopedia of Philosophy* (New York: Macmillan, 1967, 1972), vol. 7, pp. 133–134.

[40] J. Maritain, "De la métaphysique des physiciens ou de la simultanéité selon Einstein," *Revue Universelle* **10,** 426–445 (1922); "Nouveaux débats einsteiniens," ibid., **17,** 56–77 (1924). E. Morrand, "Einstein au Collège de France," *Nature* (Paris, April 1922), 315–320. Ch. Nordmann, "Einstein expose et discute sa théorie," *Revue des deux Mondes* **9,** 129–166 (1922). A. Metz and H. Bergson, "Controverse au sujet des temps réels et des temps fictifs dans la théorie d'Einstein," *Revue de Philosophie* (July–August 1924), 437–440. A. Metz, "Relativité et relativisme," Ibid. (January–June 1926), 61–87. A. d'Abro, *Bergson ou Einstein* (Paris: Gaulon, 1927); *The Evolution of Scientific Thought* (New York: Boni and Liveright; 1927; Dover Publications, 1950), chapter 15. M. Čapek, *Bergson and Modern Physics* (Dordrecht: Reidel, 1971), part 3, chapter 8.

Before we turn to more recent discussions of Einstein's proofs of the relativity of simultaneity published in his 1905 paper and in his popular 1916 exposition of the train/embankment thought experiment (as described in fig. 7.3 of chapter 7), let us discuss another, but related, effect that Einstein derived from the Lorentz transformation in both his 1905 (§ 4) and 1916 papers (§ 12). Dealing with the loss of clock synchrony it is in a certain sense the opposite of simultaneity or clock synchrony.

In his 1905 paper he described this effect as follows: "If at the points A and B of [the reference system] S there are stationary clocks which, viewed in the stationary system, are synchronous; and if the clock at A is moved with the velocity v along the line AB to B, then on its arrival at B the two clocks no longer synchronize, but the clock moved from A to B lags behind the other which has remained at B by $\frac{1}{2}\, t\, v^2/c^2$ (up to magnitudes of fourth and higher order), t being the time occupied in the journey from A to B."[41] This is, of course, the famous relativistic effect of the time retardation of a moving clock relative to a clock or clocks at rest, usually referred to as the slowing down of moving clocks and called the time dilation (or dilatation).

To understand how Einstein arrived at the time lag of $\frac{1}{2}\, t\, v^2/c^2$, by which the moving clock lags behind the clock at rest in S with which it happens to coincide, let us confine our calculation, without loss of generality, to only one spatial dimension x and standard configuration between the reference systems $S(x,t)$ and $S'(x',t')$.

Let us recall that according to the Lorentz transformation the time coordinates t and t' satisfy the equation

$$t' = \gamma(t - vx/c^2), \tag{8.1}$$

where $\gamma = (1 - v^2/c^2)^{-1/2}$ and v is the velocity of the motion of the coordinate system S' relative to [or measured in] system S.

In accordance with standard configuration, the clock U', stationary at the origin of S', reads $t' = 0$ when it passes the origin of S. Looking at U' from S at a time t later when it has moved along the x axis to the distance $x = vt$ but, being stationary at the origin of S', remains located at $x' = 0$ we find that at time t,

$$t' = \gamma(t - vx/c^2) = \gamma(t - [v/c^2]vt) = \gamma t(1 - v^2/c^2) = t(1 - v^2/c^2)^{1/2} \tag{8.2}$$

[41] A. Einstein, "Zur Elektrodynamik bewegter Körper," *Annalen der Physik* **17,** 819–981 (1905); quotation on p. 904.

or up to magnitudes of fourth and higher order

$$t' - t = \tfrac{1}{2}\, t\, v^2/c^2, \tag{8.3}$$

which shows that moving clock "goes slow" by an amount that is precisely the time dilation of the moving clock relative to clocks at rest as stated by Einstein in his 1905 relativity paper.

For the simplest proof of time dilation assume that in the inertial system S events e_1 and e_2 occur at the same place x_1 but at different times t_1 and t_2, respectively. Since according to the Lorentz transformation the times of the two events in S' (in standard configuration with S) are given by

$$t_1' = \gamma(t_1 - v\, x_1/c^2) \quad \text{and} \quad t_2' = \gamma(t_2 - v\, x_1/c^2) \tag{8.4}$$

we obtain by subtraction

$$t_2' - t_1' = \gamma(t_2 - t_1) > t_2 - t_1 \tag{8.5}$$

because $\gamma > 1$.

Einstein's prediction of this time dilation was confirmed experimentally by observations of the decay rates of fast-moving elementary particles like mesons, for example, by B. Rossi and D. B. Hall,[42] or by the use of macroscopic moving clocks.[43]

For Einstein the time dilation led to what he called "a peculiar consequence," which he described as follows: "It is at once apparent that this result [i.e. the time dilation] still holds if the clock moves from A to B in any polygonal line, and also when the points A and B coincide. If we assume that the result proved for a polygonal line is also valid for a continuously curved line, we arrive at the following result: If one of two synchronous clocks at A is moved in a closed curve with constant velocity until it returns to A, the journey lasting t seconds, then by the clock which has remained at rest the traveled clock on its arrival at A will be $\tfrac{1}{2}\, t\, v^2/c^2$ second slow."[44]

Einstein's conclusion that a clock moving in a closed curve, starting at a point A, will be found, on returning to A, to be lagging behind a clock stationary at A, was indeed "peculiar," for it led to a paradox that stirred up widespread and

[42] B. Rossi and D. B. Hall, "Variation of the rate of decay of mesotrons with momentum," *Physical Review* **59,** 223–228 (1941).

[43] J. C. Hafele and R. E. Keating, "Around the world atomic clocks," *Science* **177,** 166–170 (1972).

[44] Note 6 in chapter 7 (1905), pp. 904–905.

heated controversies among leading physicists and philosophers.[45] It became known as the "clock paradox," "the paradox of the asymmetric aging," "the paradox of Langevin's travellers," or, most frequently, "the twin paradox."

The twin paradox, which replaces the reading of clocks in the clock paradox by the aging of twins, considers two twins. One of them, called A, remains on Earth while the other twin B leaves the Earth with constant velocity v, as measured by A, to a star at a distance d from Earth and returns immediately. Since his travel time to the star is d/v, as measured by A, he will after the time interval $t_A = 2\,d/v$, as measured by A, meet at his return to Earth with A again. Hence at their reunion A has become older by t_A, whereas B has aged only by $t_A(1 - v^2/c^2)^{1/2}$. According to the relativity principle, however, one can just as well take B's point of view and regard A as the traveler and B at rest. Then at the reunion B would have aged by $t_B = 2\,d/v$ and A only by $t_B(1 - v^2/c^2)^{1/2}$. The paradox consists in the logical impossibility that at their reunion one of the twins cannot be both older and younger than the other.

We will not discuss the numerous solutions of this paradox proposed within the conceptual framework of the special theory of relativity. Nor will we explain how Albert Einstein,[46] Max Born,[47] or Christian Møller,[48] among others, invoked the general theory of relativity to resolve this paradox. Nor will we comment on Herbert Dingle's[49] claim that this paradox disproves the whole theory of relativity, or on the more recent and less radical claim by Martel Gerteis,[50] that Einstein committed an arithmetical error in his derivation of time dilation.

The clock paradox was discussed because it is intimately related to the relativity of simultaneity. That such a relation exists was recognized previously by José Alvarez López of the Universidad Católica de Córdoba, Argentina,

[45] The 105 bibliographical references to this paradox in H. Arzeliès, *Relativistic Kinematics* (Oxford: Pergamon Press, 1966) and the 250 references to it in L. Marder, *Time and the Space Traveller* (London: Allen and Unwin, 1971) are far from comprehensive even with respect to the dates of their publications.

[46] A. Einstein, "Dialog über Einwände gegen die Relativitätstheorie," *Naturwissenschaften* **6,** 697–702 (1918).

[47] M. Born, "Special theory of relativity," *Nature* **197,** 1287–1288 (1963); *Die Relativitätstheorie Einsteins* (Berlin: J. Springer, 1920, 1964), *Einstein's Theory of Relativity* (New York: Dover, 1965), p. 261.

[48] C. Møller, "On homogenous gravitational fields in the general theory of relativity and the clock paradox," *Matematisk—Fysiske Meddeelser* **20,** 1–26 (1943); *The Theory of Relativity* (London: Oxford University Press, 1952), p. 258.

[49] H. Dingle, "Relativity and space travel," *Nature* **177,** 782–784 (1956). "The case against special relativity," ibid.

[50] M. Gerteis, *Einsteins Zeitdehnung* (Osnabrück: Biblio Verlag, 1984).

when he declared that "the relativity of simultaneity . . . has not been taken into account in the definition of the intervals of length and time" and that therefore "this tacit acceptance of the absoluteness of simultaneity is the source of the 'clock paradox' of the relativity theory."[51] Ultimately the relation between the clock paradox and the relativity of simultaneity lies in the fact that both are logical consequences of the Lorentz transformation of the time variable, that is, of equation (8.1).

That this is the case for the time dilation effect and hence for the clock paradox was shown when we discussed Einstein's 1905 derivation of the time dilation. Because of its linearity equation (8.1) leads to the equation

$$\Delta t' = \gamma(\Delta t - v\,\Delta x/c^2) \tag{8.6}$$

where $\Delta t' = t_2' - t_1'$ and $\Delta x = x_2 - x_1$. Hence two spatially separated events that are simultaneous in the reference system S (i.e., $\Delta t = 0$) are not simultaneous in the reference system S' because

$$\Delta t' = -\gamma\, v\, \Delta x/c^2 \tag{8.7}$$

Clearly, both the clock paradox and the relativity of distant simultaneity are logical consequences of the linearity of the Lorentz transformation, but why must the Lorentz equations be linear? Einstein's answer to this question was: " . . . the equations must be *linear* on account of the properties of homogeneity which we attribute to space and time."[52] Although essentially correct, Einstein's answer is far from being obvious and, to be rigorously explicated, involves advanced mathematical methods, as Roberto Torretti showed in great detail.[53]

Although discussions on the clock paradox gradually abated, those on the relativity of simultaneity proliferated, confirming Einstein's prediction, published 1914 in the *Vossische Zeitung*, that the relativity of simultaneity is "the most controversial theorem of the new theory of relativity." In fact, even in 1977 Leonard Parish called Einstein's argument for the relativity of simultaneity "a most illogical deduction, . . . perhaps the most important *non-sequitur* he ever made."[54]

[51] J. A. López, "The meaning of the clock paradox," *Annales, Instituto de estudios avanzados* **8,** 7–10 (1959).

[52] "Zunächst ist klar, dass die Gleichungen *linear* sein müssen wegen der Homogenitätseigenschaften, welche wir Raum und Zeit beilegen." Note 6 in chapter 7 (1905), p. 898.

[53] R. Torretti, *Relativity and Geometry* (Oxford: Pergamon Press, 1983), section 3.6, pp. 71–76.

[54] L. Parish, *The Logical Flaws of Einstein's Relativity* (Luton: Cortney Publications, 1977), p. 21.

That the comprehension of Einstein's argumentation still causes didactical difficulties was shown recently by Rachel E. Scherr, Peter S. Shaffer, and Stamatis Vokos in their study on how far students understand the relativity of simultaneity, a study carried out at the University of Washington in Seattle. They concluded that "students at all levels often incorporate the relativity of simultaneity into their own conceptual framework which allows them to continue to believe in absolute simultaneity, so that the ideas of absolute simultaneity and the relativity of simultaneity co-exist."[55] In a subsequent paper they described the development and assessment of a curriculum designed to assist students in constructing a meaningful understanding of the relativity of simultaneity. They concluded their essay with a statement, intended for their study on the apprehension of distant simultaneity but also valid for discussions of the clock paradox: "In the traditional approach, paradoxes are often used as elicitation activities or motivational tools. However, a strategy in which the instructor elicits and exposes student beliefs to generate cognitive conflict and then resolves the paradox is inadequate. Our experience indicates that confrontation and resolution must be carried out by the students, not by the instructor, if meaningful learning is to take place. This strategy is especially crucial when the ideas are as strongly counterintuitive as in special relativity."[56]

That the relativity of simultaneity posed a problem not only for students but also for professional physicists and philosophers is evident because the logical cogency of Einstein's train/embankment thought experiment, by which he proved the relativity of distant simultaneity in his 1916 exposition of the theory of relativity,[57] has again recently become the subject of vivid debate. In an 18-page essay, published in 2003, Avi Nelson did not question the validity of the relativity of simultaneity within the special theory but claimed that Einstein's 1916 proof of this relativity by means of the train/embankment thought experiment is "incomplete and inadequate."[58] According to Nelson this thought experiment shows not the relativity of distant simultaneity but only how the observer on the embankment conceives from his point of view how the observer

[55] R. E. Scherr, P. S. Shaffer, and S. Vokos, "Student understanding of time in special relativity: simultaneity and reference frames," *American Journal of Physics, Supplement* **69,** S24–S35 (2001).

[56] R. E. Scherr, P. S. Shaffer, and S. Vokos, "The challenge of changing deeply held student beliefs about the relativity of simultaneity," *American Journal of Physics* **70,** 1238–1248 (2002).

[57] See note 60 and figure 7.3 in chapter 7.

[58] A. Nelson, "Reinterpreting the famous train/embankment experiment of relativity," *European Journal of Physics* **24,** 379–396 (2003).

in the train experiences the arrival of the two light beams. Nelson thus revived, apparently unaware, Bergson's previously mentioned 1922 argument,[59] though in a more explicit and detailed presentation.

Calling the observer on the embankment O_E and the observer on the train O_T Nelson argued as follows. He reminded us that according to Einstein's light postulate the velocity of light in the inertial reference frame of the train, as in every inertial frame, is c, and continued as follows: "At his location at M', O_T can detect the arrival of light beams and determine the directions whence they came. But he could not detect his moving toward one light beam and away from another. For if he could, he would indicate his having a different velocity with respect to one light beam versus the other. Or, equivalently, it would mean that the two light beams have different relative velocities with respect to O_T. But this would violate the requirement that the speed of light be constant throughout the inertial frame and so cannot happen. The conclusion must be, therefore, that O_T could not witness the experience O_E ascribes to him."

Nelson contended that this conclusion contradicts Einstein's result that "the observer [in the train] will see the beam of light emitted from *B* earlier than he will see that emitted from *A*." Moreover, Nelson continued, O_T will not only conclude that the lightning strokes were simultaneous, he will also observe that the light beam from the stroke at *A* reaches O_E earlier than the light beam emitted from the stroke at *B* so that O_T will declare that the strokes were not simultaneous for O_E.

Following several variations of this argument and additional explanations Nelson concluded that the solution of what he calls "the conundrum" lies in the fact "that the [thought] experiment is posited to take place in Newtonian space-time but under Einsteinian postulates. The experiment is Newtonian in that time and space are dimensionally absolute, but Einsteinian because of the second postulate and the definition of simultaneity. . . . Such a schizophrenic architecture spawns an experimental oxymoron. . . . All these contradictions and difficulties are resolved by analyzing the problem relativistically. This yields both the desired relativity of simultaneity and quantitative agreement between observers in different frames." Nelson concluded his paper with a mathematical "relativistic quantitative analysis of the train/embankment experiment."[60]

[59] See note 22.

[60] Nelson's expression "schizophrenic architecture" corresponds to what Scherr, Shaffer, and Vokos called the coexistence of the ideas of absolute simultaneity and the relativity of simultaneity.

Note that Nelson did not question the relativity of simultaneity; he only claimed that the train/embankment thought experiment cannot be used to prove the relativity of simultaneity prior to the use of the Lorentz transformation as Einstein intended to do in his popular 1916 presentation of the theory of relativity—and, we may add, as other authors of texts on relativity have done.[61]

Slightly varying Nelson's criticism one may also argue that because the train is moving with constant velocity along a straight line it constitutes an inertial system in which, according to the relativity principle, the velocity of light is in all directions equal to c. Hence the two light rays, moving in opposite directions, meet at the midpoint of the train where O_T is situated, even though O_E describes the situation differently. But since for both observers, O_E and O_T, the lightning bolts occur at the same time, in accordance with Einstein's definition, simultaneity is not a relativistic concept.

David R. Rowland criticized Nelson's claim by pointing out that the "flaw in his argument is that if, as is done in Einstein's thought experiment, point events are described in terms of what happens at the event (e.g. by describing it as the spatial coincidence of a particular particle with a particular photon of light), then such a description is in fact frame-*independent*."[62] To make this point quite clear Rowland used the following device, which had already been used by Scherr *et al.*[63]

He imagined a detector positioned at M which carries a lamp that is switched on only if the two lightning bolts arrive simultaneously at M and remains off if the two bolts do not arrive simultaneously. Because the state of the detector cannot depend on the motion of the observer of the lamp all observers must agree on whether the lightning bolts arrive simultaneously at M. Rowland thus arrived at the conclusion: "*Although the spacetime* coordinates *ascribed to point events are frame*-dependent, *point events can be described in frame*-independent *ways in terms of what happens at the event.*"[64]

Rowland ascribed to Nelson two additional logical errors, the latter one of which refers to Nelson's contention that, because the thought experiment is

[61] See, for example, P. G. Bergmann, *Introduction to the Theory of Relativity* (New York: Prentice-Hall, 1942, 1950), pp. 30–32; S. Goldberg, *Understanding Relativity* (Boston: Birkhäuser, 1984), pp. 113–115.

[62] D. R. Rowland, "Comment on 'Reinterpreting the famous train/embankment experiment of relativity,'' *European Journal of Physics* **25,** L45–L47 (2004).

[63] Note 56.

[64] Note 62, p. L46.

presented prior to a derivation of the Lorentz transformations, "Newton's concepts of absolute space and universal time *must* be applied." This conclusion, Rowland claimed, is incorrect because "the transformation laws (i.e., Galilean or Lorentzian) do not *define* the nature of spacetime, but rather *follow* from one's *postulates* about the nature of space and time." In concluding his criticism Rowland pointed out that one can derive directly not only the relativity of simultaneity but also both time dilation and length contraction merely from Einstein's postulates without any recourse to the Lorentz transformation, as P. A. Tipler showed, for example.[65]

A. John Mallinckrodt made a similar critique of Nelson's argumentation which, briefly formulated, claimed that "Nelson's contradictions are nothing more than the direct result of his attempt to perform a reconciliation that is neither possible *nor necessary* using a framework that is logically inconsistent."[66]

In his rejoinder to Rowland's and Mallinckrodt's criticisms of his reinterpretation of the train/embankment thought experiment Nelson declared that these authors ignored the fact that this experiment had been contrived only as an introductory example for the relativity of simultaneity and must therefore be set in Newtonian space–time, even though it applied Einstein's definition of simultaneity and the light postulate. "This combination of conditions, however, produces non-quantifiable contradictory results. The standard analysis evades confronting the contradictions by using a single-frame perspective, which actually does not employ the second postulate [the light postulate]. Analysing the train/embankment experiment under the conditions given, reveals not the relativity of simultaneity but, more significantly, the incompatibility of Newtonian spacetime with Einstein's postulates about light speed and simultaneity."[67] Nelson concludes therefore that, *under the conditions given,* the thought experiment does not demonstrate the relativity of simultaneity but "teaches a more profound lesson," namely, that if one accepts the second postulate the experiment reveals the logical incompatibility between this postulate and the absoluteness of time and of length.

[65] P. A. Tipler, *Modern Physics* (New York: Worth, 1978), chapter 1, section 3.

[66] A. J. Mallinckrodt, "On 'Reinterpreting the famous train/embankment experiment of relativity,'" *European Journal of Physics* **25,** L49–L50 (2004).

[67] A. Nelson, "Response to Mallinckrodt's and Rowland's comments," Ibid., **25,** L51–L56 (2004).

CHAPTER NINE

The Conventionality Thesis

The authors of the early articles and textbooks on relativity in general emphasized the importance of the concept of simultaneity for the construction of the time coordinate of an inertial reference system but rarely paid attention to the conventionality of its definition. A typical example is Max von Laue's text, *Die Relativitätstheorie,* first published in 1911. He devoted a whole section to "a complete clarification of the notion of simultaneity"[1] and its derivation from the synchronization of clocks. This synchronization, however, was obtained assuming that the one-way velocity of the propagation of light is a universal constant c. Nowhere did he mention that the thus defined simultaneity is ultimately based on the convention of the isotropy of the velocity of light.

The first author of a text on relativity to emphasize the conventionality in the definition of distant simultaneity was the Cambridge astronomer Arthur Stanley Eddington, who led the Principle Island expedition that con-

[1] "Machen wir uns den Begriff der Gleichzeitigkeit ganz klar." M. von Laue, *Die Relativitätstheorie* (Braunschweig: Vieweg, 1911, 1952), p. 30.

firmed the gravitational deflection of light in 1919, and who introduced general relativity to the English-speaking world. Whereas in his popular book *Space, Time and Gravitation*, published in 1920, he only denied the existence of an absolute simultaneity in nature,[2] in his more technical treatise *The Mathematical Theory of Relativity*[3] he stressed the conventionality in the definition of simultaneity.

In the section entitled "Simultaneity at different places," Eddington declared that two possibilities of establishing distant simultaneity exist: Einstein's light-signal method and the method of transporting clocks with "infinitesimal velocity." In either case, he claimed:

> a convention is introduced as to the reckoning of the time differences at different places; this convention takes in the two methods the alternative forms: (1) A clock moved with infinitesimal velocity from one place to another continues to read the correct time at its new station, or (2) the forward velocity of light along any line is equal to the backward velocity. Neither statement is by itself a statement of observable fact, nor does it refer to any intrinsic property of clocks or of light; it is simply an announcement of the rule by which we propose to extend fictitious time-partitions through the world. But the mutual agreement of the two statements is a fact which could be tested by observation, though owing to the obvious practical difficulties it has not been possible to verify it directly.[4]

Eddington's characterization of the time partitions, into which we divide the space–time extension of the world, as "fictitious" refers to his preceding comments in which he explained that the demand for a worldwide partition or instants arose as the result of a mistake. This mistake dates back to the times before Römer's discovery of the finite velocity of light when it was taken for granted that external events, and not only their sense impressions, take place in the time succession of our consciousness. "Physics borrowed the idea of world-wide instants from the rejected theory, and constructed mathematical continuations of the instants in the consciousness of the observer, making in this way time-partitions throughout the four-dimensional world."[5]

[2] A. S. Eddington, *Space, Time and Gravitation* (Cambridge: Cambridge University Press, 1920), p. 12.

[3] A. S. Eddington, *The Mathematical Theory of Relativity* (Cambridge: Cambridge University Press, 1923, 1952), p. 29.

[4] Ibid., p. 29.

[5] Ibid., p. 24.

In his Gifford Lectures in 1927, Eddington reiterated that the special theory of relativity does not admit the notion of an absolute simultaneity or of an absolute "Now." After describing the partition of Minkowski space–time into Here–Now, the absolute future and past, and the wedge-shaped neutral zone between the two light cones, Eddington illustrated the nonexistence of an absolute simultaneity or an absolute "Now" by the following example.

> Suppose that you are in love with a lady on Neptune and that she returns the sentiment. It will be some consolation for the melancholy separation if you can say to yourself at some—possible prearranged—moment, "she is thinking of me now." Unfortunately a difficulty has arisen because we have had to abolish Now. There is no absolute Now, but only the various relative Nows differing according to the reckoning of different observers and covering the whole neutral wedge which at the distance of Neptune is about eight hours thick. She will have to think of you continuously for eight hours on end in order to circumvent the ambiguity of "Now."[6]

Eddington then explained why this ambiguity plays no role in our ordinary experience. At the greatest possible separation on the Earth the thickness of the neutral wedge is less than a tenth of a second, so that terrestrial synchronism is not seriously interfered with. Furthermore, because events are usually not instantaneous but have a finite duration sufficient to cover the width of the neutral zone, "the event taken as a whole may fairly be considered to be Now absolutely. From this point of view the 'nowness' of an event is like a shadow cast by it into space, and the longer the event the farther will be the umbra of the shadow extend."[7] Finally, Eddington draws attention to the intimate connection between the denial of absolute simultaneity and the denial of absolute velocity by pointing out that "knowledge of absolute velocity would enable us to assert that certain events in the past or future occur Here but not Now; knowledge of absolute simultaneity would tell us that certain events occur Now but not Here."[8]

At the time Eddington presented his ideas about the conventionality of distant simultaneity, Hans Reichenbach arrived independently and for totally different reasons at similar conclusions. For whereas Eddington's approach was based on his philosophical conviction that a physical quantity should

[6] A. S. Eddington, *The Nature of the Physical World* (Cambridge: Cambridge University Press; New York: Macmillan, 1928), p. 49.

[7] Ibid., p. 49.

[8] Ibid., pp. 61–62.

not be defined "as though it were a feature in the world-picture which had to be sought out" but rather "by the series of operations and calculations of which it is the result,"[9] Reichenbach, trained in the Kantian tradition, began his study of time by examining the relation between Kant's synthetic a priori conception of time and Einstein's definition of time in his theory of relativity.

In his first major publication on relativity, his 1920 analysis of the relation between this theory and Kant's philosophy, Reichenbach pointed out that in Einstein's theory of time "there is a certain arbitrariness contained in any 'coordinate time.' This arbitrariness is reduced to a minimum if the speed of propagation of the process, the need of which for the definition of synchronicity has been pointed out by Einstein, is assumed to be constant, independent of direction, and equal for all coordinate systems."[10] Reichenbach still regarded Einstein's light principle, as we see, as a kind of selection principle for the reduction of the arbitrariness inherent in the logical construction of coordinate time. Although the notion of "arbitrariness" already contains an allusion to conventionality, Reichenbach was not yet so much concerned about the very existence of this conventionality as about restricting it to a minimum.

In the essay "On the present State of the Discussion on Relativity," written when he was appointed in 1922 as lecturer at the Technical Highschool in Stuttgart, Reichenbach was more explicit, because he said that "it would be a mistake to believe that the definition of simultaneity given in the special theory of relativity claims to be 'more correct' than any other definition of simultaneity."[11] He admitted, however, that this definition is more "advantageous" than any alternative because synchronism, thus defined, is a transitive relation, which assures that the law of causality will not be violated.

Reichenbach here referred to an issue that he discussed in greater detail in his *Axiomatization of the Theory of Relativity*[12] on which he had been working

[9]Note 3, p. 3.

[10]H. Reichenbach, *Relativitätstheorie und Erkenntnis Apriori* (Berlin: Springer, 1920), reprinted in H. Reichenbach, *Gesammelte Werke*, edited by A. Kamlah and M. Reichenbach (Braunschweig: Vieweg, 1977), vol. 3; *The Theory of Relativity and A Priori Knowledge* (Berkeley, California: University of California Press, 1965).

[11]H. Reichenbach, "Der gegenwärtige Stand der Relativitätsdiskussion," *Logos* **10,** 316–378 (1922); "The present state of the discussion on relativity," in H. Reichenbach, *Modern Philosophy of Science—Selected Essays* (London: Routledge & Kegan Paul, 1959), vol. 2, pp. 3–47.

[12]H. Reichenbach, *Axiomatik der relativistischen Raum-Zeit-Lehre* (Braunschweig: Vieweg, 1924, 1965); *Axiomatization of the Theory of Relativity* (Berkeley, California: University of California Press, 1969).

at that time and which presented his views on the role of axioms and definitions in an axiomatized formulation of relativity. The axioms of a physical theory, he declared, must not only satisfy the logical requirements of consistency, independence, uniqueness, and completeness like the axioms of mathematics, but in contrast to the latter, they must not be arbitrary but true, for they "contain the whole theory implicitly." On the other hand, "definitions are arbitrary; they are neither true nor false." In physics it is important to distinguish between *conceptual definitions*, which clarify the meaning of a concept by means of other concepts, and *coordinative definitions*, which take the meaning of a concept for granted and coordinate it to a physical thing. Just like conceptual definitions, coordinative definitions are arbitrary, and "truth" or "falsehood" are not applicable to them.

We would have to go into too much detail even to only outline the various axioms, definitions, and theorems that Reichenbach presented in his *Axiomatization* to prove the conventionality of the concept of distant simultaneity in a given reference system or, as it also has been called, of intrasystemic simultaneity. In fact, the study of the *Axiomatization* is not easy reading.[13] We shall therefore continue our outline of Reichenbach's conception of simultaneity in accordance with the less formalized presentation in his book *The Philosophy of Space and Time*[14] which has been called "the greatest work in the theory of science of the 20th century."[15]

Reichenbach began his treatment of simultaneity with the remark that there are three metrical coordinative definitions of time: the first for the unit of time, the second for the congruence of successive time intervals at the same location in space, and the third for the simultaneity of events at different locations in space. In the Newtonian theory of space–time the prob-

[13] Andreas Kamlah, one of the editors of Reichenbach's *Gesammelte Werke*, calls the *Axiomatization* "Reichenbach's genialstes Werk" (Reichenbach's most ingenious work). See A. Kamlah, "Die Analyse der Kausalrelation, Reichenbach's zweites philosophisches Hauptproblem," in H. Poser and U. Dirks (eds.), *Hans Reichenbach: Philosophie im Umkreis der Physik* (Berlin: Akademie Verlag, 1998), p. 37. On the other hand, Hermann Weyl, in his "Review," *Deutsche Literaturzeitung 1924*, p. 2122, criticized it as "wenig befriedigend, zu umständlich und zu undurchsichtig' (little satisfying. too clumsy and too obscure). Patrick Suppes, in his essay "The Desirability of Formalization in Science," *Journal of Philosophy* **65,** 651–664, also regarded it as not satisfying.

[14] H. Reichenbach, *Philosophie der Raum-Zeit-Lehre* (Berlin: W. de Gruyter, 1928); *The Philosophy of Space and Time* (New York: Dover Publications, 1958).

[15] "Dss grösste Werk der Wissenschaftstheorie des zwanzigsten Jahrhunderts," W. C. Salmon, "Introduction," in H. Reichenbach, *Gesammelte Werke* (Braunschweig: Vieweg, 1977), vol. 1, p. 25.

lem of whether two spatially separated events are simultaneous could be resolved by the transport of clocks (as we saw, for example, in chapter 4, for the determination of geographical longitude) or by use of the fact that a finite upper limit to the velocities of causal interactions or signals does not exist. According to Newton's law "force equals mass times acceleration," a force acting on a given mass during a sufficiently long time imparts to the mass an arbitrarily large velocity. In relativistic physics infinitely fast signals or causal connections do not exist. Reichenbach introduced therefore the auxiliary concept of a *first-signal,* which he defined as follows. Let A and B be two distant points in an inertial system S. "If several kinds of signals are sent from A at the same moment, they will arrive at B at different times. To order the times of their arrival at B we need only the time series at B [i.e., only one clock]. That particular signal having the earliest time of arrival at B is called the first-signal; it is therefore defined as the fastest message carrier between any two points in space."[16]

If now e_A denotes the event of the departure of a first-signal from A, e_B the event of its arrival at B and e'_A its rearrival at A after its reflection at B, then e_B is later than e_A but earlier than e'_A. But the position of any event e at A between e_A and e'_A is left "indeterminate as to time order." In Newtonian physics, where there is no limit to the speed of signals and the first-signal would have an infinite velocity, the time interval between e_A and e'_A would be reduced to a single point; and the event at this point would be the only event at A that is indeterminate as to time order relative to e_B and can therefore coordinatively be defined as the only event at A that is simultaneous with e_B. Reichenbach pointed out that modern physics provides decisive empirical evidence that light is a first-signal and that this limiting character of the finite velocity of light can be established without using the concepts of velocity or of simultaneity. Hence, e_A and e'_A are separated by a finite (nonzero) time interval and every event between these two events is indeterminate as to its time order relative to e_B. Reichenbach then stated what he called "the topological definition" of simultaneity: "Any two events which are indeterminate as to their time order may be called simultaneous."[17] In other words, the statement, that two events are simultaneous if

[16] Note 14 (1958), p. 143. The notation is changed so that it agrees with Einstein's notation as used in chapter 7.

[17] Ibid., p. 145.

and only if they cannot be connected by a causal chain that travels from one to the other in either direction, "supplies the *conceptual* definitions of simultaneity." Yet this conceptual definition does not yet coordinate to any given event a *single* event at another location that is simultaneous with the former. "The causal structure of our universe involves the consequence that exclusion of causal connection does not lead univocally to a unique simultaneity."[18]

That the meaning of the notion of simultaneity, expressed by its conceptual definition, does not uniquely determine the reference of this concept without an additional coordinative definition is shown by two arguments. First, it is obvious that there are infinitely many events between e_A and e_A' that satisfy the condition of causal nonconnectibility with e_B so that according to the conceptual definition each one of them is simultaneous with e_B. The second argument is Reichenbach's "velocity-simultaneity circle argument," which deals with the question of how to determine the time of a distant event to check whether it coincides with the time of the occurrence of a nearby event so that both events may be regarded as being simultaneous.

To perform this task we must know not only the distance between the locations of the two events but also the velocity of the signal by which the comparison of the time indications can be performed. To measure a velocity, however, one has to use synchronized clocks so that the simultaneity of distant events must already be known. "Thus we are faced with a circular argument. To determine the simultaneity of distant events we need to know a velocity and to measure a velocity we require knowledge of the simultaneity of distant events. The occurrence of this circularity proves that simultaneity is not a matter of knowledge, but of a coordinate definition, since the logical circle shows that a knowledge of simultaneity is impossible in principle."[19]

Reichenbach concluded that these considerations teach us how to understand Einstein's 1905 definition of simultaneity, $t_B - t_A = t_A' - t_B$ or equivalently $t_B = t_A + 1/2\,(t_A' - t_A)$, which defines the time of arrival of the light ray at B as the midpoint of the interval between its departure and its return at A. But this coordinative definition of distant simultaneity, Reichenbach con-

[18] H. Reichenbach, "Planetenuhr und Einsteinsche Gleichzeitigkeit," *Zeitschrift für Physik* **33,** 628–634 (1925).

[19] Ibid., pp. 126–127. The "logical circle," mentioned here by Reichenbach, had been pointed out already by Einstein in § 8 of his 1917 book (see note 35 in chapter 7).

tinues, is not the only epistemological possibility. Its generalization in the form

$$t_B = t_A + \varepsilon(t_A' - t_A) \qquad 0 < \varepsilon < 1 \tag{9.1}$$

"would likewise be adequate and could not be called false." The restriction imposed on ε, which is often called the "Reichenbach synchronization parameter," to have a numerical value in the open interval between 0 and 1 expresses the causality condition that the light pulse cannot arrive at B before it has left A and cannot return to A before it has left B ($t_A < t_B < t_A'$).

Reichenbach modified Einstein's notation by writing t_1 instead of Einstein's t_A, t_2 instead of t_B, and t_3 instead of t_A', so that equation (9.1) reads in Reichenbach's formulation

$$t_2 = t_1 + \varepsilon(t_3 - t_1) \quad 0 < \varepsilon < 1 \tag{9.1'}$$

As we will see in the sequel most authors prefer Reichenbach's notation although, according to Roberto Torretti's criticism of Reichenbach's equation, Einstein's notation, involving the symbols of the places A and B of the clocks which Einstein assumed to be at rest in an inertial frame, should be preferable. For as Torretti claims, Reichenbach's formulation cannot be accepted without qualification, because "Reichenbach's rule [of synchronisation] says nothing about the state of motion of the clocks being synchronized, while Einstein's definition presupposes that they are both at rest in the same inertial frame. In fact, unless this presupposition is made, Reichenbach's rule may not be consistently applicable. (Suppose, for instance, that the clock from which the bouncing signal is sent is affixed to the rim of a rotating disk.)"[20]

In defense of Reichenbach, however, one may say that because Reichenbach explicitly associates (9.1′) with the name of Einstein ("how to understand the definition of simultaneity given by Einstein") he also assumes *tacitly* that the definition of simultaneity (9.1′) applies only to clocks stationed in an inertial frame.

Reichenbach, of course, was aware that setting ε equal to $\frac{1}{2}$ in equation (9.1)—a choice now usually called "standard signal synchrony"—and retrieving thereby Einstein's 1905 simultaneity definition, simplifies the for-

[20] R. Torretti, *Relativity and Geometry* (Oxford: Pergamon Press, 1983), p. 224. When mentioning a rotating disk Torretti refers the reader to H. Arzeliès' *Relativistic Kinematics* (Oxford: Pergamon, 1966), pp. 217–225, where it is shown that the velocity of light on a rotating disk "depends upon the point and the direction being considered."

mulas of the theory. This choice, moreover, made simultaneity a symmetric and transitive relation for the same rule of synchronization. This simplification is, however, merely a descriptive or notational simplification and cannot be interpreted as providing a set of relations that are "more true" than those obtained for other admissible values of the Reichenbach synchronization parameter. Reichenbach warned us not to confuse "transitivity of simultaneity" with "transitivity of simultaneity according to the same rule of synchronization." Only the latter is abandoned when a nonstandard synchronization (i.e., $\varepsilon \neq 1/2$) is adopted.

In a paragraph entitled "The construction of the space-time metric" in *The Philosophy of Space and Time,* Reichenbach emphasized that the properties of symmetry and transitivity of the simultaneity relation "are by no means self-evident." He referred the reader to his *Axiomatization of the Theory of Relativity,* where he showed that a rigorous derivation of these properties requires the use of certain axioms, like the so-called "round-trip axiom," which postulates that the time interval required for round trip of a signal does not depend on the direction of its propagation.

In the sequel Reichenbach defended his thesis against possible attempts at reintroducing a uniquely determined simultaneity in the sense of singling out a particular value of ε as the only physically admissible one. The most obvious attempt to do so is, of course, to narrow the interval (t_A, t_A') and to let it degenerate to a single number by using increasingly higher signal velocities. This attempt, Reichenbach pointed out, cannot succeed because no signals can be faster than light. In making this contention, Reichenbach continued, physics does not commit the fallacy of regarding absence of knowledge as evidence for knowledge to the contrary; it is physical experience, as we know it today, that shows us that the velocity of light is the limit of the velocity of all causal propagation.

The second group of attempts listed by Reichenbach to eliminate the arbitrariness makes use of specially designed mechanisms. Reichenbach contended, though, that a closer examination would always reveal that these mechanisms presuppose in a more or less disguised manner the existence of arbitrarily high velocities of causal propagation that disqualifies them for the same reason as the attempts of the first group.

Reichenbach was aware, of course, that knowledge of the one-way velocity of light determines unambiguously the value of ε, and conversely, that knowledge of ε determines unambiguously the one-way velocity of light. This can be seen as follows. Let c_{AB} denote the one-way velocity of light in the direction from A to B and c_{BA} its one-way velocity in the opposite direction so that (see fig. 7.1 in chapter 7)

$$c_{AB} = AB/(t_B - t_A), \quad c_{BA} = AB/(t'_A - t_B), \quad \text{and } t'_A - t_A = 2\, AB/c$$

Then it follows from equation (9.1) that

$$c_{AB} = c/2\varepsilon \quad \text{and} \quad c_{BA} = c/2\,(1 - \varepsilon) \tag{9.2}$$

Hence, knowledge of a one-way velocity of light (or, as we shall see, of any object whatever) determines a unique value of ε and knowledge of ε determines the one-way velocity of light. Thus, if it were possible to measure a one-way velocity, the value of ε would be empirically determined and the conventionality thesis would be refuted.

The third group of attempts to measure distant simultaneity mentioned by Reichenbach is based on the transport of clocks. Eddington, as mentioned earlier,[21] acknowledged clock transportation (or *transport-synchronization* in Reichenbach's terminology), though only if carried out with infinitesimal velocity, as a legitimate means to establish distant simultaneity. In contrast, Reichenbach rejected such a proposal because the theory denies that two clocks, synchronized at one place, remain synchronized when brought to a different place along different paths with different velocities. Even if the theory were wrong on this point, Reichenbach added, and "the transport of clocks could be shown to be independent of path and velocity, this type of time comparison could not change our epistemological results, since the transport of clocks can again offer nothing but a *definition* of simultaneity.

Moreover, "even if the two clocks correspond when they are again brought together, how can we know whether or not both have changed in the meantime?"[22] This question is undecidable as is the question of the comparison of length of rigid rods. Again "a solution can be given only if the comparison of time is recognized as a definition. If there exists a unique transport—synchronization, it is still merely a definition of simultaneity."

[21] Note 3.
[22] Note 14 (1958), p. 133.

Reichenbach summarized the results obtained so far as follows: "The time metric depends on three coordinative definitions. The first deals with the unit of time and determines the numerical value of a time interval. The second deals with uniformity and refers to the comparison of successive time intervals. The third deals with simultaneity and is concerned with the comparison of time intervals which are parallel to each other at different point of space. These three definitions are required in order to make a time measurement possible; without them the problem of the measurement of time is logically undetermined."

Reichenbach's argumentation for the conventionality of distant simultaneity was based on his epistemological conviction that statements are conventional, as opposed to factual, if they are unverifiable in principle. It is for this reason that his refutations of all attempts to reach absolute simultaneity, as he called it, form an integral part of his doctrine of conventionality. Because his contention of the impossibility of obtaining *knowledge* of distant simultaneity was the ultimate basis of his doctrine, his approach to conventionality may be regarded as being founded on epistemological considerations.

Reichenbach's analysis of the simultaneity concept, like many of his foundational studies in other branches of physics, was motivated by his recognition that it is important to distinguish between the factual and the conventional components of a physical theory. Einstein also noted this characteristic feature of Reichenbach's work when he emphasized in his appreciative review of Reichenbach's 1927 book that "special care has been taken to ferret out clearly what in the relativistic definition of simultaneity is a logically arbitrary decree and what in it is a hypothesis, i.e., an assumption about the constitution of nature."[23]

Reichenbach was not the only one at the time who recognized and emphasized the impossibility of experimentally determining the one-way velocity of light. A few months before Einstein's visit to England in 1921 the prestigious journal *Nature* published a special issue on relativity with contributions by leading British physicists. In one of these papers[24] James Hopwood Jeans, referring to certain experiments performed by Quintilio Majo-

[23] A. Einstein, Review, *Deutsche Literaturzeitung* 1928, pp. 19–20.

[24] J. H. Jeans, "The general physical theory of relativity," *Nature* **106**, 791–793 (1921).

rana,[25] claimed that it was possible to measure the one-way velocity of light. This statement by the Cambridge astronomer and Fellow of the Royal Society was soon questioned by an unknown freelance physicist, C. O. Bartrum, who after examining Majorana's measurements wrote in a subsequent issue of *Nature*: "I venture to suggest that these experiments do not bear the interpretation that Mr. Jeans puts upon them, and that the experiment has not yet been devised that will enable a comparison to be made between the velocity of light on its outward and return journeys along the same path, or that will give a measure of the velocity on a single journey. The author of these papers makes no claim to have done this. I fear such an experiment is impossible."[26] The debate between Jeans and Bartrum, later joined by Oliver Lodge, went on for several months[27] but never touched upon any deeper methodological or epistemological issues.

Reichenbach's *Axiomatization of the Theory of Relativity*, admittedly no easy reading, was too philosophical for physicists and too physical for philosophers. Even Hermann Weyl, as mentioned earlier,[28] found it unduly complicated. The Finnish physicist Halmar J. Mellin used it to attack the theory of relativity. Among other objections he rejected Reichenbach's reformulation of Einstein's simultaneity definition on the grounds that simultaneity is absolute and can be recognized as being so by a thought experiment, which, unknown to him, resembles the above-mentioned twin-rod experiment. Imagine, he said, two rigid rods AB and $A'B'$ of equal rest length (i.e., of equal length if at rest relative to an inertial system) gliding one along the other; we then can "in arbitrarily many experiments both by visual inspection and by the sense of touch determine that when A coincides with A' also B coincides with B' *in the same moment*" (*in demselben Augenblick*).[29]

In a rejoinder Reichenbach rejected Mellin's interpretation of such an experiment for two reasons. First, Mellin ignored the causal chain inherent in

[25] Q. Majorana, "Démonstration expérimentale de la constance de vitesse de la lumière émise par une source mobile," *Comptes Rendus* **167,** 71–73 (1918); "Démonstration expérimentale de la constance de la lumière réfléchie par un miroir en mouvement," Ibid., **165,** 424–426 (1917); "On the second postulate of the theory of relativity," *Philosophical Magazine* **35,** 163–174 (1918); "Experimental demonstration of the velocity of light emitted by a moving source," Ibid., **37,** 145–150 (1919).

[26] C. O. Bartrum, "Relativity and the velocity of light," *Nature* **107,** 42 (10 March 1921).

[27] *Nature* **107,** 73 (17 March 1921); **107,** 141 (31 March 1921); **107,** 169 (7 April 1921).

[28] Note 13.

[29] H. J. Mellin, "Kritik der Einsteinschen Theorie an Hand von Reichenbach," *Annales Academiae Scientarum Fennicae* **26,** 26 (1926).

the very process of sensation, which gives rise to a temporal uncertainty of the same order of magnitude as the relativistic time difference in question. Second, Mellin's proposed experiment would only provide just another definition of distant simultaneity, which, according to relativity, depends on the relative velocity of the rods under discussion.[30]

A common objection voiced by philosophers against Einstein's 1916 proof of the relativity of distant simultaneity claimed that the simultaneity criterion, according to which two spatially separated events are simultaneous if the observer *midway* between them sees them at the same time, could not be applied, because the observer in the train sees the two light flashes coming from *A* and *B* when he is no longer at equal distances from the latter. Hans Thirring, in his lucid exposition of the fundamental ideas of the theory of relativity,[31] showed that this objection can easily be refuted by assuming that the extremities of the train themselves carry light sources, which flash at the time of the bolts.

A skillfully formulated critique of Einstein's concept of distant simultaneity and its relativity was published by the American epistemologist and historian of intellectual thought Arthur Oncken Lovejoy, professor of philosophy at Johns Hopkins University and first editory of the *Journal of the History of Ideas*. Early influenced by Bergson, Lovejoy called himself a "temporalist realist" who believes in the empirical reality of time. Einstein's notion of time, therefore, was for him an anathema that had to be disproven.

In a paper published in 1930 Lovejoy called Einstein's approach, as manifested in his analysis of the notion of simultaneity, "the radically experimental theory of the nature of meaning." "By this name," he wrote:

> I designate the following duplex thesis: any general attributive term—such as the adjective 'simultaneous' and the abstract noun 'simultaneity'—(1) has 'meaning' only if its definition formulates some practicable method by which the applicability of the term in question to a given subject of discourse can be experimentally determined, i.e., describes some event capable of being directly observed at first hand under exactly determinable conditions, which event shall serve as the criterion for such applicability; and (2) the occurrence of such event, under the conditions set forth in the definition, *is* the meaning, and the whole meaning of the term.

[30] H. Reichenbach, "Erwiderung auf eine Veröffentlichuing von Herrn Hj. Mellin," *Zeitschrift für Physik* **39,** 106–112 (1926).

[31] H. Thirring, *Die Idee der Relativitätstheorie* (Berlin: Springer, 1921), pp. 51–52.

Even if we admit the acceptability of clause (1), Lovejoy continued, clause (2), so crucial for Einstein's definition of simultaneity, is contrary to the spirit of scientific empiricism, for it declares that a relation that, under specific conditions, is actually exemplified in experience cannot exist under different circumstances because, according to the experimental theory of meaning, a diversity of operations would always imply a diversity of concepts. Using a Bergsonian argument, Lovejoy contended that, in particular, simultaneity-at-a-place and simultaneity-at-a-distance are not different relations, but even if they were, Einstein's definition of distant simultaneity would be completely arbitrary.[32]

Recalling Einstein's thought experiment of the two thunderbolts striking at points *A* and *B* and Einstein's definition of their being simultaneous if the light signals from these points reach an observer midway between *A* and *B* at the same time, Lovejoy asked: "But when must the observer be midway between them?" After quoting Einstein's answer and its implication concerning the relativity of distant simultaneity, Lovejoy continued: "It would have been equally possible to say the events at *A* and *B* are to be considered simultaneous if the light-signals from them arrive together at a point which *is* midway between *A* and *B* at the moment of their arrival; and the separateness of the moments of arrival of the signals at the position of an observer who has been but no longer is midway between *A* and *B* is not to be considered evidence of the non-simultaneity of the events." If the criterion had been formulated in the second way, the "relativity" of simultaneity would not have followed. The observer on the train in that case would simply say: These signals reach me separately; but since I am not *now* midway between *A* and *B*, that fact does not show that the events at *A* and *B* were not simultaneous. Not only would this alternative definition be equally possible, but it would obviously accord, as the other does not, with our "natural" idea of simultaneity, that is, it would express the implications, with respect to events at a distance, of that fundamental concept that we empirically acquire through the experience of simultaneity at a place. The only reason for preferring the first statement of the criterion to the second is that from it alone can the relativity of simultaneity be proved. "The logical procedure is that of

[32] A. O. Lovejoy, "The dialectical argument against absolute simultaneity," *The Journal of Philosophy* **27**, 617–633, 645–654 (1930); reprinted in R. Hazelett and D. Turner, *The Einstein Myth and the Ives Papers* (Old Greenwich, Connecticut: Devin–Adair, 1979), pp. 232–244.

choosing arbitrarily that definition of the crucial term in the argument which will permit the conclusion desired to be drawn."[33]

Lovejoy's proposal of an alternative definition of distant simultaneity rested on a misconception that he would have avoided if he had distinguished among the events e_A and e_B, the occurrences of the two thunderbolts, and the positions or locations of these events, the points A and B in the embankment system S, and A' and B' in the train system S'. These positions, perhaps marked by black spots due to the lightning strikes, are also identifiable long after the events have ceased to exist (events have only a pointlike existence both in space and time). At the moment when e_A occurs, A and A' coincide; and at the moment when e_B occurs, B and B' coincide; but thereafter A' and B' are moving to the right of A and B, respectively. Einstein assumed that the light signals, emitted by e_A and e_B, reach the observer on the embankment, placed at the midpoint between A and B and at rest relative to S (locally) at the same time, so that by his definition e_A and e_B are simultaneous events in S. Once the simultaneity of these two events relative to S is acknowledged, which also means that the coincidences of A with A' and of B with B' are simultaneous relative to S, it is meaningful to say that at the instant of these coincidences also the midpoint M of AB coincides with the midpoint M' of $A'B'$, and that thereafter M' is moving to the right of M, without ceasing to be the midpoint between A' and B', though no longer between A and B. Had Lovejoy distinguished consistently between events and their locations, he would have had to formulate his "alternative" definition with respect to the train system S' as follows: the events e_A and e_B are to be considered simultaneous if the light signals from them arrive together at a point that *is* midway between A' and B' at the moment of their arrival. But this is precisely what Einstein did! Because the observer at M' remains always at the midpoint between A' and B', Lovejoy's assumption, that the train observer "has been but no longer is midway" between A' and B' is self-contradictory. Hence, Lovejoy's additional stipulation that under these unreal conditions, the temporal separateness of the arrivals of the signals is not a sufficient condition for the nonsimultaneity of the events, is an empty statement, and the conclusions he drew from it in support of our "natural" idea of simultaneity are meaningless. A similar misconception lies at the root of another of Lovejoy's objections to the relativity of distant simultaneity, namely that, because

[33] Ibid. (1930), pp. 648–649.

the two observers perform their tests about the (locally) simultaneous arrivals of the signals at different distances from the sources, "they are not judging about simultaneity in the same sense" and therefore "the discrepancy between their conclusions does not prove the relativity of simultaneity."

To strengthen the cogency of his objections, Lovejoy also questioned Einstein's rejection of the universality of time on the following grounds: relativity tells us that "*while* such and such things are happening on *S*, such and such other things are happening on *S′*, " statements that can "have no meaning if there is no *common* duration with a common measure."

We have purposely discussed only some of Lovejoy's objections against Einstein to show how even an intellect as sharp as Lovejoy could not rid itself from the shackles of the conceptualizations of classical physics. Had he examined his statements in terms of Minkowski diagrams, he himself would undoubtedly have recognized where his classical preconceptions had led him astray.

In all these critical discussions of Einstein's conception of simultaneity and its relativity the fact that Einstein's definition of this concept was based on the "stipulation" that the one-way velocity of light from *A* to *B* equals that from *B* to *A*, or in Reichenbach's terminology, the conventional character of the simultaneity definition, was completely ignored. True, when Reichenbach's *Philosophie der Raum-Zeit-Lehre* was published it was highly praised in the professional press. Erwin Freundlich, the Potsdam astronomer and author of a book on relativity, called it "a work of greatest scientific importance" that "deserves widest attention among mathematicians, physicists and philosophers"[34]; Moritz Schlick called it "an excellent work" that can contribute much to the dissemination of important insights."[35] Nevertheless, until the mid-1950s, neither in Germany, which Reichenbach was forced to leave in 1933, nor elsewhere did his ideas, and in particular his conventionality thesis of distant simultaneity, gain any noteworthy attention.

An interesting example of this disregard is the following episode. In 1944 the renowned astronomer and physicist Herbert Dingle of London's Imper-

[34] "Ein Werk von grösster wissenschaftlicher Bedeutung, . . . das allgemeinstes Interesse in dem Kreis der Mathematiker, Physiker und Philosophen verdient." E. Freundlich, *Physikalische Zeitschrift* **29,** 590 (1928).

[35] Ein "ausgezeichnetes Werk", das "einen grossen Leserkreis verdient, denn es kann viel zur Verbreitung wichtiger Einsichten . . . beitragen." M. Schlick, *Die Naturwissenschaften* **17,** 549 (1929).

ial College of Science and Technology published a paper that had been submitted to him by Paul Arne Scott-Iversen, a resident of Leamington Spa near Birmingham, and research physicist on the staff of the Rover Company. Unfortunately, Scott-Iversen died during the war before completing the project he had had in mind. Dingle published the paper as a tribute to his memory and prefaced it with a short foreword in which he declared that the paper "affords an interesting example of the variety of aspects in which the special theory of relativity may be regarded, and so makes a contribution to the better understanding of the essential principles of that theory."[36]

Scott-Iversen's theory leads to the same observational effects as the special theory of relativity but, in contrast to the latter, eliminates the relativity of simultaneity, length, and mass for any two selected inertial systems. To this end it postulates that the round-trip velocity of light *in vacuo*, when measured over any closed path, is always the same constant c and the one-way velocity of light C is given by the ellipsoidal distribution

$$C = c/(1 - e \cos \omega), \tag{9.3}$$

where e is the eccentricity of the ellipsoid and ω is the angle between the direction of C and the major axis of the ellipsoid. For $e = 0$ the ellipsoid becomes a sphere and the one-way velocity of light is independent of its direction in space. Comparison of equation (9.3) with the Reichenbach formula (9.2), namely, $C = c/2\varepsilon$, shows that in the Scott-Iversen theory the Reichenbach synchronization parameter depends on the direction of the light ray in space according to the formula $\varepsilon = \frac{1}{2}(1 - e \cos \omega)$.

A direction-dependent synchronization parameter had been used already by Reichenbach in his *Philosophy of Space and Time*.[37] However, neither Scott-Iversen nor Dingle, who read German, make any reference to Reichenbach. Obviously, Reichenbach's work was unknown to them. In fact, until the mid-1950s Reichenbach's work and, in particular, his thesis of the conventionality of distant simultaneity received little attention from physicists. This can be seen, for example, from Henri Arzeliès's extensive bibliography in his comprehensive *Relativistic Kinematics*,[38] which contains a detailed section "On the

[36] P. A. Scott-Iversen, "Introductory notes on a reformulation of the special theory of relativity," *Philosophical Magazine* **35**, 105–120 (1944).

[37] Note 14, section 26.

[38] H. Arzeliès, *Relativistic Kinematics* (Oxford: Pergamon Press, 1966); revised and updated from the original French edition *Cinématique relativiste* (Paris: Gauthier-Villars, 1955).

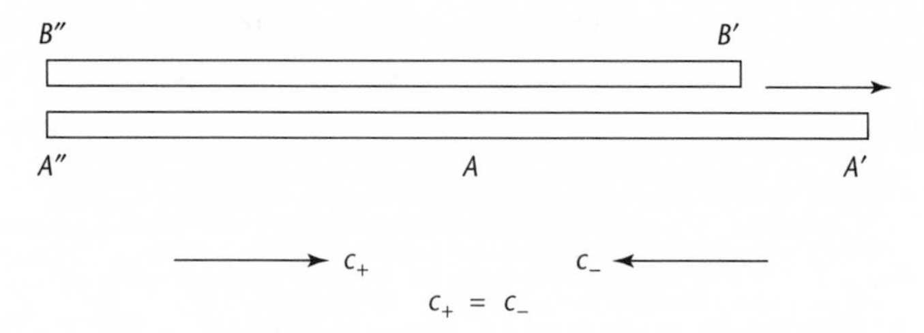

Figure 9.1

Synchronization Procedures and the Concept of Simultaneity" but does not mention even Reichenbach's name.

Scott-Iversen's proof that it is possible to eliminate the relativity of distant simultaneity for two inertial systems S_A and S_B by using an appropriate ellipsoidal velocity distribution is based on a thought experiment involving two rods r_A and r_B of equal rest length l. Rod r_A with end points A' and A'' and midpoint A is at rest in S_A, whereas rod r_B with end points B' and B'' is moving with constant velocity u along r_A in the direction from A'' to A' (see fig. 9.1) and thus is at rest in an inertial system S_B. When B'' passes A'' or when B' passes A' a light signal is emitted from these meeting points to the midpoint A. According to the special theory of relativity (or spherical velocity distribution) the length of rod r_B is contracted so that B' passes A' after B'' has passed A''; the time lapse T between the arrivals of the signals at A is obviously $T = l\,[1 - (1 - u^2/c^2)^{1/2}]/u$. In ellipsoidal velocity distribution (fig. 9.2) the two rods have equal length so that the light signals from A' and A'' are emitted simultaneously; but the direction-dependent velocity of light in the direction from A'' to A, denoted by c_+, turns out to be $c(1 - (c/u)(1 - [1 - (u^2/c^2)^{1/2}])^{-1}$, whereas in the direction from A' to A, denoted by c_-, it is given by the same expression but with the first minus sign exchanged by a plus sign. Hence the time lapse between the arrivals of the two light signals at A, given by $l(2\,c_-)^{-1} - l(2\,c_+)^{-1}$, is again precisely equal to T. Scott-Iversen concludes therefore: "We receive two signals, indicating the occurrence of two events. One signal arrives at an internal T later than the other. We know, however, that the distances and velocities involved are such that the time of travel of the last signal will be longer than that of the first by the same interval T. We conclude, therefore, that the two events must have been simultaneous in the frame of reference at rest relatively to the rod $A'A\,A''$. Owing to the complete symmetry of the two frames of reference,

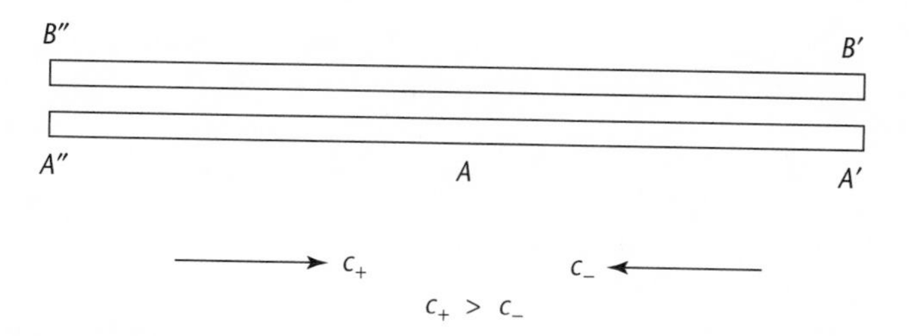

Figure 9.2

events simultaneous in one of them must also be simultaneous in the other."[39]

Had Scott-Iversen been acquainted with Reichenbach's synchronization parameter ε he would have recognized that in his theory ε is not only direction-dependent, as stated above, but is also a function of the velocity u of the reference system S_B relative to S_A. For, as he demonstrated himself, the eccentricity e in formula (9.3) is given by the expression $(c/u)\{1 - [1 - (u/c)^2]^{1/2}\}$. Hence, in the Scott-Iversen theory the Reichenbach synchronization parameter is a function both of the direction, given by ω, and of the velocity u, that is, $\varepsilon = \varepsilon(\omega, u)$.

In his paper, Scott-Iversen also eliminated from his theory the other relativistic effects between two given inertial systems without violating the observational results of the usual theory of relativity. He concluded with the statement: "The most important result is the absoluteness of simultaneity, which should facilitate the smooth blending of special relativity with quantum and wave mechanics. These changes in the expression of the special theory of relativity are brought about solely by postulating an unknowable variation in the one-way velocity of light."[40] What precisely he had in mind when he referred in this context to quantum mechanics will remain a secret forever.

In concluding this chapter, which has dealt mainly with Reichenbach's conventionality thesis of the concept of intrasystemic metrical distant simultaneity in the special theory of relativity—in contrast to the relativity of intersystemic simultaneity discussed in the preceding chapter—we should emphasize the following. It would be a mistake to believe that the importance

[39] Note 36, p. 117.
[40] Op. cit., p. 120.

of this conventionality is confined only to temporal relations or to temporal measurements. In fact, this conventionality has far-reaching effects throughout the mathematical structure of the theory as a whole but, of course, without modifying operational results. Some of these effects had already been noticed by Reichenbach himself when he discussed the notion of the length of a moving rod in classical and in relativistic kinematics. By defining such a length as "the distance between simultaneous positions of its endpoints"[41] he clearly showed the dependence of spatial measurements on the definition of simultaneity. As he subsequently explained, even "the state of a space" and "the shape of a moving object" depend on the chosen definition of simultaneity. In fact, such a dependence exists throughout the theoretical structure of relativistic physics.

The indispensability of the concept of distant simultaneity for the determination of the length of a moving rigid body, for example, a rod, as claimed by Reichenbach, has been repeatedly disputed. For example, Kenneth A. Durbin claimed that it is "possible and physically more meaningful to make such a determination nonsimultaneously by taking into account the fact that light signals require a finite time to propagate." Hence, "contrary to the usual handling of the problem, the length of the uniformly moving body need *not* be based on the ability to *simultaneously* make a measurement of the space coordinates of the ends of the moving body."[42] Durbin considered two inertial systems S and S' in standard configuration and assumed that mirrors are located at the ends of the moving rod, the first mirror being capable of both reflecting and transmitting a light signal generated in the origin O of S. Starting with the statement that "the usual relativistic expression relating the coordinate x' assigned to an event in one reference system [S'] to the coordinate x in another reference system [S] is given by $x' = \gamma(x - v\, t)$," Durbin calculates, without using the notion of simultaneity, the time lag $t_2 - t_1$ measured at O, in the return of the signals reflected from the mirrors and derives the well-known length-contraction formula without the need of measuring

[41] Note 14 (1958), pp. 155–161. Reichenbach discussed the dependence of the length of a moving rod on the choice of ε in relativistic kinematics only qualitatively. The first quantitative formula was given only in 1968 by Adolf Grünbaum in his "Reply to Hilary Putnam's 'An examination of Grünbaum's Philosophy of Geometry,' " in R. S. Cohen and M. W. Wartofsky (eds.), *Boston Studies in the Philosophy of Science* (1968), vol. 5, p. 118. Another derivation of the same formula was given in 1970 by John A. Winnie in a paper that is discussed in chapter 14.

[42] K. A. Durbin, "Nonsimultaneous measurement of the coordinates used to obtain the length of a uniformly moving body," *American Journal of Physics* **32,** 639–641 (1964) (italics in original).

simultaneously the positions of the end points of the moving rod. The hitch of this derivation lies in Durbin's use of the initial equation, that is, the Lorentz equation $x' = \gamma(x - v\,t)$, because the notion of distant simultaneity has to be used to obtain this equation.

Shortly after its publication Durbin's paper was criticized by Francis W. Sears who pointed out that "although the calculations are based on nonsimultaneous clock readings, the coordinates of the ends are in effect determined simultaneously." Sears proposed instead another method of finding the length of a moving rod, in which "the coordinates of the ends of the rod are not observed simultaneously and which does not require the use of signals or synchronized clocks." According to Sears the observer has simply to "note the times at which the ends of the rod pass him. Then, if the time interval for a rod to pass observer B is Δt_B, the length of the rod as calculated by this observer is $L_B = v\,\Delta t_B$,"[43] where v is the velocity of the rod.

The snag in this proposal is, of course, its use of the velocity v of the rod, because to determine a velocity one has to use synchronized clocks, contrary to Sears's declaration that his method "does not require the use of light signals or synchronized clocks." Hence, Sears's method of length determination also involves, even if only implicitly, the concept of distant simultaneity.

[43] F. W. Sears, "Length of a moving rod," *American Journal of Physics* **33,** 266–268 (1965).

CHAPTER TEN

The Promulgation of the Conventionality Thesis

It was only in the late 1950s that Reichenbach's philosophy of space and time, and his thesis of the conventionality of simultaneity, found the attention they deserve. An important factor in this development was the publication of his *Philosophie der Raum-Zeit-Lehre* in an English translation by his wife Maria Reichenbach and John Freund. In his introduction to the English edition Rudolf Carnap declared: "The constant careful attention to scientifically established facts and to the content of the scientific hypotheses to be analyzed and logically reconstructed, the exact formulation of the philosophical results, and the clear and cogent presentation of the arguments supporting them, make this work a model of scientific thinking in philosophy."[1]

The stimulus to this translation was given in 1955 when Adolf Grünbaum published an important essay in the widely read *American Journal of Physics* in which he stated that "my treatment of several of the issues is greatly indebted to two outstanding works on the philosophy of relativity by Hans Reichenbach,

[1] Chapter 9, note 14, p. VII.

which are not available in English."[2] It was actually a pure accident that Grünbaum became acquainted with Reichenbach's philosophy while still a Ph.D. student at Yale. Whenever they were in New York, he and his colleague and friend Robert S. Cohen would visit book stores in search of rare or out-of-print books on the philosophy of science. On one of these occasions Bob Cohen purchased a copy of Reichenbach's *Philosophie der Raum-Zeit-Lehre* and gave it to Grünbaum, who, born and educated in Cologne, had no difficulty in reading the German text. How deeply Grünbaum was influenced by this book is described in his essay entitled "Hans Reichenbach's Definitive Influence on me."[3]

In his influential 1955 paper Grünbaum argued that for a correct understanding of the special theory of relativity it is essential to note, first of all, that "the relativity of simultaneity . . . arises, in the first instance, within a *single* Galilean [inertial] frame" (called *intrasystemic simultaneity* in contrast to *intersystemic simultaneity*, which latter refers to different inertial systems as has been dealt with by Einstein in his 1905 paper). Grünbaum therefore devotes the second chapter on his paper, entitled "the relativity of simultaneity," to this subject. He began his exposition by pointing out that in a world of arbitrarily fast causal chains the concept of absolute simultaneity would have a perfectly physical meaning

> even in a temporal description of nature given by a *relational* theory of time. However, a theory, like the special theory of relativity, that denies the existence of an infinitely fast causal chain, deprives the concept of absolute simultaneity of its physical meaning even *within* a *single* inertial system. . . . But since the metrical concept of velocity presupposes that we know the meaning of a transit time and since such a time, in turn, depends on a prior criterion of clock synchronization or *simultaneity*, we must first formulate the limiting property of electromagnetic chains [the fastest causal chain] *without* using the concept of simultaneity of noncoincident events.[4]

To define intrasystemic simultaneity Grünbaum offered the following *nonmetrical* formulation of Einstein's limiting postulate: "No kind of causal chain (moving particles, radiation), emitted at a given point P_1 together with a light

[2] A. Grünbaum, "Logical and philosophical foundations of the special theory of relativity," *American Journal of Physics* **23,** 450–464 (1955); reprinted in A. Danto and S. Morgenbesser (eds.), *Philosophy of Science* (Cleveland, Ohio: The World Publishing Company, 1960), pp. 399–434.

[3] In H. Reichenbach, *Selected Writings, 1909–1953* (Dordrecht: Reidel, 1978), vol. 1.

[4] Note 2, p. 453.

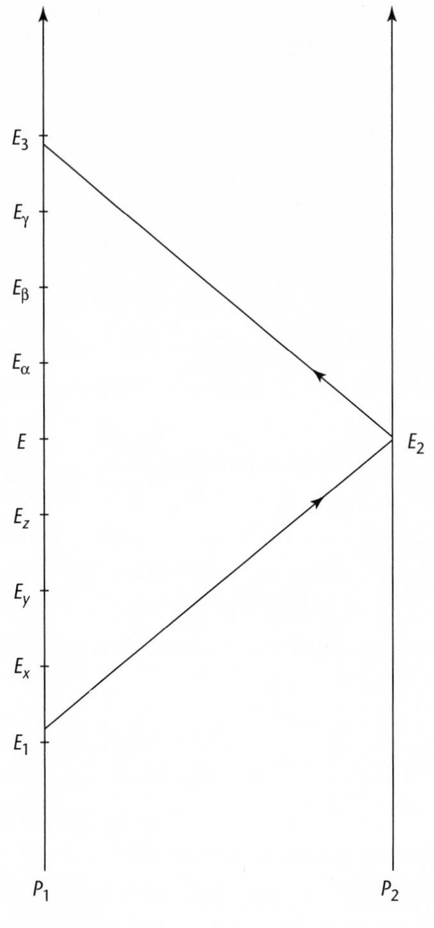

Figure 10.1

pulse can reach any other point P_2 *earlier*—as judged by a *local* clock at P_2 which merely orders events there in a metrically arbitrary fashion—than this light pulse."[5]

Applying this postulate to the case of two points, P_1 and P_2, fixed in a reference frame S and connected by light signals as shown in figure 10.1, we see that:

> Instead of E being the only event at P_1 which is neither earlier nor later than E_2 at P_2, as in the Newtonian world, on Einstein's hypothesis each one of the en-

[5] Ibid., p. 454.

tire superdenumerable infinity of point-events in the *open* interval between E_1 and E_3 at P_1 fails to have a *determinate* time-relation to E_2. For none of these events at P_1 can then be said to be either earlier or later than E_2: no signal originating at any of these events can reach P_2 soon enough to coincide there with E_2, and no causal chain emitted at P_2 upon E_2's occurrence can reach P_1 prior to the occurrence of E_3. But to say that none of these events at P_1 is either earlier or later than E_2 is to say that no one of them is *objectively* any more entitled to be regarded as simultaneous with E_2 than is any of the others. It is therefore only *by definition* that some *one* of these events comes to be simultaneous with E_2. Unlike the Newtonian situation, in which there was only a *single* event E which could be significantly held to be simultaneous with E_2, the physical facts postulated by relativity require the introduction, *within a single* inertial frame S, of a *convention* stipulating which *particular* pair of causally *non*-connectible events will be *called* "simultaneous." This relativity of simultaneity prevails within a single inertial system, because the simultaneity criterion of the system is relative to the *choice* of a particular numerical value between t_1 and t_3 as the temporal *name* to be *assigned* to E_2. Accordingly, depending on the particular event at P_1 that is chosen to be simultaneous with E_2, upon the occurrence of E_2 we set the clock at P_2 to read

$$t_2 = t_1 + \varepsilon (t_3 - t_1),$$

where ε has the particular value between 0 and 1 appropriate to the choice we made.[6]

Any event at P_1 between E_1 and E_3 is "topologically simultaneous" with E_2 for not being connectible with E_2 by any physical causal (signal) chain. Grünbaum has been criticized for calling this relation a *topological* simultaneity because it is not a topological invariant of Minkowski space. But he only followed Reichenbach who in paragraph 22 of *Philosophy of Space and Time* wrote: "We define: *any two events which are indeterminate as to their time order may be called simultaneous*. This topological definition would be sufficient for a unique definition of simultaneity in the classical theory of time."

Grünbaum then pointed out that if, for example, ε is so chosen that E_y becomes simultaneous with E_2, then all events between E_1 and E_y become *definitionally* earlier than E_2 and all events between E_y and E_3 become *defini-*

[6] Ibid., p. 454.

tionally later than E_2. Furthermore, the claim that, since only for $\varepsilon = \frac{1}{2}$ the velocity of light becomes isotropic, this value of ε is more "true" is a profound error, because no statement about velocities derives its meaning from mere facts. The apparent simplicity obtained by choosing $\varepsilon = \frac{1}{2}$ is not "simpler in the inductive sense of assuming less in order to account for our observational data, but only in the *descriptive* sense of providing a *symbolically* simpler representation of these data which expresses itself also by "assuring that synchronism will be both a symmetric and a transitive relation upon using *different* clocks in the same system,"[7] as Hans Reichenbach has shown.[8] "These considerations," Grünbaum continues, "enable us to see that physical facts which are independent of descriptive conventions do *not* dictate discordant judgments by different Galilean observers concerning the simultaneity of given events." Grünbaum then shows how the freedom of choosing the value of ε makes it possible to eliminate the relativity of the simultaneity of distant events, a task which, as we know, has been carried out already in 1944 by Scott-Iversen.[9]

To understand Grünbaum's strategy of achieving this aim we have to mention some simple mathematical consequences of equation (9.1) for the one-way velocity of light c_+ in the direction from A to B and the one-way velocity of light c_- in the opposite direction from B to A (see fig. 7.1 in chapter 7).

The light postulate or the postulate that the round-trip velocity of light equals c implies that $t'_A - t_A = 2d/c = t'_A - t_B + t_B - t_A = (d/c_+) + (d/c_-)$ or $2/c = (1/c_+) + (1/c_-)$. But $c_+ = d/(t_B - t_A) = c/2\varepsilon$ and $c_- = \frac{1}{2} c/(1 - \varepsilon)$. Hence, if $\varepsilon = \frac{1}{2}$ (standard synchrony) then $c_+ = c_-$ and if $\varepsilon \neq \frac{1}{2}$ (nonstandard synchrony), then $c_+ \neq c_-$.

Grünbaum applied these results to the thought experiment of the train moving along the embankment (fig. 7.3 of chapter 7), which served Einstein to prove the relativity of distant simultaneity. Einstein used standard synchrony ($\varepsilon = \frac{1}{2}$), however, when he showed that the two light bolts struck the end points of the train simultaneously for an observer stationary on the embankment but *not* simultaneously for an observer in the train. Grünbaum now asked whether "the train observer's observations of the lightning flashes *compel* him to say that the bolts did *not* strike simultaneously? Decidedly not!"

[7] Ibid., p. 455.
[8] Chapter 9, note 12.
[9] Chapter 9, note 36.

For, as we have seen, the one-way velocities of light, that is, in this case the velocities of the light signals emitted from the end points of the train, depend on the direction of their propagation in the case of nonstandard synchrony. These light pulses will always meet on the moving train at some point D' different from the midpoint M' of the train but lying between A' and B'. Grünbaum could therefore conclude that to define the two flashes as having occurred simultaneously, the train observer need only decide to define the ratio of the velocity of the light coming from the left to the velocity of the light coming from the right as the ratio of the distance $A'D'$ to the distance $D'B'$. Because "time = distance/velocity," $A'D'/c_+ = D'B'/c_-$, which means that the light signals began their transmission at the same time or, finally, the lightning bolts occurred also simultaneously for the observer in the train. "It follows that it is the relativity of simultaneity *within* each inertial system which allows each Galilean observer to choose his own value of ε so as to agree with other observers on simultaneity *or* so as to disagree."

The critical reader will have noticed that Grünbaum's elimination of distant simultaneity resembles Scott-Iversen's method insofar as both are based on the same two principles: (1) the transit times of two light signals are equal if the faster signal covers a proportionally larger distance, and (2) two light signals, having spent the same transit times until they meet, must have been emitted simultaneously. Scott-Iversen never realized that his method depends on the freedom of choosing different synchronization parameters in accordance with Reichenbach's conventionality thesis; Grünbaum, on the other hand, at least when writing his 1955 paper, was unaware of Scott-Iversen's essay. It is, therefore, no exaggeration to say that what Scott-Iversen had done without understanding what he had done was understood by Grünbaum without knowing that it had already been done.

In his book[10] Grünbaum offered a more detailed exposition of his derelativization of distant simultaneity. He studied, in particular, how the clocks at A' and B' have to be synchronized with a clock at the midpoint M' by

[10] A. Grünbaum, *Philosophical Problems of Space and Time* (chapter 7, note 65), pp. 360–368. H. Reichenbach, in *Philosophy of Space and Time* (note 1 in chapter 9; [1928], p. 180; [1958], p. 146), anticipated this elimination of relativity when he wrote: "We could arrange the definition of simultaneity of a system K in such a manner that it leads to the same results as that of another system K' which is in motion relative to K; in K', ε would not be equal to $1/2$ in the definition of simultaneity, but would have another value. It is a serious mistake to believe that if the state of motion is taken into consideration, the relativity of simultaneity is necessary." But Reichenbach did not elaborate this issue mathematically.

means of their synchronization parameters $\varepsilon_{A'}$ and $\varepsilon_{B'}$, correspondingly, so that in accordance with Einstein's simultaneity criterion the lightning flashes will be simultaneous also in the train system. The condition is simply that the difference between these synchronization parameters is equal to v/c.

In his book, and in his 1955 essay in the *American Journal of Physics*, Grünbaum also discussed the philosophical status of the concept of simultaneity and its definition. As the intrasystemic relativity of simultaneity shows, "in the first instance, it is the limiting character of the velocity of light and not the relative motion of inertial systems which gives rise to the relativity of simultaneity." This limiting property of light is an objective property of the causal structure of the physical world and independent of human measuring activities. The impossibility of operations that would define absolute simultaneity is not the result of our inability to carry out such measuring operations but

> a *consequence* of the more fundamental impossibility of the required causal relations between physical events. To be sure, operations of measurement are indispensable for *discovering* or *knowing* that particular physical events can or cannot sustain the causal relations which would define relations of temporal succession or of ordinal simultaneity between them. But the actual or physically possible causal relations in question are or are not sustained by physical events quite apart from *our* actual or hypothetical measuring operations and are *not* first *conferred* on nature by our operations. In short, it is because no relations of absolute simultaneity *exist* to be measured that measurement cannot disclose them; it is *not* the mere failure of measurement to disclose them that *constitutes* their non-existence, much as that failure is *evidence* for their non-existence. Only a philosophical obfuscation of this state of affairs can make plausible the view that the relativity of simultaneity . . . leads support to the subjectivism of homocentric operationism or of phenomenalist positivism.[11]

These statements were directed against the operationism and, in particular, against the philosophical ideas of its foremost contender Percy W. Bridgman, for whom a scientific "concept is synonymous with the corresponding operations."[12] Concerning the concept of simultaneity Bridgman wrote in 1927:

[11] Note 2 (1955), p. 456.
[12] P. W. Bridgman, *The Logic of Modern Physics* (New York: Macmillan, 1927, 1958), p. 5.

> Before Einstein the concept of simultaneity was defined in terms of properties. It was a property of two events, when described with respect to their relation in time, that one event was either before the other, or after it, or simultaneous with it. Simultaneity was a property of events alone and nothing else, either two events were simultaneous or they were not. . . . Einstein now subjected the concept of simultaneity to a critique, which consisted essentially in showing that the operations which enable two events to be described as simultaneous, involve measurements on the two events made by an observer, so that 'simultaneity' is therefore, not an absolute property of the two events and nothing else, but must also involve the relation of the events to the observer. Until therefore we have experimental proof to the contrary, we must be prepared to find the simultaneity of two events depends on their relation to the observer, and in particular on their velocity. Einstein, in thus analyzing what is involved in making a judgment of simultaneity, and in seizing on the act of the observer as the essence of the situation, is actually adopting a new point of view as to what the concepts of physics should be, namely, the operational view.[13]

Bridgman's claim that the special theory of relativity lends support to the operational interpretation of scientific concepts and, in particular, of the concept of distant simultaneity was criticized by Grünbaum in his essay *Operationism and Relativity.*[14] This debate between Grünbaum and Bridgman continued, so to speak, even after Bridgman's death in 1961, for in an epilogue to Bridgman's book *A Sophisticate's Primer of Relativity*, published posthumously in 1962, Grünbaum criticized Bridgman's statement that, in opposition to Reichenbach, "causal connectedness of distant events is not necessarily connected with order in time, but may just as well be correlated with directional effects in space. There are methods of setting distant clocks which do not involve causal propagation; and, in particular, clocks may be set consistently by transport, contrary to Reichenbach's explicit statement, and with no direct involvement with the properties of light."[15] Grünbaum also criticizes in this epilogue Bridgman's proposal of obtaining distant simultaneity

[13] Ibid., pp. 7–8; reprinted in R. Boyd, P. Gasper, and J. D. Trout (eds.), *The Philosophy of Science* (Cambridge, Massachusetts: MIT Press, 1991, 1993), pp. 59–60.

[14] A. Grünbaum, "Operationism and relativity," *The Scientific Monthly* **79,** 228–238 (1954); reprinted in P. Frank (ed.), *The Validation of Scientific Theories* (Boston: Beacon Press, 1957), 84–94.

[15] P. W. Bridgman, *A Sophisticate's Primer of Relativity* (Middletown, Connecticut: Wesleyan University Press, 1962; London: Routledge & Kegan Paul, 1962), p. 147.

by the use of superlight velocities produced by light signals that emanate from a rotating searchlight and reach distant clocks that are to be synchronized.

We will not discuss these methods and their critiques here in any detail because we will deal with them later on in chapter 12.

Instead, we question whether Reichenbach's and Grünbaum's conceptions of the conventionality of distant simultaneity differ from each other. Michael Friedman[16] and Michael Redhead[17] answered this question in essentially the same way. Both contended correctly, as it seems, that the conventionalist positions of Reichenbach and Grünbaum differ from a semantical point of view.

> Reichenbach argues from an epistemological point of view; he argues that certain statements are conventional as opposed to "factual" because they are unverifiable in principle. Grünbaum argues from an ontological point of view; he argues that certain statements are conventional because there is a sense in which the properties and relations with which they purportedly deal do not really exist, they are not really part of the objective physical world. Thus, Reichenbach's and Grünbaum's arguments depend on two different characterizations of the difference between conventional and "factual" statements. According to Reichenbach, the "factual"/conventional distinction is just the verifiable/unverifiable distinction. According to Grünbaum, the "factual"/conventional distinction rests on a prior distinction between properties and relations that are objective constituents of the physical world and those that are not.[18]

In short, for Reichenbach the criterion of the conventionality of a statement was the unverifiability of its possession of a truth value, whereas for Grünbaum it was the lack of an objectively existent relation in physical reality that corresponds to the contention of the statement.

[16] M. Friedman, "Simultaneity in Newtonian Mechanics and Special Relativity," in J. Earman, C. Glymour, and J. Stachel (eds.), *Foundations of Space-Time Theories* (Minnesota Studies in the Philosophy of Science, vol. 8) (Minneapolis: University of Minnesota Press, 1977), pp. 403–432. Quotation on p. 426.

[17] M. Redhead, "The Conventionality of Simultaneity," in J. Earman, A. I. Janis, G. J. Massey, and N. Rescher (eds.), *Philosophical Problems in the Internal and External World: Essays in the Philosophy of Adolf Grünbaum* (Pittsburgh: University of Pittsburgh Press, 1993), pp. 103–128.

[18] Note 16, p. 426.

CHAPTER ELEVEN

Symmetry and Transitivity of Simultaneity

As stated in chapter 7, in his 1905 paper on relativity, Einstein "assumed" that clock synchronization or simultaneity are symmetric and transitive relations,[1] and he defined the "time" of a reference system in terms of the notion of simultaneity. Because of the importance of the relation between these two notions, the "time of a reference system," briefly denoted by t, and "simultaneity," denoted by σ, it would be useful to recall the following set-theoretical definitions.

(I) A binary relation R on a set S is called "reflexive" if, for every element a of S, the proposition $a\ R\ a$ is true, that is, if every element of S stands in relation R to itself; (II) R is said to be "symmetric" in S if, for any two elements a and b of S, $a\ R\ b$ implies $b\ R\ a$; (III) R is said to be "transitive in S if, for any three elements a, b, c of S, $a\ R\ b$ and $b\ R\ c$ imply $a\ R\ c$; (IV) a relation R is said to be an "equivalence" relation if it is reflexive, symmetric, and transitive; (V) a collection of subsets of S is a "cover" of S if their (set-theoretical) sum is S; (VI) a cover of S is a "partition" of S if its members are pairwise disjoint.

[1] See chapter 7, note 34.

With these definitions it is easy to prove the following theorem: Any equivalence relation in S leads to a unique partition of S and, conversely, any given partition of S defines an equivalence relation on S.

If R is an equivalence relation on S and a any fixed (but arbitrary) element of S, then the set of all elements x of S that satisfy the condition $x\ R\ a$ is called the *equivalence class of a* and denoted by $[a]$; hence, symbolically,

$$[a] = \{x \mid x \varepsilon S \text{ and } x\ R\ a\}$$

Finally, the collection of all equivalence classes generated in S by an equivalence relation R is called the *quotient set* of S *modulo R* (or *induced by R*) and denoted by S/R.

If S denotes the set of all events, R the (standard) simultaneity relation σ, and t, the *time* of a reference system, then Einstein's definition of *time* (*Edt*) as presented at the end of § 1 of his 1905 paper on relativity can be expressed symbolically by

$$t =_{df} S/\sigma$$

This definition says that the *time* of a reference system is the quotient set of all events induced by the (standard) simultaneity relation.

True, Einstein never published such a set-theoretical formulation of his definition of *time*, but he would have undoubtedly endorsed it because it faithfully expresses what he had in mind. To substantiate this claim let us recall that in his Kyoto lecture, as mentioned in chapter 7, Einstein declared that "an analysis of time was my solution," whereas he should have said "an analysis of (the concept of) simultaneity was my solution." His referral to t, instead of to σ, suggests that he had *Edt* in his mind. This suggestion is strongly supported also by Wertheimer's report of his discussions with Einstein, according to which Michelson's famous experiment or other experimental discoveries did not lead to the genesis of the theory of relativity, but the fact that "it occurred to Einstein that time measurement involves simultaneity." That Einstein early distinguished between distant simultaneity, as in *Edt*, and local simultaneity of events was also recorded by Wertheimer when he quoted Einstein as having said: "If two events occur in one place, I understand what simultaneity means. . . . But am I really clear about what simultaneity means when it refers to events in two places? What does it mean to say that this event occurred in my room at the same time as another event in some distant place? Surely I can use the concept of simultaneity for different places in the same way as for one and the same place—but can I? Is it as clear to me in the former as it is the latter case? . . . It is not!"[2]

Einstein's early awareness of the difference between local and distant simultaneity or consequently, in accordance with *Edt*, between local time and the time t of a reference system, has been quoted purposely for the following reason. We may object that the concept of an *event* already implies the concept of *time*, which according to *Edt* presupposes the notion of *event*, which in turn involves the concept of time so that *Edt* is based on a vicious circle. Even if we adopt Minkowski's definition of the term "event" as "a physical occurrence which has no . . . duration in time,"[3] the very condition of "having no duration in time" may be claimed to imply the use of the concept of time.

We avoid this circularity, however, if we distinguish between two different meanings of *time* (or of simultaneity) as Einstein did: (i) "time" in the sense of a coordinate on the time axis of a reference system or briefly t as in *Edt*, and (ii) "time" in the sense of local time, corresponding to the concept of local simultaneity and independent of any coordination in a reference system. If the term "time of an event," when used prior to *Edt*, is understood in the sense (ii) and not in the sense (i), Einstein's definition of *time* as S/σ is not vitiated by any vicious circularity.

Note, however, that Einstein's definition of *time* can be maintained only if σ (or simultaneity), as defined by him, is an equivalence relation. Einstein, it will be recalled, only *assumed* that this condition is satisfied.[4] The problem to be faced, therefore, is whether, instead of "assuming" this fact, it could not be logically derived from the definition of simultaneity.

That the simultaneity relation σ is a reflexive relation, that is, that every event occurs simultaneously with itself, is trivially true, of course. But, according to Einstein's definition of standard simultaneity, that σ is a symmetric relation is not at all self-evident and must be proved. Einstein, it will be recalled, simply *assumed* in his 1905 paper this symmetry when he wrote that "(1) if the clock at *B* synchronizes with the clock at *A*, the clock at *A* synchronizes with the clock at *B*." He also *assumed* the transitivity for he wrote: "(2) If the clock at *A* synchronizes with the clock at *B* and also with the clock at *C*, the clocks at *B* and *C* also synchronize with each other."[5]

[2] See chapter 7, note 36, pp. 218–219.
[3] Chapter 1, notes 10 and 11.
[4] Chapter 7, note 34.
[5] See chapter 7, note 34.

Strictly speaking, Einstein's formulation of the transitivity of σ presupposes its symmetry. For, denoting the clock at A by U_A etc., Einstein defined the transitivity of σ by demanding that $U_A\sigma U_B$ and $U_A\sigma U_C$ imply $U_B\sigma U_C$ and $U_C\sigma U_B$, whereas according to the above-mentioned definition he should have defined it by demanding that $U_A\sigma U_B$ and $U_B\sigma U_C$ imply $U_A\sigma U_C$. If, however, σ is also symmetric, then $U_A\sigma U_B$ can be replaced by $U_B\sigma U_A$ and hence $U_B\sigma U_A$ and $U_A\sigma U_C$ imply $U_B\sigma U_C$, as required.

In a footnote of *Axiomatization*[6] Reichenbach maintained that Einstein apparently regarded the symmetry and transitivity of clock synchronization, and hence of simultaneity, as facts of experience (Erfahrungstatsachen), in contrast to his view that synchronization itself is a matter of definition, and that Einstein "was therefore the originator of this important distinction." This statement by Reichenbach may be challenged by pointing out that such a differentiation had already been made before Einstein, for example, by Poincaré, as we saw in chapter 6.

An apparently most straightforward proof of the symmetry of σ was proposed by Arthur I. Miller[7] in his profound analysis of Einstein's 1905 paper on relativity by contending that this symmetry "follows directly from the second postulate," according to which the velocity of light in empty space is a universal constant.

Although apparently incontestable, Miller's argument cannot be applied in the context of Einstein's logical construction of the theory for the following reason. Because the notion of velocity presupposes the concept of *time*, namely, the time of the reference system S, and because this concept of time is defined as S/σ, where σ is supposed to be an equivalence relation, and hence a symmetric relation, Miller's argumentation involves a vicious circle, for his proof of the symmetry of σ is based on the assumption that σ is an equivalence relation and, as such, is a symmetric relation.

Miller also offers a proof of the transitivity of σ as follows. "Assume that a clock at A is synchronous with clocks at B and C. From the definition of synchronization we have: (A) $t_B - t_A = t'_A - t_B$ and (B) $t_C - t_A = t'_A - t_C$. Eqs. (A) and (B) are consistent only if $t_B = t_C$. Therefore the clocks at B and C are synchronous."[8] Miller's argument holds only for clocks equidistant from the

[6] Chapter 9, note 12, § 10, (1924), footnote 2; (1969), footnote 17.

[7] A. I. Miller, *Albert Einstein's Special Theory of Relativity* (Reading, Massachusetts: Addison-Wesley Publishing Company, 1981), p. 197.

[8] Ibid., p. 199.

clock at A because in the general case the t'_A in (A) differs from the t'_A in (B), it being assumed that the light signals emitted from A to B and C are emitted at the same moment t_A.

The earliest attempt to prove the transitivity of the simultaneity relation was probably made by von Laue in the first edition of his previously mentioned text.[9] Based on the assumption that "signals propagating in opposite directions are equivalent," as Einstein expressed it in a footnote of his 1912 manuscript[10] the proof starts with the algebraic identity (both sides of the *equation* being zero):

$$(t_A - t_0) + (t_B - t_A) + (t_0 - t_B) = (t'_B - t'_0) + (t'_A - t'_B) + (t'_0 - t'_A)$$

and continues, according to Einstein, as follows: "But since [by assumption] $t_A - t_0 = t'_0 - t'_A$ and $t'_B - t'_0 = t_0 - t_B$, therefore also $t_B - t_A = t'_A - t'_B$." Hence if the clocks at A and at B are individually synchronized with the clock at 0, they are also synchronized among themselves.

Ludwik Silberstein, then a professor at the University of Rome, made an interesting attempt to prove the symmetry of the simultaneity relation in 1914. Silberstein presupposed "that the time employed by the light-signal to pass from A to B is *always the same*." Denoting by a_d, a_a, and a_r, respectively, the instants of *d*eparture, *a*rrival, and *r*eturn, of a light signal at the clock at A, and correspondingly by b_d, etc., the corresponding instants at the clock at B, Silberstein argued as follows: If the clock at B is synchronous with that at A, then,

> We have by [Einstein's] definition [of simultaneity] $b_a = \frac{1}{2}(a_d + a_r)$, or
>
> $$b_a - a_d = a_r - b_a = a_a - b_a,$$
>
> because the return at A may be equally well considered as an arrival at that place. Now, if at the instant a_a the flash be sent again towards B, where it arrives at the instant b_r, we have, by our above requirement;
>
> $$b_a - a_d = b_r - a_a$$
>
> and, by the last equation,
>
> $$a_a - b_a = b_r - a_a.$$

[9] Chapter 7, note 92.

[10] Chapter 7, note 75, p. 23. The fact that Einstein mentioned von Laue's proof of the transitivity of σ only in a footnote explains why we have called Einstein's appeal to the transitivity of simultaneity "almost tacitly."

But here b_a is identical with the instant of departure b_d, and, consequently, $a_a = \frac{1}{2}(b_a + b_r)$, i.e. the clock placed at A is synchronous with that placed at B. Q.E.D.[11]

Ingenuous as Silberstein's proof of the symmetry of simultaneity seems to be, it cannot replace Einstein's "assumption" that simultaneity is a symmetric relation, because it would be vitiated, just like Miller's above-mentioned much later attempt, by the same vicious circularity. In fact, Silberstein's presupposition "that the *time* employed by the light-signal to pass from A to B is *always the same*" employs the notion of *time*, which, as explained earlier, is defined by Einstein as S/σ, where σ is an equivalence and hence a symmetric relation.

Reichenbach was probably the first who realized the danger of committing vicious circles in an axiomatic construction of the special theory of relativity. In *Axiomatization of the Theory of Relativity*[12] he emphasized the need to develop "the theorems in our presentation in such a way that a theorem never presupposes axioms that follow it in the exposition."[13] Reichenbach even warned, in particular, that the danger of such a vicious circularity could easily occur in the context of dealing with the notion of simultaneity because "the measurement of simultaneity presupposes the knowledge of velocity. On the other hand, the measurement of velocity presupposes the knowledge of the simultaneity, because time measurements at two different places are required."[14]

To avoid such inadvertencies Reichenbach's constructive axiomatization of Einstein's theory of relativity deviated in certain points from Einstein's 1905 approach, although it led, of course, to the same conclusions. Thus, although Reichenbach followed Einstein in defining that a clock at B is said to be synchronized with a clock at A if, in the notation of chapter 7, $t_B = \frac{1}{2}(t_A + t_A')$. But, in contrast to Einstein, Reichenbach proved the symmetry of simultaneity by using a light signal that "travels back and forth several times; its times at A are t_A, t_A', t_A'' and at B, correspondingly, t_B, t_B'. According to [the just-quoted] definition, $t_B = \frac{1}{2}(t_A + t_A')$ and $t_B' = \frac{1}{2}(t_A' + t_A'')$." Referring now to Axiom IV, 1, from which it follows that $t_A' - t_A = t_A'' - t_A'$, Reichenbach

[11] L. Silberstein, *The Theory of Relativity* (London: Macmillan, 1914, 1924), p. 95.

[12] Chapter 9, note 12.

[13] "Wir . . . wählen die Darstellung so, dass ein Satz niemals Axiome voraussetzt, die ihm in der Darstellung folgen." Ibid., § 6.

[14] Ibid., § 2.

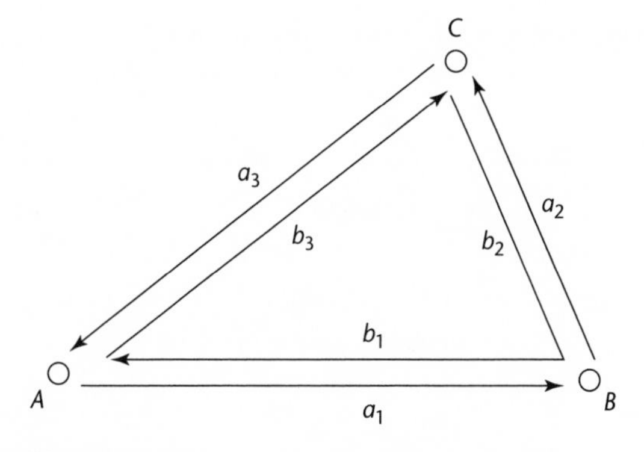

Figure 11.1

continued: "There are three equations for the five variables $t_A \ldots t''_A$; if t_A and t''_A are eliminated, $t'_A = \frac{1}{2}(t_B + t'_B)$ remains as the relation between $t_B\ t'_A\ t'_B$. This means that clock *A* simply shows the times according to which it would have to be set if it were to be made synchronous with *B*. Since *B* was synchronized with *A*, *A* is also synchronous with *B*."[15]

Reichenbach's method of demonstrating the symmetry of simultaneity is physically identical with the method used ten years earlier by Silberstein,[16] but it has the advantage of exposing ostensibly the assumptions underlying the proof.

Reichenbach's proof of the transitivity of simultaneity is also based on an axiom, the "round-trip axiom," which states that "if from a point *A* . . . two light signals are sent in opposite directions along a closed triangular path *ABCA*, they will return simultaneously."[17] To prove that if a clock at *A* is synchronous with a clock at *B* and also with a clock at *C*, the clocks at *B* and *C* are synchronous Reichenbach considered the following procedure (see fig. 11.1).

> A light signal is sent in the direction *ABCA*, the second signal, traveling in the opposite direction, is not sent simultaneously but so that it will arrive at *C* at the same time as the first signal. The intervals $\overline{AB}$, $\overline{BC}$, and $\overline{CA}$ [where the bar over *AB*, for example, denotes the time of the signal to proceed from *A* to *B* as measured at *A*] are $a_1\ a_2\ a_3$, and $b_3\ b_2\ b_1$ in the other direction. Because of the

[15] Loc. cit., § 10.
[16] Note 11.
[17] Chapter 9, note 12, § 11.

synchronization from A $a_1 = b_1$ and $a_3 = b_3$. According to the round-trip axiom, $a_1 + a_2 + a_3 = b_1 + b_2 + b_3$ and, therefore, $a_2 = b_2$. These two time intervals were chosen so that they would correspond to a signal reflected directly at C, because the signal CB leaves C the moment BC arrives. The formula $a_2 = b_2$ says that this moment at C has a time index corresponding to half the time interval between the time of departure and the time of arrival of the signal BCB; therefore, C is synchronous with B. If the time of departure of the second signal is chosen so that it will arrive at B simultaneously with the first signal, then B is synchronous with C. Therefore, transitivity holds.

As the preceding considerations confirm, the statement that the standard synchrony σ is an equivalence relation is not a logical consequence of the definition of this relation but requires some additional assumptions. Hence, because, according to Einstein's 1905 definition, the *time* of the reference system S of all events is defined as S/σ or a partition of S modulo σ, where σ is an equivalence relation, it is clear that Einstein's conception of *time* is also based on these additional assumptions. Einstein's admission, "we assume" (wir nehmen an) that simultaneity is an equivalence relation, is therefore an integral part of his conception of *time*.

From the historical point of view it is interesting to compare Einstein's conception of *time* with that of other philosophers. It is tempting, for example, to identify the Aristotelian "*now*" (νῦν) with an equivalence class of σ. However, whereas according to Einstein every such equivalence class may well be regarded as synonymous with a "now" and "*time*," as the set of such equivalence classes, as composed of such "nows," according to Aristotle, as stated in chapter 2, "time is not composed of atomic 'nows.' "[18] On the other hand, Leibniz's causal theory of time, which defines "*time*" as "the order of existence of those things which are not simultaneous"[19] and ascribes therefore logical priority to the notion of "simultaneity" over that of "time," may well be regarded, at least in this respect, as an anticipation of Einstein's approach.

So far only the properties of the standard light-signal synchronization have been studied.

Turning now to the study of the analogous properties of the nonstandard signal synchronization it is useful to employ the following notation. Let U_B denote a clock at rest at point B of an inertial system and U_A another clock

[18] Chapter 2, note 32.
[19] Chapter 5, note 42.

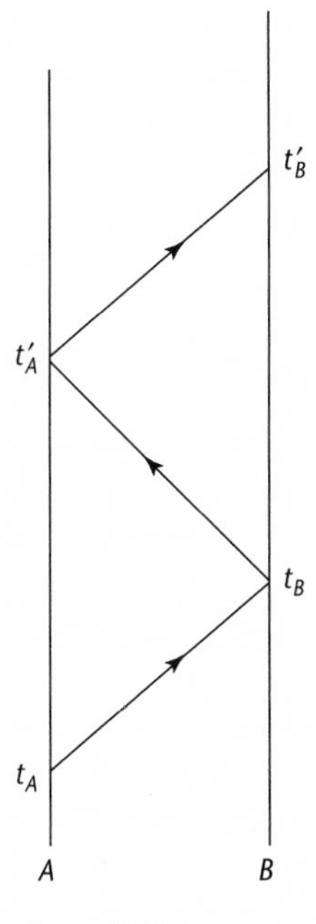

Figure 11.2

at rest at point *A* of the same system. The fact that U_B is nonstandard synchronized with U_A by setting U_B to agree with U_A, so that U_A serves as the master clock, is expressed, following Grünbaum,[20] by the formula

$$U_B \text{ syn } (\varepsilon_{AB})\, U_A, \tag{11.1}$$

where ε_{AB} is the Reichenbach synchronization parameter used in this operation. If U_A is set to agree with U_B so that U_B serves now as the master clock, then, of course, letters *A* and *B* will have to be permuted and (11.1) will be replaced by the expression

$$U_A \text{ syn } (\varepsilon_{BA})\, U_B. \tag{11.2}$$

Because the operations described by (11.1) and (11.2) differ, there is no reason to assume that $\varepsilon_{AB} = \varepsilon_{BA}$.

If, as in figure 11.2, t_A denotes the moment of the emission of a light signal from *A*, t_B its reception at *B*, from where it is immediately reflected to

[20] Chapter 7, note 65 (1973), p. 671.

reach A at t'_A and once again reflected to reach B at t'_B, then according to the empirically verifiable two-way light principle:

$$t'_B - t_B = t'_A - t_A \tag{11.3}$$

Grünbaum now pointed out that

The reading t'_A of U_A on the arrival of the light ray does not *automatically* qualify U_A to be synchronized from the points of view of U_B or of some third clock U_C. It is true that U_A's reading t'_A (along with its reading t_A) was already invoked to impart a setting t_B to U_B which satisfies the condition (11.1). And the principles asserted thus far permit us to *tamper* with the otherwise existing setting of U_B in order to satisfy the latter condition of synchronism. But these principles also allow us to "*correct*" computationally, if necessary, the reading t'_A of U_A to assure the fulfillment of the condition (11.2) for any chosen ε_{BA} between 0 and 1. It is to be understood that computational correcting of the reading of U_A does *not* involve physically tampering with the setting of U_A. If the latter kind of "correcting" is to be disallowed, we must introduce the following restriction of governing any clock X whose readings have already been used to synchronize some other clock Y with X via some ε_{XY} between 0 and 1: the setting of any such clock X *must* be accepted as automatically qualifying X to be synchronized from the point of view of Y *and* of any other clock Z. This restriction will be called *the rule of committed synchronism*, and abbreviated to "RCS."[21]

If this rule and hence the t'_A reading of U_A are accepted then clearly the value ε_{BA} is uniquely determined by the equation

$$t'_A = t_B + \varepsilon_{BA}\, T, \tag{11.4}$$

where, in accordance with (11.3), $T = t'_B - t_B = t'_A - t_A$. Using the arithmetical identity

$$t_B - t_A + t'_A - t_B = t'_A - t_A = T \tag{11.5}$$

we obtain $\varepsilon_{AB}T + \varepsilon_{BA}\, T = T$ or finally

$$\varepsilon_{AB} + \varepsilon_{BA} = 1. \tag{11.6}$$

Equation (11.6) shows that, in contrast to standard synchrony or standard simultaneity ($\varepsilon = 1/2$), nonstandard synchrony or nonstandard simultaneity

[21] Ibid., p. 672.

($\varepsilon \neq 1/2$) is not only not a symmetric, but even an asymmetric relation, because if $\varepsilon_{AB} \neq 1/2$ then also $\varepsilon_{BA} = 1 - \varepsilon_{AB} \neq 1/2$ and vice versa.

As a matter of fact, equation (11.6) had already been published in June 1967 by Brian Ellis and Peter Bowman,[22] then of the University of Melbourne, two years before Grünbaum presented it in "A Panel Discussion on Simultaneity"[23] in March 1969. Ellis and Bowman claimed that equation (11.6) is a logical consequence of the two-way light principle. Denoting the distance between A and B by d, the velocity of light in the direction from A to B by c_{AB} and in the opposite direction by c_{BA}, they argued as follows. According to the two-way light principle,

$$(d/c_{AB}) + (d/c_{BA}) = 2d/c, \tag{11.7}$$

where $c_{AB} = d/(t_B - t_A)$ and $c_{BA} = d/(t_A' - t_B)$. But in $c_{AB} = d/(t_B - t_A)$ the quantity t_B serves to qualify U_B as being synchronized with U_A. True, according to the two-way light principle, $T = 2d/c$. But, said Grünbaum, "the conjunction of the arithmetical identity (11.5) with this principle cannot yield

$$(d/c_{AB}) + (d/c_{BA}) = 2\ d/c \tag{11.8}$$

unless *RSC* is involved to *accept* the reading t_A' as qualifying U_A to be synchronous with U_B and write

$$c_{BA} = d/(t_A' - t_B) \tag{11.9}$$

Yet Ellis and Bowman use (11.8) to deduce (11.6) via the relations $\varepsilon_{AB} = c/(2\ c_{AB})$ and $\varepsilon_{BA} = c/(2\ c_{BA})$. Since the *two*-way light principle cannot itself entail any relation between the two distinct *one*-way synchronisms specified by ε_{AB} and ε_{BA}, the two-way principle cannot entail (11.6)."

The proven asymmetry of a relation, like that of nonstandard simultaneity, does not enforce the relation to be either intransitive, nontransitive, or transitive. Thus, for example, the order relation among the natural numbers, "*a is smaller than b*," or symbolically $a < b$, is an asymmetric and *transitive* relation, asymmetric because, if $a < b$, then it is not the case that

[22] B. Ellis and P. Bowman, "Conventionality in distant simultaneity," *Philosophy of Science* **34,** 116–136 (1967).

[23] A. Grünbaum, W. C. Salmon, B. C. van Fraassen, and A. I. Janis, "A Panel Discussion of Simultaneity by Slow Clock Transport in the Special and General Theories of Relativity," *Philosophy of Science* **36,** 1–105 (1969). Grünbaum's contribution to this Panel Discussion is reprinted in chapter 7, note 65, pp. 670–708.

$b < a$, and *transitive* because, if $a < b$ and $b < c$, then $a < c$. In contrast, the family relation among men "*a is the son of b*," or symbolically $a \pi b$, although also asymmetric, is an *intransitive* relation, asymmetric because, if $a \pi b$, then it is not the case that $b \pi a$, and *intransitive* because, if $a \pi b$ and $b \pi c$, then it is not the case that $a \pi c$, for then a is the grandson and not the son of c. Because every nonempty intransitive relation is ipso facto also a *nontransitive* relation it is clear that asymmetric relations that are *nontransitive* exist.

Turning, therefore, to the question of whether nonstandard light-signal synchronization or, equivalently, nonstandard distant simultaneity is a transitive relation we will see that the answer depends on the choice of the synchronization procedure.

Let us begin with a review of Grünbaum's proof of the intransitivity of nonstandard synchronisms.[24] It is based on a generalized version of RCS, hence called GRCS, which states that "at most one act of setting is permissible per clock to achieve the *mutual* synchronism of any two or more clocks. Thus, if a clock U_Y has been synchronized from the point of view of a clock U_X, then the resulting setting of U_Y must be accepted as a basis for synchronizing any other clock U_Z with U_Y, and the setting of a clock U_Z resulting from using U_Y to synchronize U_Z must automatically be accepted as rendering U_Z synchronous with U_X."[25]

Let three clocks U_A, U_B, and U_C be stationed at, respectively, the three vertices A, B, and C of a triangle in an inertial system (see fig. 11.3) and let AB have the length l, BC the length m, and AC the length n. The proof is based on the following three postulates: (1) the two-way light principle according to which the round-trip velocity of light *in vacuo* has the numerical value c. (2) The generalized rule of committed synchronism (GRCS) is valid, according to which at most one act of setting is permitted per clock to obtain mutual synchronism. (3) The transitivity of synchronism requires the equality of the three synchronization parameters: $\varepsilon_{AB} = \varepsilon_{BC} = \varepsilon_{AC}$, which may therefore be denoted collectively by ε. Let two light rays be emitted from A at the time when U_A indicates $t_A = 0$. One ray transverses AC, the other first the segment AB and, immediately afterward, the segment BC; it will reach C by a time difference Δ later than the first ray. The value of Δ, being measurable

[24] Ibid., pp. 674–678.
[25] Ibid., p. 673.

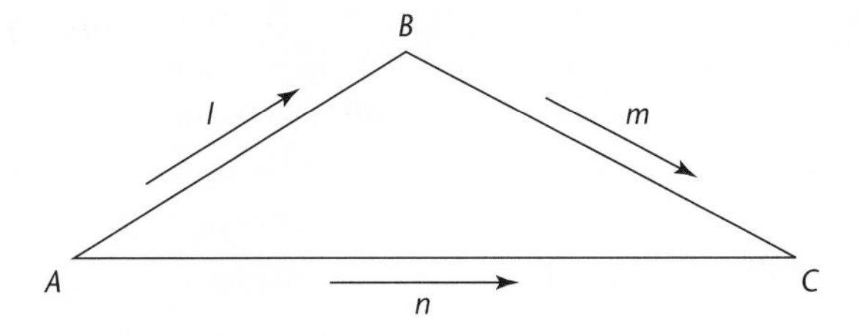

Figure 11.3

by only U_C, can therefore be determined by standard synchronization which yields

$$\Delta = (l + m - n)/c \tag{11.10}$$

If T_1, T_m, and T_n denote the round-trip times of the light rays *ABA*, *BCB*, and *ACA* as measured by U_A, U_B, and U_C, respectively, the two-way light principle tells us that

$$T_l = 2l/c \quad T_m = 2m/c \quad T_n = 2n/c \tag{11.11}$$

The direct ray reaches C at the time $\tau_1 = \varepsilon_{AC}\, T_n$, whereas the other rays arrive at B at the time $t_B = \varepsilon_{AB} T_1$ and then at C at the time $\tau_2 = \varepsilon_{AB}\, T_l + \varepsilon_{BC}\, T_m$. Implementing GRCS we see that it suffices that the synchronization parameters are so chosen that at most one setting of any one clock is needed to satisfy the equation

$$\tau_1 + \Delta = \tau_2 \tag{11.12}$$

Because $\Delta = \tau_2 - \tau_1$, using (11.10) and (11.11) yields

$$\tfrac{1}{2}\,(l + m - n) = l\,\varepsilon_{AB} + m\,\varepsilon_{BC} - n\,\varepsilon_{AC} \tag{11.13}$$

As Grũnbaum now points out, fulfillment of condition (11.13) implements for clock U_C the requirement of GRCS and also shows that even though there are nonstandard synchronisms $0 < \varepsilon < 1$ satisfying GRCS, any such synchronism is intransitive. Clearly, imposing upon (11.13) the transitivity requirement $\varepsilon_{AB} = \varepsilon_{BC} = \varepsilon_{AC} = \varepsilon$ yields

$$\tfrac{1}{2}\,(l + m - n) = \varepsilon\,(l + m - n) \tag{11.14}$$

Because $l + m - n \neq 0$, however, we obtain $\varepsilon = \tfrac{1}{2}$ and conclude that a transitive synchronization must be a standard synchronization or, in other words, that standard synchronism is a *necessary* condition for the transitivity of synchronisms, which proves the nontransitivity of nonstandard synchronizations. To prove also their intransitivity we see that equation (11.13) implies that if $\varepsilon_{AB} = \varepsilon_{BC} = \varepsilon_k$ such that $\varepsilon_k \neq 1/2$, then also $\varepsilon_{AC} \neq \varepsilon_k$. Hence, if $\varepsilon_{AB} = \varepsilon_{BC} = \varepsilon_k$ (11.13) can be written in the form

$$(l + m)(1/2 - \varepsilon_k) = n(1/2 - \varepsilon_{AC}) \qquad (11.15)$$

which, because $l + m \neq n$ and neither side of this inequality is zero, can hold only if the second factors on each side are equal to zero, which is incompatible with the hypothesis of *non*-standard synchronization $\varepsilon_k, \neq 1/2$ or $1/2 - \varepsilon_k \neq 1/2 - \varepsilon_{AC}$. We thus obtain $\varepsilon_k \neq \varepsilon_{AC}$, which shows that the nonstandard synchronism $\varepsilon_k \neq 1/2$ is intransitive.

In 1974 Philip L. Quinn of Brown University, a former doctoral student of Grünbaum, pointed out that "Grünbaum's argument, although mathematically flawless, does not prove quite what he has claimed."[26] In fact, Grünbaum's proof hinged on the condition that $l + m \neq n$, which fails to be satisfied if A, B, and C are collinear in this order. Quinn demonstrated that Grünbaum's contention remains valid also for three collinear points, provided the optical synchronization procedure that was implicit in the derivation is used throughout the argument. However, application of an alternative synchronization procedure leads to the conclusion that all nonstandard synchronisms are transitive. As Grünbaum pointed out, "the moral of Quinn's paper is that (1) there are at least two procedures for synchronizing triplets of clocks in an inertial system; (2) these alternative procedures impose different constrains on the transitivity properties of non-standard synchronisms, so that (3) assertions about the transitivity properties of non-standard synchronisms need to be relativized to the synchronization procedure(s) for which they hold."[27]

In the following presentation of Quinn's proof that a one-signal nonstandard synchronization of three clocks, U_A, U_B, and U_C, stationed at the collinear points A, B, and C (in this order) is transitive we will use the notation of equation (11.1). Thus, a subindex (e.g., B in t_B) indicates by what clock the time coordinate t is measured (e.g., by U_B); and equation $t_B = t_A + \varepsilon_{AB}(t'_A - t_A)$ for the synchronization of clock U_B with U_A corresponds to the expression U_B syn(ε_{AB}) U_A.

[26] P. L. Quinn, "The transitivity of non-standard synchronisms," *The British Journal for the Philosophy of Science* **25**, 78–82 (1974).

[27] Chapter 7, note 65 (1973), pp. 852–853.

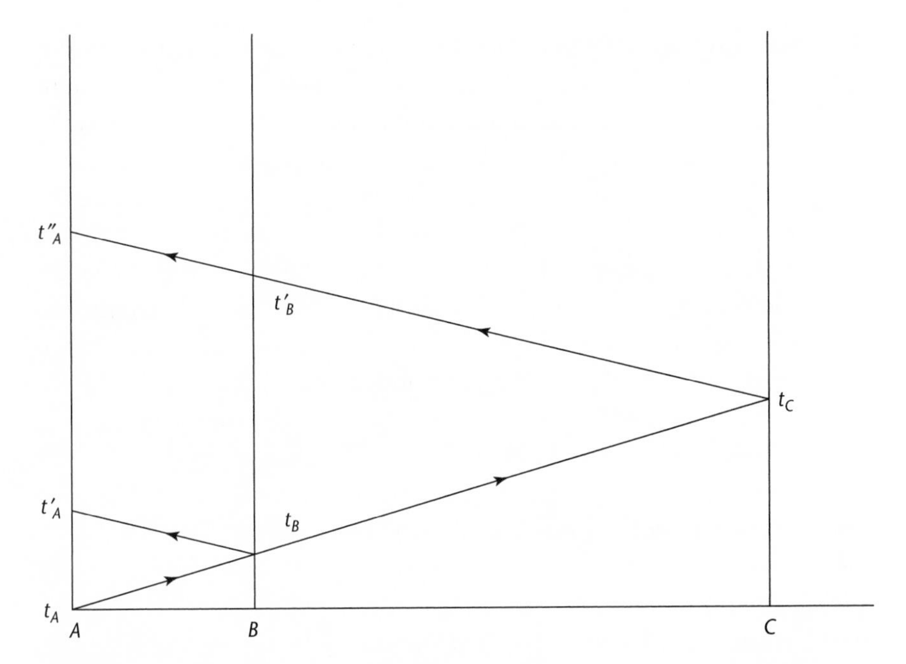

Figure 11.4

A light signal is emitted from A at the time $t_A = 0$ via B to C where it arrives at time $t_C = \varepsilon_{AC}\, 2\,(l + m)/c$, l being the distance between A and B and m the distance between B and C. Let U_C be synchronized with U_B by the same signal by which U_B is synchronized with U_A; this ray is partially reflected by B at t_B and reached A again at t'_A; it is also partially transmitted to reach C at t_C; after reflection at C it passes B again at t'_B and reaches A at t''_A (see fig. 11.4). This synchronization procedure, therefore, leads to the equation $(\varepsilon_{AB}2l/c) + (\varepsilon_{BC}\, 2\, m/c) = \varepsilon_{AC}\, 2\,(l + m)/c$, or $\varepsilon_{AB}l + \varepsilon_{BC}\, m = \varepsilon_{AC}(l + m)$. Synchronization of U_C with U_B by the same rule as U_B with U_A (so that $\varepsilon_{BC} = \varepsilon_{AB} =$ [say] ε_k) gives for $\varepsilon_k = \varepsilon_{AB} = \varepsilon_{BC} = 1/2$ and, because $l + m = 0$, $\varepsilon_k = \varepsilon_C$. Hence, $\varepsilon_{AB} = \varepsilon_{BC} = \varepsilon_{AC}$. Hence, U_C syn (ε_{BC}) U_B and U_B syn (ε_{AB}) U_A imply $U_C\, \varepsilon_{AC}\, U_{AC}$, which shows that this nonstandard synchronization is a transitive synchronization.

In 1975 Ian Walter Roxburgh of the University of London argued that Quinn's proof for the transitivity of nonstandard synchrony is flawed because "he has not, in fact, tested his procedure for transitivity, only applying it to one degenerate case" and used different rules for synchronization in so far as only one of the light rays passes through point B, whereas the other ray is reflected by B. If only one synchronization is used, Roxburgh claims, every nonstandard synchrony is intransitive.[28] In Quinn's case to set the clock at C, the light ray goes through B just as the ray reflected to A. To adhere to one and the same synchronization definition Roxburgh introduces an additional clock U_D at a point D and calls the distance between A and B, B and C, and B and D, respectively, l, p, and q. Roxburgh then shows that t_1, the time when the ray leaves D, and t_3, the time when it reaches B, satisfy the equations $t_3 = t_1 + 2\,\varepsilon\, p/c + (1 - \varepsilon)\, 2\, q/c$, whereas the synchronization of U_D with U_C leads to $t_3 = t_1 + \varepsilon(t_5 - t_1) = t_1 + \varepsilon(2p + 2q)/c$, where t_5 denotes the time when the ray, returning from B, reaches A. Clearly, these equations are compatible if either $\varepsilon = 1/2$ or $q = 0$. In Quinn's proof $q = 0$ so that U_D is identical with U_A. But, in general, the synchronization is transitive only if $\varepsilon = 1/2$, which proves Roxburgh's contention.

Roxburgh concluded his essay by pointing out that this must be the case, for if U_C is synchronized with U_A, and U_D with U_A, then replacing U_C by U_A and U_D by U_B implies U_A is also synchronized with U_B, and the synchronism is symmetric.

The problem of the symmetry and transitivity or intransitivity of nonstandard synchronizations, which originated from Reichenbach's thesis of the conventionality of distant simultaneity and its promulgation and elaboration by Grünbaum, engaged in its subsequent development not only the attention of Quinn and his commentator Roxburgh, but attracted the attention of numerous philosophers and physicists. We would have to go into too much technical detail to describe the distinctive features of each of these further contributions to the problem. Instead we refer the interested reader primarily to the contributions by Jarrett Leplin[29] of the University of North Carolina and by Russell Francis[30] of the University of Toronto and to the detailed paper by Leo Karlov[31] that deals with the nonstandard synchronization even for sets of infinitely many clocks.

We draw attention to only one point in Leplin's essay which seems to be of philosophical importance. Analyzing Quinn's procedure Leplin asks:

[28] I. W. Roxburgh, "The transitivity of non standard synchronisms," *The British Journal for the Philosophy of Science* **26,** 47–49 (1975).

[29] J. Leplin, "Synchronization rules and transitivity," *The British Journal for the Philosophy of Science* **27,** 399–402 (1976).

[30] R. Francis, "On the interpretation and transitivity of non-standard synchronisms," *The British Journal for the Philosophy of Science* **31,** 165–173 (1980).

[31] L. Karlov, "Clocks in nonstandard synchrony," *General Relativity and Gravitation* **19,** 455–464 (1987).

> Allowing that it does issue in the transitivity of non-standard synchronisms, is it correct to regard it as a legitimate synchronization rule? . . . It is natural to regard Quinn's method not as a synchronisation rule at all, but only as an arbitrary restriction on the procedure for comparing clocks set independently according to some common choice of ε . . . The natural view would be that an ε choice is intransitive if it requires a different ε choice for the identification of resultant readings on clocks compared directly, or, perhaps, *via* any path. It is possible to declare that the 'natural view' is merely a convention, since in principle we are free to choose any signal path, including Quinn's. Quinn's accomplishment might then be represented as the revelation of a new level of conventionality—the conventionality of transitivity.

If we accept this point of view and make the following statement: Event e is simultaneous with event e' and event e' is simultaneous with event e'', therefore event e is simultaneous also with event e'', then we use two different categories of conventions. The term "simultaneous" is subject to the Reichenbach thesis of the conventionality of distant simultaneity, and the term "therefore" and its sequence is subject to what may be called the "Leplin thesis of the conventionality of the transitivity of simultaneity."

As shown in this chapter, the question of whether standard simultaneity is a symmetrical and transitive relation has been the subject of extensive research. The more fundamental problem concerning the very possibility of this relation, however, or, in other words, the problem of what are the sufficient and necessary conditions for the possibility of establishing this relation or, in brief, what are the conditions that clocks *can be standard-synchronized* at all, has been generally ignored.

Recall that Einstein, after having stated his definition of standard synchrony in 1905, declared that "we *assume* that this definition of synchronism is free from contradictions and possible for any number of points," and that synchrony, thus defined, is a symmetric and transitive relation.[32] Because physicists and philosophers subsequently thought it necessary, as we have seen, to *derive*, rather than merely to *assume*, the symmetry and transitivity of this relation, we should expect that they likewise should have thought it necessary to replace Einstein's *assumption* of the possibility of establishing this relation by a *logical proof* of this possibility.

[32] "Wir *nehmen an*, dass diese Definition des Synchronismus in widerspruchsfreier Weise möglich sei . . . " (italics added). See chapter 7, note 6.

However, the important problem of what the necessary and sufficient conditions are, that clocks in an inertial system can be synchronized, has been ignored, in general. A noteworthy exception is an article by Alan Lang Macdonald[33] of Luther College in Decorah, Iowa, which with slight modifications can be summarized as follows.

In accordance with Einstein's 1910 definition of standard simultaneity[34] a clock U_A at point A in an inertial system S is synchronized with a clock U_B at point B in S if for a light signal emitted from A at time t_A, as indicated by U_A, and received at B at time t_B, as indicated by U_B, and for a light signal emitted from B at t'_B, as indicated by U_B, and received at A at t'_A, as indicated by U_A

$$t_B - t_A = t'_A - t'_B. \tag{11.16}$$

The following two conditions are now shown to be necessary and sufficient that clocks in an inertial system can be synchronized:

Condition (i): If light signals are emitted from point R at times t_R^1 and t_R^2 according to a clock U_R at R and arrive at point S correspondingly at times t_S^1 and t_S^2 according to a clock U_S at S, then

$$t_S^1 - t_R^1 = t_S^2 - t_R^2. \tag{11.17}$$

Condition (ii): The time required for light to traverse a triangle is independent of the direction taken around the triangle (round-trip axiom).

Condition (i) is necessary, because if the clocks can be synchronized, (11.17) must hold in order that they remain synchronized (date-independence). Conversely, if (11.17) holds and the signal from B to A is the reflection of the signal from A to B, that is, $t'_B = t_B$ (as in Einstein's 1905 definition of simultaneity), then (11.16) reads

$$t_B = t_A + T \tag{11.18}$$

where $2\,T = t'_A - t_A$, which shows that the light signal arrives at B in half the time it takes for the round trip. Because by definition $2\,T = (t'_A - t_B) + (t_B - t_A)$ equation (11.17) shows that T does not depend on the date of the emission of the signal.

[33] A. L. Macdonald, "Clock synchronization, a universal light speed, and the terrestrial redshift experiment," *American Journal of Physics* **51**, 795–797 (1983).

[34] See figure 7.2 in chapter 7.

Assuming that all clocks have been thus synchronized with a clock at A, we now prove that the round-trip axiom (ii) is a necessary and sufficient condition that if clocks located at B and C are synchronized with a clock at A they are synchronized with each other. Assume a light signal, transmitted around the triangle ABC, passes A at local time t_A, B at local time t_B, C at local time t_C, and returns to A at local time t_D. Analogously let the corresponding times for a signal sent around in the opposite direction be t'_A, t'_B, t'_C, and t'_D. Obviously

$$t_D - t_A = (t_D - t_C) + (t_C - t_B) + (t_B - t_A) \tag{11.19}$$

$$t'_D - t'_A = (t'_D - t'_B) + (t'_B - t'_C) + (t'_C - t'_A). \tag{11.20}$$

But because the clock at A is synchronized with the clocks at B and C,

$$t_B - t_A = t'_D - t'_B \quad \text{and} \quad t_D - t_C = t'_C - t'_A. \tag{11.21}$$

Hence, the middle terms on the right-hand sides of (11.19) and (11.20) are equal if and only if the left-hand sides are equal. This means that the clocks at B and C are synchronized if and only if the round-trip condition (ii) is satisfied for the triangle ABC.

CHAPTER TWELVE

Arguments against the Conventionality Thesis

The thesis of the conventionality of intrasystemic distant simultaneity states that there are no logical or empirical reasons to prefer any particular value, or range of values, of the Reichenbach synchronization parameter ε, except, of course, the condition that it is confined (for causality reasons) to the open interval between 0 and 1. The thesis has been challenged by numerous philosophers and physicists, in fact, by so many that it would require a separate volume to give a comprehensive, let alone exhaustive, account of their arguments. We will therefore review only the historically most important or methodologically most significant examples and ignore technical details.

To refute the conventionality thesis it suffices to show that it is possible by a convention-free method, that is, by a method that, in particular, does not presuppose standard synchrony, to establish either (1) distant simultaneity or (2) synchronization of distant clocks or alternatively to measure (3) the one-way velocity of light or (4) that of any physical object of nonzero mass. Each of these methods, if successful, would single out a unique value of ε. The validity of the last two methods is because, in the notation of chapter 9 and equation (9.2), knowledge of the one-way velocity of light c_{AB} in the direction from point A to point B equals $c/2\,\varepsilon$ and c_{BA} in the opposite direction equals $c/[2(1 - \varepsilon)]$, where c denotes the round-trip light velocity, which is measurable by only a single clock. From equation (9.1) it also follows that the one-way velocity v_{AB} of any physical object equals $cv/[c + v(2\varepsilon - 1)]$ and v_{BA} in the opposite direction equals $c\,v/[c - v(2\varepsilon - 1)]$, where v is the velocity of the object measured in standard signal synchrony.[1] In the case $v = c$ the last two expressions are, of course, identical with the preceding two expressions.

A supposedly convention-free experimental verification of even only a qualitative anisotropy of the velocity of light, such as suggested in 1904 by Wilhelm Wien[2] and by Alfred Schweizer,[3] or with technically more advanced instrumentation, like electronically synchronized Kerr cells by Maurice Jacob[4] in 1927 or Alessandro Amerio[5] in 1947, would already be a serious challenge. It would not only invalidate standard signal synchrony, based on Einstein's "equal-time stipulation," it would also invalidate the conventionality thesis according to which any value of ε, including ½, should be acceptable. Formally, this is because, if the one-way velocity of light in one direction, say c_{AB}, differs from c_{BA}, their ratio and therefore also $\varepsilon/(1 - \varepsilon)$ would differ from 1. But this implies that $\varepsilon \neq \frac{1}{2}$ and excludes standard synchronization.

Reichenbach himself, shortly after having asserted the earlier-quoted "simultaneity–velocity circle," discussed an experimental proposal to establish convention-free distant simultaneity by the so-called "galvanometric method." This proposal was made by Friedrich Adler, formerly a fellow student with Einstein at the ETH in Zürich. It consists of a galvanometer connected at both ends via switches with an electric battery so that to close the

[1] The last two equations can be derived as follows. Let a light pulse and a particle be emitted (locally) simultaneously from A toward B. The difference between their arrival times at B, measurable by a local clock, must be the same whether calculated in standard or in nonstandard synchrony. Hence $(AB/c) - (AB/v) = (AB/c_{AB}) - (AB/v_{AB})$. Substituting $c_{AB} = c/2\varepsilon$ yields the above-mentioned value for v_{AB}. An analogous calculation yields the above-stated value for v_{BA}.

[2] W. Wien, "Über einen Versuch zur Entscheidung der Frage, ob sich der Lichtäther mit der Erde bewegt oder nicht," *Physikalische Zeitschrift* **5,** 585–586, 604–605 (1904).

[3] A. Schweizer, "Über die experimentelle Entscheidung der Frage, ob sich der Lichtäther mit der Erde bewegt oder nicht," *Physikalische Zeitschrift* **5,** 809–811 (1904).

[4] M. Jacob, "Procédé expérimental permettant de comparer, à un instant donné, la vitesse de la lumière dans un sense et dans le sense opposé," *Comptes Rendus de l'Académie des Sciences* **184,** 1432–1434 (1927).

[5] A. Amerio, "Un' esperienza sulla teoria della relatività," *Atti della Accademia Nazionale dei Lincei* **2,** 736–739 (1947).

circuit the two instants of momentarily closing the switches must be simultaneous. Adler therefore claims that a deflection of the galvanometer needle indicates distant simultaneity.[6] In his rebuttal Reichenbach pointed out that, if one describes the process not in terms of a primitive and only approximately correct theory of electric circles, as Adler did, but in terms of the full-fledged Maxwellian theory of the electromagnetic field, he will recognize that the "entire arrangement is really nothing but a disguised signalling process" and the arbitrariness in the determination of simultaneity is just as large as in the case of signal synchronization. Ultimately, the difference between Adler and Reichenbach is this: whereas Adler admits the physical reality of instantaneous actions at a distance, Reichenbach admits only actions at contact, which exclude the possibility of an infinitely fast transfer of an effect over any finite arbitrary distance.

The denial of actions at a distance, which were admitted in Newtonian physics, also denies the existence of perfectly rigid rods. The definition of perfect rigidity, as preservation of geometrical shape, implies that if one end of such a rod would be set into motion, the other end would instantaneously start moving as well, so that the rod could serve as a generator of actions at a distance. If perfectly rigid rods existed, the problem of distant synchronization could, of course, be solved simply by coupling clock mechanisms by such rods. In fact, numerous synchronization procedures, proposed to disprove the conventionality thesis, are but more or less disguised versions of such coupling proposals. Typical examples are the thought experiments called "shaft synchronizations."

One of the earliest proponents of shaft synchronization was the Swedish physicist Carl Benedicks, Director of Stockholm's Institute of Metallography. Although not denying the heuristic value of the theory of relativity, he felt confident that its results could be obtained also without posing, as he put it in his book *Rum och Tid*, "a challenge against the modes of thought applied in all previous physical reasoning."[7] The basic idea of his approach to the problems of space and time is what he called "the principle of the solid body,"[8] which in his view plays in "quantitative natural philosophy" the

[6] F. Adler, *Ortszeit, Systemzeit, Zonenzeit und das ausgezeichnete Bezugssystem der Elektrodynamik* (Vienna: Wiener Volksbuchhandlung, 1920), pp. 81–82.

[7] C. Benedicks, *Rum och Tid* (Stockholm: 1922); *Raum und Zeit* (Zürich: Füssli, 1923); *Space and Time* (London: Methuen, 1924; New York: Dutton, 1925), p. 79.

[8] Ibid. (1924), p. 9.

same role as "the principle of identity" in "exact scientific thinking." By means of the postulated immutability of solids he claimed to be able to construct geometry in a way similar to the program of the protophysicists like Hugo Dingler. Benedicks also used this principle to define distant simultaneity as follows: "We say that two distant clocks are synchronous, provided that their hands are moving as though their axles were connected by one rigid axle, consisting of an absolutely solid body." And he continued: "This definition of synchronism is precise, and has no ambiguity. It is founded only upon the fundamental basis of all measurement of time—the accepted unchangeability of the rotation process chosen as standard—and upon pure geometry—the fundamental basis of which is the existence of the absolute solid body"[9] and this method of synchronization "is independent of a supposed motion of the observer."[10]

Another widely discussed example of the shaft synchronization was proposed in 1938 by the mathematician Albert Eagle of the University of Manchester. Denoting by S'' a certain inertial system he declared: "The only correct method of synchronizing S'' clocks which I can see is as follows: two clock dials, at $x = 0$ and at $x = 1$, in planes normal to the x-axis, must be connected by a shaft and the hands set to exactly the same readings. These dials must then be driven by a master clock, at $x = 1/2$, driving the shaft at its exact centre through a gear-wheel or crank."[11] Even the mathematical corrections that he subsequently introduced in treating the shaft as a thin-walled cylinder or spindle of a given radius and given angular velocity, traveling with a given velocity in the direction of its length, and taking account of the Lorentz–FitzGerald length contraction, does not save his proposal from joining the long list of instrumental simultaneity constructions incompatible with the theory of relativity.

A similar proposal was made by Eugene Feenberg of Washington University, St. Louis, Missouri, who, in a paper, published in 1974, described several thought experiments in which a rotating shaft is used to define distant synchrony for a system of clocks along the shaft and declared: "In this context Einstein's definition of distant simultaneity is seen as based on a physi-

[9] Ibid., p. 26.

[10] Ibid., p. 27.

[11] A. Eagle, "A criticism of the special theory of relativity," *Philosophical Magazine* **26,** 410–414 (1938). "Note on synchronizing 'clocks' in a moving system by a connecting spindle," Ibid., **28,** 592–595 (1939).

cal assumption (and not merely on an overwhelmingly sensible choice in a range of conventional possibilities)."

Among a number of "shaft synchronizations" Feenberg described the following adaptation to the measurement of the one-way velocity of light of a well-known experiment performed by Hippolyte Louis Fizeau in 1849 to measure the average two-way velocity of light. "Two opaque disks, each containing one narrow radial slit, are mounted on the shaft at distance L apart. The slits are displaced in angle by φ radians. A pencil of light parallel to the axis of the shaft can pass through both slits if the angular velocity has the optimum value given by $\omega = \nu\, \varphi/L$. Notice that this structure transmits light in one direction, but not in the other."[12] Another contrivance proposed by Feenberg to measure the one-way velocity of light consisted of a long straight shaft with a circular cross section, mounted on frictionless bearings. After the shaft is set into rotation by a torque and all torsional vibrations have damped out, the shaft is supposed to be rotating uniformly with an angular velocity ω. Two counting devices, placed at points A and B alongside the shaft, are supposed to record the number of rotations. Their countings N_A and N_B are translated into clock readings by $T_A = 2\pi N_A/\omega$ and $T_B = 2\pi N_B/\omega$, respectively. "Thus the counters serve as clocks and, in fact, as synchronized clocks" contends Feenberg and adds that this shaft synchronization agrees with standard signal synchronization and that "with supplementary clocks at each position the possibility of a dependence of synchrony on the direction of rotation can be tested."

That Feenberg's shaft synchronizations were not independent of standard signal synchrony and hence did not refute the conventionality thesis was shown convincingly by Peter Øhrstrøm of the University of Aarhus in Denmark. Øhrstrøm[13] declared that every method by which the one-way velocity of light is supposedly measured depends, implicitly at least, on the method of standard signal synchronization, a statement that obviously affirms the validity of the conventionality thesis of distant simultaneity.

To strengthen his point Øhrstrøm also analyzed the much-discussed synchronization procedure proposed by Frank Jackson and Robert Pargetter[14] of

[12] E. Feenberg, "Conventionality in distant simultaneity," *Foundations of Physics* **4,** 121–126 (1974).

[13] P. Øhrstrøm, "Conventionality in distant simultaneity," *Foundations of Physics* **10,** 333–343 (1980).

[14] F. Jackson and R. Pargetter, "Relative simultaneity in the special relativity," *Philosophy of Science* **44,** 464–474 (1977).

La Trobe University, Australia. To synchronize clocks U_A and U_B stationary at points A and B, respectively, they constructed an axis XY perpendicular to AB through the midpoint C of AB. They then considered a rigid straight rod $A'B'$ with midpoint C' and length equal to the length of AB and declare: "Move $A'B'$ with uniform velocity such that C' travels along XY towards C, and $A'B'$ is perpendicular to XY (i.e., parallel to AB); then if the readings (as noted by an observer at A) on U_A just when A' coincides with A is the same as the reading (as noted by an observer at B) on U_B just when B' coincides with B, clocks U_A and U_B will be synchronous." As with respect to Feenberg's experiment, Øhrstrøm proved that, on closer analysis, the Jackson–Pargetter method also depends ultimately on standard signal synchronization.

Jackson and Pargetter were aware, of course, that a crucial point in their argument is the question of how to make sure that the rod $A'B'$ remains perpendicular to the XY line all through its motion. To ensure this perpendicularity they assumed that the rod is electrically conductive and carries at its midpoint a galvanometer, whereas its end points A' and B' glide along conductive wires of a uniformly increasing potential, and that this potential is equal at any two points that are connectable by a normal between the two lines. Hence, it was claimed, if the rod is perpendicular to the line XY and therefore parallel to AB, no current would be indicated by the galvanometer, whereas if the rod were tilted at some angle there would some current indicated by the galvanometer.

The Jackson–Pargetter proposal evoked numerous criticisms precisely with respect to this perpendicularity criterion. Thus Carlo Giannoni showed that "zero current does not in general imply the perpendicularity of the conducting rod, and *ipso facto* the conducting rod cannot be used to give an absolute synchronization to clocks."[15] Also Roberto Torretti rejected the Jackson–Pargetter synchronization proposal by showing that "the requirement that $A'B'$ remain at all times parallel to AB has no meaning unless the simultaneity of distant events is already defined in the rest-frame of AB. For as $A'B'$ travels toward AB, B' will sooner or later cross all the parallels to AB that are crossed by A'. But we shall not say that $A'B'$ remains parallel to AB unless B' crosses each such parallel *at the very time* that A' is crossing it," a statement which obviously presupposes a criterion of distant simultaneity.[16] As to the

[15] C. Giannoni, "Comment on 'Relative simultaneity in the special theory of relativity,' " *Philosophy of Science* **46,** 306–309 (1979).

[16] R. Torretti, "Jackson and Pargetter criterion of distant simultaneity," *Philosophy of Science* **46,** 302–305 (1979).

control device by an electrostatic field of constant nonzero gradient perpendicular to *AB*, Torretti pointed out that such an arrangement relies on the laws of electrodynamics in their standard form, based on standard signal synchrony, and that a differently defined coordinate time in the rest frame of *AB* would imply a deflection of the galvanometer even if $A'B'$ remains parallel to *AB* throughout its motion.

Note that such a relatively simple instrumental arrangement gave rise to profound and far-reaching considerations. Burke Townsend's 10-page critical examination of the Jackson–Pargetter proposal is a good illustration of this. We would have to go into too much detail even only to summarize Townsend's paper, which concludes with the statement that "the Jackson–Pargetter proposal is another in a now lengthy list of attempts to show that the thesis of the conventionality of distant simultaneity in the Special Theory of Relativity must run afoul of experimental facts."[17] Their proposal was also criticized severely by Herman Erlichson of the College of Staten Island, New York, because in testing for bending of the rod a light beam is sent from A' to B'. "This can be used as a test for bending in the frame $A'B'$ but not in the frame *AB*. To test for bending of $A'B'$ relative to the frame *AB* you would need previously synchronized clocks in the *AB* frame which are not available. Only with such clocks could the straightness of the rod *in the AB frame* be ascertained." Also their statement (made on p. 471) that the rod is traveling with "uniform velocity" presupposes the availability of synchronized clocks in the frame *AB*.[18]

Optical measurements have been quoted to refute the conventionality of distant simultaneity more frequently and perhaps more thought provokingly than these mechanical proposals. The earliest quantitative optical measurement in astronomy, Olaf Roemer's famous 1676 determination of the velocity of light,[19] has often been mentioned as a refutation of the conventionality thesis. Not involving any reflections and based on the use of only a single clock, an earth-bound chronometer, Roemer's method is apparently a convention-free measurement of the one-way velocity of light. It is not surprising, therefore, that Zdzislaw Augustynek, for example, in his "Critique of

[17] B. Townsend, "Jackson and Pargetter on distant simultaneity," *Philosophy of Science* **47,** 646–655 (1980).

[18] H. Erlichson, "The conventionality of synchronism," *American Journal of Physics* **53,** 53–55 (1985).

[19] See chapter 4, note 9.

the conventionalistic Interpretation of the Definition of Simultaneity," rejected Reichenbach's conventionality thesis because "Roemer's measurement does not deal with the round-trip of the light but with the propagation of light in *one direction*—from Jupiter to Earth."[20] In a similar vein G. Burniston Brown declared that "Roemer's . . . method nullifies Einstein's contention, repeated by Eddington and others, that we only know the out-and-return velocity, not the one-way velocity, so that the time of arrival of a signal at a distant point is never known from observation but can only be a convention."[21] A few years later Brown wrote: "One of Einstein's most extraordinary errors was the statement that we only know the mean two-way velocity of light (i.e., up to a mirror and back) so light might, for instance, go quicker one way than the other. The fact is, however, that we have known the one-way velocity of light since the very first time this velocity was measured, in 1676, by the Danish astronomer Roemer."[22]

In the sequel of this paper Brown declared that "there is no difficulty in synchronizing clocks at rest at a distance: a flash is sent at t_1 which returns at t_2; its interaction at the distant reflector was therefore at $(t_1 + t_2)/2$. This formula is not just a convention as Einstein tried to maintain: it is standard practice." These words prompted Hermann Erlichson to declare that "in order for Brown's statement to be correct one would have to have established, in advance of the synchronization, that the speed of light is the same in the forward and backward direction. Failing this knowledge, the setting of the distant clock at $(t_1 + t_2)/2$ is only a convention; i.e., the Einstein synchronization convention."[23] But this argument did not convince Brown and in his reply[24] he referred to a terrestrial model of Roemer's measurement of the velocity of light which uses only one clock stationed at rest with respect to the source of light.

Similarly, Louis Essen, the noted British expert on atomic clocks, emphasized repeatedly that Roemer's method makes it possible to measure "the velocity of light in a single direction."[25]

[20] Z. Augustynek, "Kritik der konventionalistischen Interpretation der Definition der Gleichzeitigkeit," *Naturwissenschaft und Philosophie* (Berlin: Akademie Verlag, 1960), pp. 177–182.

[21] G. B. Brown, "What is wrong with relativity?," *Bulletin of the Institute of Physics* **18,** 71–77 (1967).

[22] G. B. Brown, "Experiment versus thought-experiment," *American Journal of Physics* **44,** 801–802 (1976).

[23] H. Erlichson, "Einstein synchronization," *American Journal of Physics* **46,** 1017 (1978).

[24] G. B. Brown, "Reply by the author," Ibid., pp. 1071–1072.

[25] K. D. Froome and L. Essen, *The Velocity of Light* (London: Academic Press, 1969), p. 3.

Not surprisingly, the earliest critical analysis of Roemer's measurement from the point of view of the conventionality thesis was a 1925 paper by Reichenbach.[26] Reichenbach found it "strange that here a velocity is measured in *one* direction; for this would require a comparison of time indication at one extremity of the earth's orbit with that at the other extremity; but such a comparison is impossible unless simultaneity has been already defined." Asking himself what definition of simultaneity lies at the root of Roemer's method and noting that time is measured here solely by a clock attached to the earth, Reichenbach concluded that the definition of simultaneity in this case was that provided by transport of clocks. And because astronomical clocks are regulated in accordance with the rotation of the Earth, the Earth itself is the transported clock that transmits the synchronism.

It has rarely been noticed that in 1944 Scott-Iversen also analyzed Roemer's method by comparing its formulation in spherical and in elliptical coordinates and concluded that "Roemer's measurements and his method of calculation do not give the one-way velocity of light; they give only the there-and-back velocity."[27]

An important approach to clarify the problem is to view Roemer's method as an application of the Doppler effect. Textbooks on optics or essays on the measurement of the velocity of light rarely, if ever, emphasize the essential identity between Roemer's method and the Doppler effect. Exceptions are Arnold Sommerfeld's textbook *Optics*,[28] an essay by O. Costa de Beauregard,[29] and Mogens Pihl's book on Danish contributions to classical physics.[30] But many explanations of Roemer's determination of the velocity of light, like Max Born's in his book on relativity, are actually explanations of the Doppler effect.[31]

[26] H. Reichenbach, "Planetenuhr und die Einsteinsche Gleichzeitigkeit," *Zeitschrift für Physik* **33,** 628–634.

[27] Chapter 9, note 36, p. 111.

[28] A. Sommerfeld, *Optik–Vorlesungen über theoretische Physik* (Leipzig: Akademische Verlagsgesellschaft, 1943), vol. 4; *Optics—Lectures on Theoretical Physics* (New York: Academic Press, 1954), vol. 4, 60–63.

[29] O. Costa de Beauregard, "De la mesure de la vitesse de la lumière sur un parcours aller simple," *Bulletin Astronomique* **15,** 159–162 (1950).

[30] M. Pihl, *Betydningsfulde Danske Bidrag til den klassiske Fysik* (Copenhagen: Dag, 1972).

[31] M. Born, *Die Relativitätstheorie Einsteins und ihre physikalischen Grundlagen* (Berlin: Springer, 1920); *Einstein's Theory of Relativity* (London: Methuen, 1924; New York: Dover Publications, 1962, 1965).

Because an interpretation of Roemer's determination of the velocity of light as a measurement of a one-way velocity would be fatal to the conventionality thesis of distant simultaneity, we draw the interested reader's attention to some recent literature on this issue. In 1970 the Australian physicist Leo Karlov published a mathematical analysis of Roemer's procedure in defense of the conventionality thesis. He concluded "that the method of Roemer (as indeed *all* observational method of finding the speed of light) can only supply an objective average speed in a closed path."[32] Reza Mansouri and Roman U. Sexl criticized Karlov's work as incorrect because, in their view, Roemer determined "the one-way velocity of light . . . with the help of clocks synchronized by slow clock transport."[33] Karlov[34] later published a modified version of his argument, which, in turn, was censured by Louis Essen,[35] who claimed that "Roemer's was a straight forward measurement of the time taken by light to travel in one direction across the diameter of the Earth's orbit round the Sun. It seems to me that this is the fact and any attempt to interpret the result as giving a two-way value is an illusion." In 1991 three Yugoslavian physicists, Babović, Davidović, and Aničin, identified Roemer's determination of the speed of light as a Doppler effect which "can be well fitted and properly modified to suit the requests of a student laboratory curriculum" and mentioned the papers of Karlov and of Essen, but without stating who of them is right.[36]

In 1958, James H. Shea presented a profound analysis of Roemer's calculations that showed that "although the method Roemer conceived is unquestionably valid, his original and only paper on the subject left out much of the detail necessary to determine whether his measurements were adequate to the task of demonstrating the effect he claimed to have observed. . . . Mathematical analysis of the dynamics of the Earth/Jupiter synodic system allows a more thorough analysis of Roemer's work than has previously been made."[37]

[32] L. Karlov, "Does Roemer's method yield a unidirectional speed of light?," *Australian Journal of Physics* **23,** 243–258 (1970).

[33] R. Mansouri and R. U. Sexl, "A test theory of special relativity II: first order test," *Vienna Reports on Gravitation and Cosmology*, 1976; *General Relativity and Gravitation* **8,** 515–524 (1977).

[34] L. Karlov, "Fact and Illusion in the speed-of-light determinations of the Roemer type," *American Journal of Physics* **49,** 64–66 (1981).

[35] L. Essen, "Fact and Illusion in the speed-of-light determinations of the Roemer type," Ibid., 620 (1981).

[36] V. M. Babović, D. M. Davidović, and B. A. Aničin, "The Doppler interpretation of Roemer's method," *American Journal of Physics* **59,** 515–519 (1991).

[37] J. H. Shea, "Ole Roemer, the speed of light, the apparent period of Io, the Doppler effect, and the dynamics of Earth and Jupiter," *American Journal of Physics* **66,** 561–569 (1998).

After this historical introduction let us investigate whether Roemer's determination of the velocity of light does really not involve any convention as has been claimed. It is based on the observations of the emersions of Jupiter's satellite Io, that is, the moments when Io leaves the shadow of Jupiter. These emersion processes are regarded as emission processes of a constant frequency ν_0 which if, observed from the Earth at a changing distance L from Jupiter or Io, experiences a Doppler shift to ν. If T_0 denotes the proper period of the satellite's revolution around Jupiter (which is assumed to be stationary) and ΔL is the change in distance between the Earth (which recedes from, or approaches, Jupiter) and Jupiter during T_0, then the time interval between two consecutive emersions as observed from the Earth is $\Delta t = T_0 + (\Delta L/c)$, where c is the velocity of light to be measured. The time interval $\Delta_k t$ between k consecutive observed emersions is then given by $kT_0 + (\Delta_k L/c)$, where $\Delta_k L$ is the total change in L during $\Delta_k t$.

To find the unknowns T_0 and c we consider two boundary situations: (1) let m be the number of emersions observed in the time interval $\Delta_m t$ during which the Earth returns to its initial position so that $\Delta_m L = 0$; then $T_0 = \Delta_m t/m$; (2) let n be the number of emersions observed in the time interval $\Delta_n t$ during which the Earth passes from conjugation to opposition so that $\Delta_n L = D$, where D is the known diameter of the Earth's orbit, then $\Delta_n t = nT_0 + (D/c)$ or $c = D/(\Delta_n t - n\,T_0)$. Setting $\Delta_n t = n\,T$ so that T is the average time interval between consecutive emersions, and $D = \mathrm{v}\,\Delta_n t$ so that v is the average velocity of the Earth along D, we obtain $T = T_0/(1 - \mathrm{v}/c)$. Finally, introduction of the frequencies $\nu = 1/T$ and $\nu_0 = 1/T_0$ yields the well-known formula of the Doppler effect $\nu = \nu_0(1 - \mathrm{v}/c)$.

We thus have to conclude that contrary to cosmology where one applies the Doppler effect to measure the velocity of a light source, assuming that c is known, Roemer applied the formula to measure the one-way velocity of light, assuming that v is known. The Doppler effect transforms or reduces, so to speak, the measurement of the one-way velocity of light c to a measurement of the one-way velocity v of the observer which, like every one-way velocity, involves for its determination synchronized clocks and is therefore not convention-free as long as no convention-free method for clock synchronization has been proved to exist.

Apart from the astronomical (Roemer and Bradley) and terrestrial (Fizeau, Foucault, and Michelson) methods, since World War II, a third method of measuring the velocity of light or electromagnetic radiation has existed, which, like Roemer's, is based on a one-clock operation: the cavity microwave resonance method.[38] Essen,[39] Bol,[40] and others obtained results by its high precision. In this method one uses a metallic cavity of known dimensions that contains electromagnetic radiation; if the cavity's dimensions are equal to an integral number of half wavelengths a resonance state of standing waves is generated; knowing the frequency ν (a one-clock operation) of the enclosed radiation and measuring its wavelength yields the velocity without apparently any need of a conventional stipulation of distant simultaneity. It seems natural to argue, therefore, as Benjamin Liebowitz did in 1952, that this method makes it possible to investigate the isotropy of light propagation "in full conflict with $\varepsilon \neq 1/2$ and yet independent of stipulations of simultaneity."[41]

In his rebuttal of this argument Grünbaum pointed out that, although for nonstanding waves the very notion of wavelength, that is, the distance between simultaneous like phases, has, so to speak, a built-in simultaneity convention, for standing waves the length of a wave is independent of any simultaneity criterion. What invalidates Liebowitz's argument, however, is his "using the number ν of vibrations per unit of *local* time on the clock at the transmitter to compute the distance traversed per unit of one-way *transit* time. Even in the simple case of the formula $v = \nu\lambda$, the velocity of one-way transit can be obtained from the frequency based on the transmitter's clock, only because a prior choice of $\varepsilon = 1/2$ renders the *emission* of the tail of the last of ν waves *simultaneous* with the *arrival* of the head of the first of these ν waves at a point whose distance from the transmitting point is $\lambda\nu$. In other words, it is only because this particular pair of events is *decreed* to be simultaneous by the choice of $\varepsilon = 1/2$ that the arrival of the first wave can be said to occur one unit of *transit* time after that wave's own emission."[42]

Wesley C. Salmon also defended the conventionality thesis against Liebowitz's challenge on similar grounds. He called into question the tacit

[38] For details see, for example, R. P. Feynman, R. B. Leighton, and M. Sands, *The Feynman Lectures on Physics* (Reading, Massachusetts: Addison-Wesley, 1964), vol. 2.

[39] L. Essen and A. C. Gordon-Smith, "The velocity of propagation of electromagnetic waves derived from resonant frequencies of a cylindrical resonator," *Proceedings of the Royal Society A* **194,** 348–361 (1948).

[40] K. Bol, "A determination of the speed of light by resonant cavity method," *Physical Review* **80,** 298 (1950).

[41] B. Liebowitz, "Relativity of simultaneity in a single Galilean frame," *American Journal of Physics* **24,** 587–588 (1956).

[42] A. Grünbaum, "Relativity of simultaneity within a single Galilean frame: a rejoinder," *American Journal of Physics* **24,** 588–590 (1956).

assumption that the wavelength of cavity radiation is equal to the wavelength of one-way propagation of radiation of the same frequency through free space; and he argued that the assumed equality of the one-way velocities, in opposite directions, of the waves generating the pattern of standing waves can be verified only on the basis of a prior simultaneity convention.[43]

The last point was beautifully worked out by Shing-Fai Fung and K. C. Hsieh in their discussion of a one-dimensional cavity of length L. Supposing from the start that the waves, forming the standing-waves pattern, propagate inside the cavity in opposite directions with velocities c_+ and c_- and wavelengths λ_+ and λ_-, correspondingly, they showed that the wavelength λ, obtained by the measurement, is indeed an average wavelength given by $\lambda = 2(\lambda_+{}^{-1} + \lambda_-{}^{-1})^{-1} = 2L/n$, where n is an integer. They concluded that "the above argument precludes any possibility of determining the anisotropy in the speed of light by means of interference between a light wave and its reflection, thus giving support to the conventionality thesis." Nevertheless, they showed that on closer analysis certain difficulties arose. As stated at the beginning of this chapter, the one-way velocities of any object in opposite directions are ε dependent. Because there is no a priori reason to assume that ε has the same value for all directions, it is conceivable that two particles of equal mass and equal round-trip velocity will travel with different one-way velocities when traveling in different directions. Assume that their kinetic energies are converted into heat in a calorimeter. Because the temperature increase would not be the same, energy and temperature would depend on direction. These and other more technical arguments, involving two thought experiments, lead the authors to the conclusion that the "one-way speed of light can be measured and that Einstein's postulate can be tested."[44]

The invention of the maser and laser at about 1960 and the discovery of the Mössbauer effect made it possible to observe frequency shifts with great precision. Martin Ruderfer[45] and Christian Møller[46] were among the first to draw attention to the feasibility of utilizing the new techniques for the ex-

[43] W. C. Salmon, "The philosophical significance of the one-way speed of light," *Nous* **11,** 258–292 (1977).

[44] S.-F. Fung and K. C. Hsieh, "Is the isotropy of the speed of light a convention?," *American Journal of Physics* **48,** 654–657 (1980).

[45] M. Ruderfer, "First-order terrestrial ether drift experiment using the Mössbauer radiation," *Physical Review Letters* **5,** 191–192 (1960).

[46] C. Møller, "New experimental tests of the special principle of relativity," *Proceedings of the Royal Society of London A* **270,** 306–311 (1962).

perimental study of a possible anisotropy of light propagation. Similarly, commending the new techniques "for a direct test of Einstein's concept of simultaneity," Pascal M. Rapier of Richmond, California, proposed to synchronize locally two identical maser-controlled clocks (to an accuracy of 10^{-9} sec) before transporting one of them a distance of 18.8 miles. Assuming the velocity of the solar system in the universe to produce a first-order effect of v/c on the time required for light to pass in the two directions Rapier concluded that a null effect would experimentally confirm Einstein's postulate and "relative space would have been proved isotropic to the propagation of light" whereas a non-null effect would completely discredit the axiomatic foundations of the theory of relativity.[47] Mansouri and Sexl claimed that the experiment could be performed without transporting any clock by merely comparing two separated clocks, one in Europe and the other in the United States, with the help of radio signals, due to the effect of the terrestrial rotation on the directional sensitivity of the one-way velocity of light. "Two such clocks synchronized, for example, at noon will be out of synchronization by midnight if anisotropies are actually present."[48]

The preceding discussion does not exhaust the long list of interferometric, or more generally, optical or electromagnetic experiments proposed to measure the one-way velocity of light, or to detect, at least, an anisotropy of the velocity of light. Some physicists like Stefan Marinov[49] of Sofia, Bulgaria, claim to have found incontrovertible evidence for such an anisotropy. Astrophysicists like T. W. Cole[50] of Sydney, Australia, or C. Bare et al.[51] of Greenbank, West Virginia, believe that the so-called method of very-long-baseline interferometry could yield a positive result. T. Chang[52] of Peking claimed it possible to detect by a precision measurement of the angle of stellar aberration an anisotropy of the one-way velocity of light. But at the same time

[47] P. M. Rapier, "A proposed test of the constancy of the velocity of light," *Proceedings of the IRE* **49,** 1322 (1961).

[48] Note 33, pp. 519–520.

[49] S. Marinov, "The velocity of light is direction-dependent," *Czechoslovakian Journal of Physics B* **24,** 965–970 (1974); "A proposed experiment to measure the one-way velocity of light," *Journal of Physics A* **12,** L99–L101 (1979).

[50] T. W. Cole, "Astronomical tests for the presence of an ether," *Monthly Notices of the Royal Astronomical Society* **175,** 93P–96P (1970).

[51] C. Bare et al., "Interferometric experiment with independent local oscillators," *Science* **157,** 189–190 (1967).

[52] T. Chang, "A suggestion to detect the anisotropy of the one-way velocity of light," *Journal of Physics A* **13,** L207–L209 (1980).

K. Ruebenbauer[53] of the University of Krakow, Poland, quoted experimental data obtained by Mössbauer measurements as evidence for the isotropy of the light velocity. Moreover, T. Sjödin[54] of Brussels and M. F. Podlaha[55] of Munich claimed to have proved the impossibility of one-way light-velocity measurements. Even disregarding the nontechnical literature of that time, such as the books by Arthur S. Otis[56] or S. P. Gulati and S. Gulati[57] or the publications by Gotthard Barth,[58] we see that the conventionality thesis concerning distant simultaneity continued to be an issue of heated debate.

In 1977 Wesley C. Salmon published a very-well-written analysis of about a dozen thought experiments designed to ascertain the one-way velocity of light and thus to refute the conventionality thesis of distant simultaneity. He showed that none of them presented a convention-free method and concluded that "the evidence, thus far, favors those who have claimed that the one-way speed of light unavoidably involves a non-trivial conventional element."[59]

An interesting example of those proposals, not mentioned so far in our survey of such thought experiments, is the one reported by Salmon but originally conceived by Henry Hill of the University of Arizona and Jerry D. Long, Salmon's student.[60]

A beam of light is split by a half-silvered mirror at point A so that one part of it travels along an equilateral triangle ABC clockwise and the other half travels counterclockwise. Because according to the theory of relativity the round-trip speed of light over any closed route *in vacuo* is a constant, the two beams at their arrival back at A will not exhibit any interference fringes. Let us now place a piece of glass in the path between B and C so that the two beams travel through it in opposite directions and let us assume that the velocity of light traveling from B to C is not equal to that from C to B. Then,

[53] K. Ruebenbauer, "Isotropy of the velocity of light," *International Journal of Theoretical Physics* **19,** 217–219 (1980).

[54] T. Sjödin, "Synchronization in special relativity and related theories," *Nuovo Cimento* **51B,** 229–246 (1979).

[55] M. P. Podlaha, "On the impossibility to measure one-way velocity of light," *Lettere al Nuovo Cimento* **28,** 216–221 (1980). See also M. Podlaha, B. Althaus, and H. Baensch, "Zur Problematik der Geschwindigkeitsmessung," *Philosophia Naturalis* **16,** 315–317 (1977).

[56] A. S. Otis, *Light, Velocity and Relativity* (Yonkers-on-Hudson, New York: Burckel, 1863).

[57] S. P. Gulati and S. Gulati, *A Big Howler* (New Delhi: Delta Publications, 1982).

[58] G. Barth, *Die Relativitätstheorie* (Zwingendorf, Austria: 1985); *Licht aus den Atomen* (Gehrden, Germany: Raum und Zeit Verlag, 1985).

[59] Note 43.

[60] Ibid., pp. 280–281.

because the velocity of light in a refractive medium is proportional to that *in vacuo*, interference fringes should appear at A. Because no such fringes were detected in the performance of this experiment it was concluded that the velocity of light in opposite directions is the same, a conclusion that, of course, would refute the conventionality thesis.

Salmon showed that this conclusion is unwarranted for it "makes factual assumptions not warranted by the experimental evidence. When Foucault measured the speed of light through refractive media, he used essentially the same apparatus as he had used to measure the speed *in vacuo*. He therefore measured the round-trip speed of light in refractive media and compared it with the round-trip speed *in vacuo*. The result, carefully stated, is this: the ratio of the round-trip speed of light in a refractive medium is equal to the index of refraction. It does not follow from this statement that the same relation obtains between the respective one-way speeds. It was this latter statement, however, which was used in the analysis of the foregoing experiment."[61]

Apparently unaware of Salmon's paper, George Stolakis of the University of Warwick published a slightly modified version of the same experiment twelve years later. He imagined three collinear points ACB with C as the midpoint of the line AB, dividing AB into two equal segments each of length l. Two light signals are emitted simultaneously from C, one to A and the other to B through the same medium of refractive index n. After their reflection at A and B they return through a vacuum to C. If c_- denotes the velocity of light *in vacuo* in the direction from C to A and c_+ the velocity of light *in vacuo* in the opposite direction, then the duration of the whole journey along AC is $t_A = ln/c_- + l/c_+$ and along BC $t_B = ln/c_+ + l/c_-$. The time difference Δt of their arrival times at C is therefore given by $t_A - t_B$ which equals $l(n-1)(c_- - c_+)$. Because $l \neq 0$ and $n \neq 1$ the experimental result $\Delta t = 0$ obviously implies $c_+ = c_-$. Stolakis concludes therefore that any "spatial anisotropy of light propagation can be experimentally detected."[62]

In 1989 Robert K. Clifton presented a profound analysis of this problem and went further than Salmon by giving a nonstandard derivation of Snell's Law from Fermat's Principle, which obviously plays a key role in this argu-

[61] Loc. cit.

[62] G. Stolakis, "Against conventionalism in physics," *The British Journal for the Philosophy of Science* **37**, 229–232 (1986).

ment. Clifton showed that, in general, experiments "involving regions of space with varying refractive indices, cannot 'single out' any factual value of the Reichenbach–Grünbaum ε factor thus posing no threat to the conventionalist thesis."[63]

Like Salmon in his thought experiment, R. de Ritis and S. Guccione of the University of Naples also used an equilateral triangle *ABC* but supposed that at each of its vertices there is an optical device splitting a light ray into one returning along the direction of its emission and one proceeding to the next vertex. They assumed that at *A* at the *A* time t_A (i.e., at the time measured by a clock located at *A*) a ray leaves *A* toward *B*, where it arrives at *B* time t_B and after reflection arrives back at *A* at *A* time t_A'. Then $t_B = t_A + \varepsilon_1(t_A' - t_A)$ with $0 < \varepsilon_1 < 1$. At t_B at *B* time the unreflected part of the original ray leaves *B* for *C*, where it arrives at *C* time t_C, so that $t_C = t_B + \varepsilon_2\,(t_B' - t_B)$, with $0 < \varepsilon_2 < 1$, where t_B' is the *B* time of this ray's arrival back at *B*. Now, by transitivity of fulfilment of simultaneity criterion, clocks at *A* and *C* must also be synchronized. Hence $t_A'' = t_C + \varepsilon_3(t_C' - t_C)$ with $0 < \varepsilon_3 < 1$, where t_C' is the *C* time of the arrival at *C* of the ray reflected from *A*. Now, according to the round-trip principle $t_B' - t_B = t_C' - t_C = 2\,T$ and $t_A'' - t_A = 3\,T$, where $T = \frac{1}{2}(t_A' - t_A)$. Addition of the first three equations yields $t_A'' - t_A = 2\,T\,(\varepsilon_1 + \varepsilon_2 + \varepsilon_3)$. But $t_A'' - t_A = 3\,T$ implies $\varepsilon_1 + \varepsilon_2 + \varepsilon_3 = 3/2$. Evidently it is easy to find $\varepsilon-$ values (each between 0 and 1) that do not satisfy this equality. But, of course, it is satisfied if each ε equals $\frac{1}{2}$. De Ritis and Guccione claimed therefore to have refuted the conventionality thesis.[64] The critical reader will have noticed that their refutation rests on the transitivity of nonstandard synchronization which, as shown earlier, does not hold.

Clearly, the fourth equation is satisfied by standard simultaneity ($\varepsilon_1 = \varepsilon_2 = \varepsilon_3 = \frac{1}{2}$) but, as De Ritis and Guccione claimed, it falsifies the conventionality thesis, because it is easy to find ε_1, ε_2, and ε_3, each between 0 and 1, which do not satisfy the fourth equation.

Note that the argument, as presented by these authors, is based on the assumption that nonstandard synchronization is a transitive relation, which, as we know, does not hold. Another more profound reason to reject the validity of De Ritis and Guccione's argumentation was pointed out by Peter

[63] R. K. Clifton, "Some recent controversy over the possibility of experimentally determining isotropy in the speed of light," *Philosophy of Science* **56,** 688–696 (1989).

[64] D. De Ritis and S. Guccione, "Can Einstein's definition of simultaneity be considered a convention?," *General Relativity and Gravitation* **17,** 595–598 (1985).

Havas.[65] Although he agreed that the fourth equation is correct, for "light should take 3/2 as long to go round the triangle than back and forth along one side," the assumption of a complete arbitrariness in choosing the value of each ε is unwarranted. As Havas showed in a generally covariant formulation of the special theory of relativity, the synchronization parameters ε, in general, are direction dependent (or even position and time dependent), so that $\varepsilon = {}^1\!/_2(c^{-1}\, g_{io}n^i + 1)$, where g_{io} is a component of the metrical tensor g_{mn} and n^i is a unit vector that determines the direction. Hence, $\varepsilon_1(n_1^i) + \varepsilon_2(n_2^i) + \varepsilon_3(n_3^i) = {}^1\!/_2[c^{-1}\, g_{io}(n_1^i + n_2^i + n_3^i) + 3]$. Since the path is closed $n_1^i + n_2^i + n_3^i = 0$, which leads to the fourth equation. At the end of his paper Havas presented an example of a position- and direction-dependent definition of simultaneity and concludes that "there would be little point in trying to summarize a definition of simultaneity in the form $t_B = t_A + \varepsilon(t_A' - t_A)$, since its apparent simplicity would hide a time, position, and direction dependence much better summarized by the metric tensor" of the space–time under discussion.

In 1983 Burke Townsend[66] criticized the Fung and Hsieh[67] contention of the measurability of the one-way velocity of light and a similar argument proposed by Charles Nissim-Sabat,[68] which was inspired by a thought experiment suggested by P. Kolen and D. G. Torr.[69] He showed their contention to be incapable of invalidating the conventionality thesis of distant simultaneity.

That Townsend's confutation of the opponents of the conventionality thesis was fully justified not only for technical or purely physical reasons was contended in 1983 by Robert Weingard on the grounds that "a change from a standard to a nonstandard synchronization, relative to an inertial frame, is a change in the coordinate system we are using to coordinize space-time" and

[65] P. Havas, "Simultaneity, conventionalism, general covariance, and the special theory of relativity," *General Relativity and Gravitation* **19,** 435–453 (1987). For the mathematics involved see C. Møller, *The Theory of Relativity* (Oxford: Clarendon Press, 1951, 1972), § 89.

[66] B. Townsend, "The special theory of relativity and the one-way speed of light," *American Journal of Physics* **51,** 1092–1096 (1983).

[67] Note 44.

[68] C. Nissim-Sabat, "Can one measure the one way velocity of light?," *American Journal of Physics* **50,** 533–536 (1982). See also his later article "A Gedankenexperiment to measure the one way velocity of light," *The British Journal for the Philosophy of Science* **35,** 62–64 (1984) which was retracted three years later, ibid., **38,** 75 (1987).

[69] P. Kolen and D. G. Torr, "An experiment to measure the one-way velocity of propagation of electromagnetic radiation," *Foundations of Physics* **12,** 401–411 (1982).

because "the laws of physics are covariant and thus valid when written relative to any coordinate system."[70]

In an earlier essay Weingard offered a new argument for the thesis, announced in 1967 by Hilary Putnam[71] but conceived much earlier, that in a relativistic space–time, events, whether in the past, present, or future, are equally real. Although accepting the thesis Weingard rejects Putnam's argumentation for it on the grounds that it ignores the fact that in special relativity temporal relations between events at a point *P* and events outside of *P*'s lightcone "are conventional, depending not only on the frame of reference chosen but also on the *definition of simultaneity adopted*."[72] It is precisely this convention, combined with the contention that all things or events that exist now are real and that if an event can be considered real it must be real, and the conclusion that all events outside the lightcone of "me-now" are real, on which Weingard based his proof.

Weingard's association of the notion of simultaneity with that of reality prompted Vesselin Petkov of the Institute of Philosophy in Sofia to combine a refutation of Stolakis's above-mentioned experiment[73] against the conventionality thesis with a philosophical chapter entitled "On the essence of the conventionality of simultaneity."[74]

The declared purpose of this chapter is to demonstrate that (1) the simultaneity of distant events must be a matter of convention and (2) "to elucidate the essence of this convention." To combine these two issues Petkov identified the concept of simultaneity with the concept of *simultaneity at the present moment of time*. In prerelativistic theories of time only the present exists. Because the present is the set of simultaneous events that occur at the present moment, the simultaneity of events is an objective fact. "Therefore, according to the classical view of reality, simultaneous events at the present moment of time (i.e. present events) are *objectively* privileged in comparison with past and future events (since only present events are considered as existing), which means that simultaneity is not conventional." According to

[70] R. Weingard, "Remark on 'The special theory of relativity and the one-way speed of light,' " *American Journal of Physics* **53,** 492 (1985).

[71] H. Putnam, "Time and physical geometry," *Journal of Philosophy* **64,** 240–247 (1967).

[72] R. Weingard, "Relativity and the reality of past and future events," *The British Journal for the Philosophy of Science* **23,** 119–121 (1972).

[73] Note 62.

[74] V. Petkov, "Simultaneity, conventionality and existence," *The British Journal for the Philosophy of Science* **40,** 69–76 (1989).

the theory of relativity, in contrast, the fact that observers in relative motion assign simultaneity to different classes of events shows that not a single class of events is objectively privileged. Petkov thus arrived at the conclusion that "all events are equally real, which means that it really is a question of convention which events should be considered as *simultaneous* for a given observer."[75]

That the relativity of simultaneity implies the reality of all events, whenever they occur, whether in the past, the present, or the future, has been argued already by C. W. Rietdijk,[76] H. Putnam,[77] and others. For if a given event is in the present for one observer and hence real for him, the fact that according to the relativity of simultaneity it is in the past or in the future for another observer cannot deprive it of its ontological status of being real.

These arguments support the idea of what Francis Herbert Bradley called a "block universe" in which past, present, and future events possess the same degree of reality and which was described most vividly by Hermann Weyl: "The objective world simply *is*, it does not *happen*. Only to the gaze of my consciousness, crawling upward along the life-line of my body, does a section of the world come to life as a fleeting image in space which continuously changes in time."[78] Even Einstein was interpreted as having sympathized with this idea. For example, when informed of the death of his lifelong friend Michele Angelo Besso he wrote on 21 March 1955—almost exactly one year before his own decease—in a condolence letter to Besso's widow: "Now he has gone a little ahead of me in departing from this quaint world. This means nothing. For us faithful physicists, the separation between past, present and future has only the meaning of an illusion, though a persistent one."[79]

[75] Loc. cit., p. 74.

[76] C. W. Rietdijk, "A rigorous proof of determinism derived from the special theory of relativity," *Philosophy of Science* **33,** 341–344 (1966).

[77] H. Putnam, "Time and physical geometry," *The Journal of Philosophy* **64,** 240–247 (1967); also in H. Putnam, *Mathematics, Matter and Method* (Cambridge: Cambridge University Press, 1975), pp. 198–205.

[78] H. Weyl, *Philosophy of Mathematics and Natural Science* (Princeton: Princeton University Press, 1949), p. 116.

[79] "Nun ist er mir auch mit dem Abschied von dieser sonderbaren Welt ein wenig vorausgegangen. Dies bedeutet nichts. Für uns gläubige Physiker hat die Scheidung zwischen Vergangenheit, Gegenwart und Zukunft nur die Bedeutung einer wenn auch hartnäckigen Illusion." Einstein Archive, reel 7-245. See also *Albert Einstein–Michele Besso Correspondence 1903–1955* (Paris: Hermann, 1972), pp. 537–538.

CHAPTER THIRTEEN

Clock Transport Synchrony

The use of the transport of clocks for the establishment of distant simultaneity is as old as the invention of portable timepieces. As we saw in chapter 4, transported clocks had been used in the sixteenth century, for example, by the Flemish cartographer Reinerus Gemma, for the determination of the geographic longitude of a certain location. Still in 1904 Henri Poincaré, in his *The Value of Science*,[1] discussed in detail the use of transported clocks for the determination of distant simultaneity and geographical longitude without making any reference to the relativistic retardation involved.

In prerelativistic physics, in fact, the transport of clocks poses no problem because according to classical physics motion does not affect the time indication of a clock. In relativistic physics, however, the so-called time dilation, an experimentally well-confirmed effect, impairs the use of moving clocks for the establishment of distant simultaneity unless this time dilation is taken into account. The simplest way to confirm this effect is to synchronize locally two clocks, say U_1 and U_2, and to move one of them, say U_2, away from

[1]H. Poincaré, *La Valeur de Science* (Paris: Flammarion, 1904); *The Value of Science* (New York: Dover Publications, 1958), p. 35.

its partner and after some time to bring them together again. It will be seen that the two clocks are no longer in synchrony. The well-known "twin paradox" is a famous example of this effect. Only obstinate opponents of the special theory of relativity, like Hugo Dingler,[2] who claimed that motion cannot affect the rate of a clock, saw no problem in the transport of clocks for the establishment of distant synchronization.

The retardation of a moving clock U relative to a clock U_B, which is at rest in an inertial system S, that is, the amount of time by which U, after leaving U_B, lags behind U_B when it meets again with U_B, had already been calculated by Einstein in § 4 of his 1905 relativity paper.

Because of the importance of this effect for our present discussion a presentation of its mathematical derivation is not out of place. Let U_A be a clock located at the origin A of an inertial system S and U_B a clock located at the point B on the x axis of S and synchronized with U_A, and let d denote the distance between these two clocks. Then in S a third clock U, which leaves A at the time $t = 0$ and moves with constant velocity v from A to B, reaches U_B at the time

$$t = d/v. \tag{13.1}$$

In the inertial system S', in which U is at rest and which is in standard configuration with S, the time required by U to reach B is according to the Lorentz transformation given by

$$t' = (1 - v^2/c^2)^{-1/2}\,(t - vd/c^2). \tag{13.2}$$

By substitution of (13.1) in (13.2) we obtain

$$t' = t(1 - v^2/c^2)^{1/2} \tag{13.3}$$

Hence,

$$t' < t, \tag{13.4}$$

which means that the arrival time as indicated by U is less than the arrival time as indicated by U_B and the difference or retardation of U is given by the expression

$$t - t' = t - t(1 - v^2/c^2)^{1/2} = t[1 - (1 - v^2/c^2)^{1/2}]. \tag{13.5}$$

[2] H. Dingler, "Kritische Bemerkungen zu den Grundlagen der Relativitätstheorie," *Physikalische Zeitschrift* **21,** 668–675 (1920).

Neglecting magnitudes of fourth and higher order we conclude that the moving clock lags behind the stationary clock by $\frac{1}{2}(v^2/c^2)$ seconds per second.[3]

As this mathematical analysis reveals, the retardation of a moving clock can be reduced arbitrarily by diminishing sufficiently the velocity v of the moving clock. It was for this reason that Joseph Winternitz, a former student of Philipp Frank at the University of Prague, in his critique of Dingler's rejection of the theory of relativity pointed out that clock transport can serve as a method of synchronization because "one can arbitrarily decrease the retardation effect by sufficiently diminishing the velocity of the clock."[4]

At the same time also Eddington declared in his previously mentioned treatise[5] that there are two equivalent methods of establishing distant simultaneity: "(1) A clock moved with infinitesimal velocity from one place to another" and (2) "the forward velocity of light along any line is equal to backward velocity." "Neither statement," Eddington continued, "is by itself a statement of observable fact, nor does it refer to any intrinsic property of clocks or light; it is simply an announcement of the rule by which we propose to extend fictious time partitions through the world. But the mutual agreement

[3]"If at the points *A* and *B* of *S* there are stationary clocks which, viewed in the stationary system, are synchronous; and if the clock at *A* is moved with the velocity v along the line *AB* to *B*, then at its arrival at *B* the two clocks no longer synchronize, but the clock moved from *A* to *B* lags behind the other which has remained at *B* by $\frac{1}{2}tv^2/c^2$ (up to magnitudes of fourth and higher order), t being the time occupied in the journey from *A* to *B*. It is at once apparent that this result still holds good if the clock moves from *A* to *B* in any polygonal line, and also when the points *A* and *B* coincide. If we assume that the result proved for a polygonal line is also valid for a continuously curved line, we arrive at this result. If one of two synchronous clocks at *A* is moved in a closed curve with constant velocity until it returns to *A*, the journey lasting t seconds, then by the clock which has remained at rest the traveled clock at its arrival at *A* will be $\frac{1}{2}tv^2/c^2$ seconds slow." See chapter 7, note 6 (1905), pp. 904–905. Although Einstein called this result a "peculiar consequence" (eine eigentümliche Konsequence; p. 904), he did not regard it as a paradox. It was called a paradox only after P. Langevin, in his essay "L'évolution de l'espace at du temps," *Scientia* **10,** 31–45, illustrated it by replacing the clocks by two twins. Since then it has been called "the twin paradox" or "the relativistic clock paradox" and became the topic of countless disputes; even today opinions still differ as to whether general relativity is required for a satisfactory solution. A detailed bibliography of this effect can be found in H. Arzeliès, *Relativistic Kinematics* (Oxford Pergamon Press, 1966), pp. 191–195, or in L. Marder, *Time and the Space-Traveller* (London: George Allen & Unwin, 1971), pp. 185–200. For a historically interesting essay on this topic, with special reference to Herbert Dingle's use of the clock paradox as an argument against the validity of the special theory of relativity, see Hasok Chang, "A misunderstood rebellion—The twin paradox and Herbert Dingle's vision of science," *Studies in the History and Philosophy of Science* **24,** 741–790 (1993).

[4]J. Winternitz, *Relativitätstheorie und Erkennislehre* (Leipzig: Teubner, 1923), p. 85.

[5]Chapter 9, note 3, p. 29.

of the two statements is a fact which could be tested by observation, though owing to the obvious practical difficulties it has not been possible to verify it directly."

Neither Winternitz nor Eddington seems to have noticed that by referring to v as a one-way velocity they involve themselves in a logical circle, for the determination of such a velocity, as Reichenbach at the same time clearly recognized, presupposes the synchronization of distant clocks. The French philosopher-physicist André Metz, who was cursorily mentioned earlier,[6] and who published a book[7] on the theory of relativity, in which he systematically rebutted misconceptions or misrepresentations of this theory, seems to have been aware of this logical flaw. Searching for a definition of distant simultaneity that resembles that of classical physics but nevertheless agrees with Einstein's synchrony he defined what he called the "rapidity" (rapidité) of a moving clock as the ratio between the distance traversed and the proper time, the time indicated by the moving clock itself. He then proposed to synchronize two distant clocks, say, U_A at point A and U_B at point B, by means of a third clock U', which is first synchronized with U_A, then transported from A to B, where U_B is then synchronized with it, but with the proviso that the rapidity of U' tends to zero.[8]

Metz proved the equivalence of this definition with Einstein's definition of simultaneity by showing that the difference between the two readings of the clocks at A and B is given by $dr/[(l + r^2/c^2)^{1/2} - 1]$ or, neglecting higher powers, by $d\,r/2\,c^2$, where d denotes the distance between A and B and r is the rapidity of the transported clock. Clearly, this expression becomes zero if r tends to zero.

On 15 June 1925, Emile Borel presented Metz's new definition of distant simultaneity to the Paris Academy of Sciences in these terms: "The time at a point B is synchronous, relative to a reference frame S, with the time of a clock at point A, if it is the time indicated at B by a transported auxiliary clock which has initially been synchronized with a clock at A, provided the rapidity of the transported clock tends to zero. *Simultaneity* relative to S is the agreement in the readings of clocks thus synchronized relative to S."[9]

Thus twenty years, almost to the day, after Einstein had published his light-signal definition of distant simultaneity in June 1905, a different syn-

[6]Chapter 8, note 25.

[7]A. Metz, *La Relativité* (Paris: Chiron, 1923).

[8]A. Metz, "Une définition relativiste de la simultanéité," *Comptes Rendus de l'Académie des Sciences* (Paris) **180,** 1827–1829 (1925).

[9]Ibid., p. 1828.

chronization method had been shown to exist consistently and to yield identical results. All earlier proposals of slow clock transport synchronizations were conceptually flawed by using, as explained previously, the concept of a one-way velocity the definition which presupposes distant simultaneity. Nor can the ideas of the Dieppe mathematical physicist Ernest Maurice Lémeray, who six months after the publication of Metz's paper claimed to have anticipated its contents by several years, be said to have provided a consistently elaborated new method of synchronization. The optical method used by him was admittedly merely a variation of Einstein's signal synchronization.[10]

Metz's notion of rapidity and similar concepts, like Evander Bradley McGilvary's notion of "heterogeneous velocity,"[11] were not widely accepted. Relating a quantity (the distance) measured in one inertial reference system to a quantity (the time interval) measured in another inertial reference system, the system in which the measuring clock is at rest, they were often regarded almost contemptuously as "hybrids" or even as "bastard" conceptions.

It was only about 1948 that Herbert Eugene Ives, the Philadelphia-born physicist and ingenious inventor of numerous optical and electronic devices, made use of them again, independently of Metz or McGilvary, when he conceived the idea of a "one-clock one-way velocity" or as he called it a "self-measured velocity," which is defined as the ratio between the distance traversed by a body and the time interval measured by a clock comoving with, or carried by, the moving body.

Ives himself says: "This is one of the commonest ways of measuring velocity. It is used by the mariner with his chronometer, by the automobilist in traversing the 'measured mile,' and by the train traveller who counts telegraph poles in the interval given by the watch in his hand."[12]

Despite the fact that the famous canal-ray experiment, which Ives performed in 1938 with his colleague George R. Stillwell at Bell Telephone Lab-

[10] M. Lémeray, "Sur une définition relativiste de la simultanéité," *Comptes Rendus de l'Académie des Sciences* (Paris) **81,** 770–772 (1925).

[11] E. B. McGilvary, "Newtonian time and Einsteinian times," *Proceedings of the Sixth International Congress of Philosophy,* Harvard University, September 13–17 (1926); (New York: Longmans, Green and Co., 1927), pp. 47–53. See also his *Times, New and Old* (Berkeley, California: University of California Publications, vol. 6, no. 5 (1928).

[12] H. E. Ives, "The measurement of the velocity of light by signals sent in one direction," *Journal of the Optical Society of America* **38,** 879–884 (1948); reprinted in D. Turner and R. Hazelett, *The Einstein Myth and the Ives Papers* (Old Greenwich, Connecticut: Devin-Adair Company, 1979), pp. 136–141.

oratories, New York, is still one of the most direct empirical proofs of the relativistic slowdown of moving clocks, Ives interpreted it as a verification of the Larmor–Lorentz ether theory and remained a vigorous antirelativist until the end of his life. He criticized Einstein's definition of distant simultaneity as a "pseudo operational procedure," since "the assignment of a definite value to an unknown velocity, by fiat, without recourse to measuring instruments, is not a true physical operation, it is more properly described as a ritual."[13]

Ives was not interested in employing his "self-measured velocity" for the formulation of a new criterion of distant simultaneity. He also regarded the idea of a slow clock transport with a velocity approaching zero as self-defeating, for a clock "transported" with zero velocity would never arrive at its destination, "since an infinite time would be required."[14] It was Bridgman,[15] Ives paragon of scientific methodology, who now employed Ives's "self-measured velocity" to obtain a clock transport synchronization procedure without recourse to an asymptotical zero velocity which for Ives was an impasse on this road.

Bridgman's operationalism became the ultimate source also for a philosophically interesting and widely discussed clock transport synchronization proposed in 1967 by Brian Ellis and Peter Bowman, then of the University of Melbourne, Australia. As a matter of fact, their work had been prompted by that of Bridgman on two counts. First, their slow clock synchronization procedure is basically a modification of Bridgman's approach, but also the feature by which it differs from the latter, namely the replacement of "self-measured velocities" by what Ellis and Bowman call "intervening velocities," can be traced back to Bridgman.

To understand this connection we must briefly digress into the history of modern psychology. It is well known that the school of behaviorism, founded by John B. Watson in the early twentieth century, confined itself to the study of behavioral responses as functions of environmental stimuli without recourse to the notions of mind and consciousness. One of its leaders was Edward Chase Tolman of the University of California, who deliberately used to call himself an "operational behaviorist," "to indicate a certain general positivistic attitude now being taken by many modern physicists and philoso-

[13] H. E. Ives, "Revisions of the Lorentz transformations," *Proceedings of the American Philosophical Society* **95,** 125–131 (1951); reprinted in D. Turner and R. Hazlett, op. cit., pp. 154–160.

[14] H. E. Ives, "Extrapolation from the Michelson-Morley experiment," *Journal of the Optical Society of America* **40,** 185–191 (1950); reprinted in D. Turner and R. Hazlett, op. cit., pp. 147–153.

[15] See chapter 10, note 15.

phers and for which Professor Bridgman has selected this word 'operational.' "[16] Following Bridgman in defining concepts "in such a manner that they can be stated and tested in terms of concrete repeatable operations by independent observers," Tolman found it necessary to introduce logical constructs, which he called "intervening variables," for theorizing about the relations between stimuli and resulting behavior.

To cut a long story short, we will review how two students of Tolman, Kenneth MacCorquodale and Paul E. Meehl, characterized "intervening variables"[17] in a paper that attracted the attention of Ellis and Bowman. They distinguished between "hypothetical constructs," which involve the supposition of unobservable entities or processes, and what they called "intervening variables," which are quantities obtained by a specific manipulation of empirical variables and do not involve any hypothesis about the existence of unobserved entities or processes and no words that are not definable either explicitly or by reduction sentences in terms of empirical variables.

To understand how Ellis and Bowman[18] applied such "intervening variables" for the definition of distant simultaneity, consider the following physical scenario. U_A and U_B are two clocks at rest at points A and B and in standard synchrony in an inertial system S. A clock V, initially synchronized with U_A at A, leaves A at U_A – time t_A through the distance s for B with velocity v, determined by measurements made on U_A and U_B. When V arrives at B at B time t_B it indicates the time T_B. According to the relativistic time dilatation

$$T_B - t_A = (t_B - t_A)(1 - v^2/c^2)^{1/2} = (t_B - t_A)(1 - \tfrac{1}{2}v^2/c^2). \tag{13.6}$$

The retardation of V relative to U_B is therefore

$$t_B - T_B = \tfrac{1}{2}(t_B - t_A)v^2/c^2 = v\ s/2\ c^2. \tag{13.7}$$

So far Ellis and Bowman had summarized only familiar results of the special theory of relativity.

[16] E. C. Tolman, "Operational behaviorism and current trends in psychology," *Proceedings of the 25th Anniversary Celebration of the Inauguration of Graduate Studies at the University of Southern California* (Los Angeles, California: University of Southern California, 1936), pp. 89–103; reprinted in E. C. Tolman, *Collected Papers in Psychology* (Berkeley, California: University of California Press, 1951), pp. 115–129.

[17] K. MacCorquodale and P. E. Meehl, "Hypothetical constructs and intervening variables," *Psychological Review* **55,** 1–41 (1948); reprinted in H. Feigl and M. Brodbeck, *Readings in the Philosophy of Science* (New York: Appleton-Century-Croft, 1953), pp. 596–611.

[18] B. Ellis and P. Bowman, "Conventionality in distant simultaneity," *Philosophy of Science* **34,** 116–136 (1967).

They now assumed that an additional clock U'_B is at rest at B and *set arbitrarily*. It indicates the time t'_B at the arrival of V at B. The "intervening velocity" v' is defined by the ration $s/(t'_B - t_A)$, an expression that involves time readings of different clocks. Because all clocks are assumed to run isochronically (with the same rate), however, $t'_B - t_B = k$, a constant depending on the setting of U'_B. As $v = s/(t_B - t_A) = s/(t_B - t_A - k)$ and $v' = s/(t'_B - t_A)$ it follows that

$$v = v'/[1 + (k\, v'/s)]. \tag{13.8}$$

This equation shows that if v' tends to zero, v also tends to zero. Hence, even if k is completely unknown we can determine a limit as v approaches zero by determining the limit as v' approaches zero. Now, we know from equation (13.7) that $\lim_{v\to 0}(t_B - T_B) = 0$. Hence, we also know that $\lim_{v'\to 0}(t_B - T_B) = 0$. But $t_B = t'_B - k$. Therefore

$$\lim_{v'\to 0}\,(t'_B - T_B) = k. \tag{13.9}$$

> Hence, to synchronize clocks by slow transport, we may adopt the following procedure: Take two clocks U_A and U'_B at the places A and B, and given them any arbitrary settings. Synchronize a clock V locally with U_A and move it with the intervening "velocity" v' from A to B, where v' is determined by measurements made on U_A and U'_B. Repeat this procedure several times using different transport "velocities," and extrapolate to find the limit, as v' approaches zero, of the difference between the reading on the clock U'_B and that on the clock V on its arrival at B. Suppose that this limiting difference is found to be k. Set the clock U'_B back by the amount k. The clocks U_A and U'_B will then be in slow transport synchrony in S.

Ellis and Bowman pointed out that their synchronization procedure does not involve the circularity noted by Reichenbach and therefore falsifies his statement that "to determine the simultaneity of distant events we need to know a velocity." They summarized their paper on slow transport synchronization with the remark that "if the empirical predictions of the Special Theory regarding clock transport are correct, a slow transport definition of simultaneity can be constructed that is logically independent of any signal definition, but is in fact equivalent to standard signal definition."[19] Note that, in contrast to Bridgman, who did not regard his clock transport synchronization, which as he showed agrees with the standard signal synchronization, as a refutation

[19] Op. cit., p. 131.

of the conventionality thesis, Ellis and Bowman claimed that their synchronization method renders the concept of simultaneity nonconventional.

The approaches of Bridgman and Ellis and Bowman were not the only slow clock-transport synchronizations proposed,[20] but the Ellis–Bowman proposal, probably because of its almost provocative anticonventionalist tenor, attracted more than any other proposed of its kind the attention of many philosophers of science. Among them, of course, were the proponents of conventionalism like Adolf Grünbaum, Wesley C. Salmon, Bas C. van Fraassen, and Allen I. Janis, who devoted to it a special panel discussion.[21] Its purpose was to prove that slow clock transport, although fully qualified to establish distant simultaneity, is not less nontrivially conventional than Einstein's standard signal synchrony.

We would have to stray too far from our main subject, the concept of simultaneity, to review in detail the epistemological issues raised in this panel discussion, such as, for example, Grünbaum's contention that the very choice of a time metric for the assignment of the measure of an interval to pairs of time coordinate already involves a conventional ingredient. Let it suffice to mention in brief how Ellis summarized this debate. He argued that according to Grünbaum, Salmon, and van Fraassen

> There is no absolute distinction between conventional and empirical statements in science. For they all agree that it is often a matter of convention whether a given statement is conventional. As I understand it, a statement occurring in a theory is conventional if and only if there exists an empirically equivalent theory in which a contrary of that statement occurs, and in which the basic semantic commitments of the terms used in making that statement are preserved. In this sense the statement that the one-way velocity of light is constant and independent of direction is conventional, and hence we may say that distant simultaneity is conventional. But having agreed that distant simultaneity is conventional in this sense, it does not follow that we cannot have good physical reasons for preferring some criteria for distant simultaneity to others.

It is with respect to this point, in particular, that Grünbaum and his colleagues differed from Ellis and Bowman, who contend that "there are indeed

[20] See, for example, S. J. Prokhovnik, "Slow transport as a criterion for synchronizing clocks," *Journal and Proceedings, Royal Society of New South Wales* **106,** 111–114 (1973).

[21] "A panel discussion of simultaneity by slow clock transport in the special and general theory of relativity," *Philosophy of Science* **36,** 1–81 (1969).

good physical reasons for preferring some criteria for distant simultaneity to others."[22]

Wesley Charles Salmon of Indiana University at Bloomington, who participated at the panel discussion while on a visiting professorship at Grünbaum's department in Pittsburgh, presented a new method of synchronizing spatially separated clocks to the panelists. It offered a unique determination of distant simultaneity by clock transport but without the need of using concepts like "infinitely slow transport," "self-measured," or "intervening velocities." As usual, let U_A and U_B denote clocks stationed at points A and B, respectively, in an inertial system and let each of a series of clocks C_1, C_1', C_1'', . . . be synchronized with U_A when leaving A at T_1 in inertial motion toward U_B. At B each of these clocks meets its mate C_2, C_2', C_2'', . . . moving inertially toward A, and is synchronized with it at B. U_B is set arbitrarily to record the arrival times of the clocks C_1, C_1', . . . *without any assumption about its synchrony with any other clocks*. Let T_3 denote the common arrival time at A, as measured by U_A, of the clocks C_2, C_2', . . . so that all clocks make the round-trip at the same average speed. Denote the retardations which C_2, C_2', . . . show with respect to U_A by R, R', R'' . . . , which differ from each other. "We can then take the event of the arrival at B of the pair which minimizes the total retardation for the round-trip at a fixed average speed as the event at B which is to be simultaneous with the midpoint T_2 between the departure and return times at A, i.e. $T_2 = \frac{1}{2}(T_1 + T_3)$, and reset U_3 accordingly. This definition of simultaneity is tantamount to the stipulation that the one-way velocities of the pair of clocks with minimal retardation are equal in magnitude."[23]

Salmon also pointed out that, if the factual assertions of the special theory of relativity are correct, his synchronization method is identical with the standard signal synchronization and with the slow transport synchronization of either the Bridgman or the Ellis–Bowman variety.

Furthermore, at this panel discussion, Salmon presented a criterion for the nontriviality of conventions and showed, as he later summarized it, that

> This conventional element in clock transport synchrony does not depend upon the fact that synchrony can be destroyed by relative motion of clocks. Even if, contrary to fact, the clock is transported from A to B and back to A were always

[22] B. Ellis, "On conventionality and simultaneity—a reply," *Australian Journal of Philosophy* **49,** 177–204 (1971).

[23] W. C. Salmon, "The conventionality of simultaneity," *Philosophy of Science* **36,** 44–63 (1969).

> in agreement with the clock that remained at A, clock transport synchrony would still involve the same kind of conventionality. Consequently, the fact that the retardation can be made arbitrarily small by transporting clocks slowly enough has no bearing upon the conventionality of distant simultaneity. The crucial basis for the conventionality of simultaneity in special relativity is not the time dilation phenomenon, but rather, the limiting character of the speed of light. This feature of special relativity constitutes one of its most fundamental departures from classical mechanics. If distant simultaneity is conventional, it is so because of a pervasive fact about the physical world, namely, that light is a first signal.[24]

The Ellis and Bowman paper also elicited numerous other responses of far-reaching importance. Thus, Grünbaum[25] argued that the very choice of a time metric for the assignment of the measure of an interval to pairs of time coordinates involves a conventional ingredient. Salmon, in addition to his above-mentioned contributions, also referred to Winnie's ε-Lorentz transformation, which we will soon discuss, and claimed that it implies the equivalence between slow clock transport and standard signal synchrony and that, therefore, "Roemer's method does not constitute an independent method for ascertaining the one-way speed of light within the special theory. It shows that, whatever value we assign to ε, slow clock transport synchrony must agree with standard signal synchrony."[26]

Winnie demonstrated the compatibility of slow clock-transport synchrony with any synchrony choice and argued therefore that it cannot be used to single out any particular choice of synchronization as the correct one.[27] Friedman showed that *if* the necessary condition "for something being a representation of physical time is that it agree with slow transport simultaneity," then the conventionalist is refuted only "if he concedes that this requirement—agreement with slow-transport simultaneity—is not itself conventional. And the conventionalist does not have (nor does he in fact) concede this."[28]

[24] W. C. Salmon, "Clocks and simultaneity in special relativity or, which twin has the timex?," in P. K. Machamer and R. G. Turnbull, *Motion and Time—Space and Matter* (Columbus, Ohio: Ohio State University Press, 1976), pp. 508–545. Quotation on pp. 524–525.

[25] Note 21.

[26] W. C. Salmon, in note 24, p. 508.

[27] I. A. Winnie, "Special relativity without one-way velocity assumptions: Part II," *Philosophy of Science* **37,** 233–238 (1970).

[28] M. Friedman, "Simultaneity in Newtonian mechanics and special relativity," in J. S. Earman, C. N. Glymour, and J. Stachel (eds.), *Foundations of Space-Time Theories—Minnesota Studies in the Philosophy of Science* (Minneapolis, Minnesota: University of Minnesota Press, 1977), vol. 8, pp. 403–432.

CHAPTER FOURTEEN

Recent Debates on the Conventionality of Simultaneity

An important publication supporting the conventionality thesis of distant simultaneity is W. F. Edwards's 1963 paper on the special theory of relativity based on nonstandard synchrony.[1] It shows that the Lorentz transformations can be generalized by admitting anisotropic light propagation and yet be observationally equivalent to those conventionally constructed. Since spatial anisotropy of the propagation of light corresponds to nonstandard synchrony and the Lorentz transformations comprise the whole special theory of relativity, Edwards's paper implies that no experiment in the special theory could ever disprove the conventionality thesis. It proves that the equivalence of different forms of the Lorentz transformations, obtained by different definitions of simultaneity, implies that direct observables, such as the readings on a single clock, would not be affected by adopting a different form of the Lorentz equations.

> But such quantities as time lapse involving two clocks separated in distance with respect to the observer, are not directly observable and involve the definition of

[1] W. F. Edwards, "Special relativity in anisotropic space," *American Journal of Physics* **31,** 482–489 (1963).

simultaneity which is, to some extent, arbitrary. . . . One may well argue about the *reality* of time dilation, or the relativity of simultaneity or other indirect observables, but in the final analysis, the argument is academic unless a two-way signal velocity greater than c is discovered outside of electromagnetism. As far as any measurement is concerned, one can adopt any view he wishes consistent with the fact that the circulation speed of light is c (if, indeed, it is). For most problems the most convenient assumption to make is still that of isotropic space.

Edwards's essay attracted little favorable attention at the time it was published. In fact, it was even rejected by Martin Strauss, who, by the way, was an ardent admirer of Reichenbach. In a contribution to an international seminar on problems of relativistic physics, organized by the University of Jena in 1965, Strauss declared, with explicit reference to Edwards's paper, that the admission of anisotropic propagation of light, within the limits of light propagation, "does *not* give a kinematics physically equivalent to that of special relativity (contrary to what is claimed by the author) but a kinematics either contradicting special relativity . . . or else essentially poorer in physical content."[2] One of the reasons of this deficiency is the declared unavoidable interrelation between the notions of simultaneity and velocity that led to the conceptual circularity stressed by Reichenbach. But this circularity follows only "if the traditional operational definition of velocity in terms of length and time is adopted. Yet we are free to use any other operational definition such as the definition by means of the Doppler effect, or no operational definition at all but an implicit definition by a set of axioms." Strauss thus proposed to "give up trying to define velocity in terms of length and time and, instead, *take the relation of constant velocity as a primitive notion to be characterized ('defined implicitly') by a set of axioms*. This, by the way, is the normal procedure in axiomatics."[3]

Compared with Edwards's paper, another synchrony-free derivation of the Lorentz transformations, published seven years later, gained a much more favorable reception. In fact, the paper published in 1970 by John A. Winnie,[4]

[2] M. Strauss, "The Lorentz group: axiomatics–generalizations–alternatives," *Wissenschaftliche Zeitschrift der Friedrich-Schiller-Universität* **25,** 109–118 (1966); English translation in M. Strauss, *Modern Physics and Its Philosophy* (Dordrecht: Reidel, 1972), pp. 130–151. See also M. Strauss, "Ist die Isotropie der Lichtausbreitung in einem Inertialsystem eine Konvention?" *Monatsberichte der Deutschen Akademie der Wissenschaften* **7,** 626–627 (1965).

[3] Op. cit., p. 139.

[4] J. A. Winnie, "Special relativity without one-way velocity assumptions," *Philosophy of Science* **47,** 81–99, 223–238 (1970).

then of Hawaii University, which derived the kinematics of the special theory of relativity without any assumption concerning the one-way velocity of light or, equivalently, concerning the numerical value of the synchronization parameter ε, was hailed by the conventionalists as an important contribution to the philosophy of time. Winnie's approach was based on the simple argument that if the value of ε is indeed merely a matter of convention then it ought to be possible to derive all experimental results of the special theory of relativity without choosing a numerical value for ε; and conversely, if it is possible to derive all these results without choosing any numerical value for ε, then the value of ε must be a matter of convention. Because the Lorentz transformations provide the theoretical explanation of all observable effects in relativistic kinematics, Winnie concluded that a derivation of the Lorentz transformation without choosing a special numerical value for ε and thus obtaining what he called the "ε-Lorentz transformations" suffices to prove the validity of the conventionality thesis.

Winnie realized that to accomplish this task he had to prove that the ε-Lorentz transformations can be obtained without any hidden one-way velocity assumptions and that these transformations yield all the observational results of the special theory of relativity.

The three postulates that Winnie assumed for the derivation of the ε-Lorentz transformations are: (1) *The round-trip light principle* according to which "the average round-trip speed of any light-signal propagated (*in vacuo*) in a closed path is equal to a constant c in all frames of reference." (2) *The principle of equal passage times* which says: "Let K and K' be two inertial frames in relative motion, and let A and A' be arbitrary points on the x axes of K and K', respectively. Let Δt be the time interval in K of the passage of a rod at rest in K' of rest length s past the point A in K, and let $\Delta t'$ be the time interval in K' of the passage of a rod at rest in K of rest length s past the point A' in K'. Then $\Delta t = \Delta t'$." (3) *The principle of linearity*: "For any two inertial frames K and K' *regardless of the choice of ε in K and K', any point P* in constant straight-line motion with respect to K is also in constant straight-line motion with respect to K', and conversely."

Note that although these principles confine the choice of the one-way velocity of light to be compatible with the average round-trip velocity c and refer to the notion of a one-way velocity they involve only a single clock in each reference frame and do not involve any *petitio principii.* Winnie could therefore conclude that

> any kinematical experiment which results in the disconfirmation of the ε-Lorentz transformations, when a particular value of ε is chosen, disconfirms all permissible choices of ε, including ε = ½. There can be no question, therefore, of somehow determining the unique "correct" value of ε. The consequences of the ε-Lorentz transformations, for arbitrary choices of ε, are simply *one-one translatable* into consequences of the standard Lorentz transformations. . . . This result thus demonstrates that different choices of ε result in *kinematically equivalent* versions of the Special Theory, and this is precisely the claim made by the thesis of the conventionality of simultaneity.

Other deviations of synchrony-free so-called "ε-Lorentz transformations" (which do not imply Einstein's isotropy convention) have been proposed by Peter Mittelstaedt,[5] Michael Friedman,[6] Carlo Giannoni,[7] and Abraham Ungar.[8] Mittelstaedt and Friedman challenged the coventionality thesis of distant simultaneity, Mittelstaedt on the grounds that "this conventionality has been shown to be almost equivalent to the free choice of the coordinate system K' in an inertial system" and Friedman on the grounds that the conventionalists, like Grünbaum, have "given us no reason to accept the view that the only objective temporal relations are constituted by causal relations."[9] Giannoni generalized Winnie's "ε-Lorentz transformation" to form a group by dispensing with the causality condition $0 < \varepsilon < 1$. Ungar presented "an ε-Lorentz transformation group that is harmonious with Reichenbach's ε-signal synchrony, assigns no preferred status to ε = ½, obeys the causality condition, and does not embody Einstein's isotropy convention."[10] It imposed, however, on each reference frame the same synchrony choice in contradiction to the spirit of the conventionality thesis.

[5] P. Mittelstaedt, "Conventionalism in special relativity," *Foundations of Physics* **7,** 573–583 (1977).

[6] Chapter 10, note 16, p. 420. See also M. Friedman, *Foundations of Space-Time Theories* (Princeton, New Jersey: Princeton University Press, 1983), chapter iv, section 7.

[7] C. Giannoni, "Relativistic mechanics and electromagnetics without one-way velocity assumptions," *Philosophy of Science* **45,** 17–45 (1978); "Clock retardation, absolute space, and special relativity," *Foundations of Physics* **9,** 427–444 (1979).

[8] A. Ungar, "The Lorentz transformation group of the special theory of relativity without Einstein's isotropy convention," *Philosophy of Science* **53,** 395–402 (1986).

[9] Note 5, p. 582; note 6, p. 430.

[10] Note 8, p. 396.

Just as many commentators interpreted Winnie's 1970 paper as having vindicated the conventionality thesis of distant simultaneity, seven years later they regarded David Malament's paper[11] as having completely reversed the situation by its proof that the standard simultaneity relation is the only non-trivial simultaneity relation that is definable in terms of the causal structure of the Minkowski space–time of special relativity.

Malament presumed that the conventionality thesis of distant simultaneity, as maintained by Grünbaum, was based on the following two assertions: "(1) The relation is not uniquely definable in terms of the relation of causal connectibility. (2) Temporal relations are non-conventional if and only if they are so definable." Malament claimed that only assertion (2) had so far been the object of critical objections to the conventionality thesis. Although expressing some sympathy with those who claimed that there is no reason why we must adopt just a causal theory of time, as presumed in assertion (2), Malament confined himself to proving that assertion (1) is false, that is, to proving that the simultaneity relation is uniquely definable in terms of the causal connectibility relation, which he denoted by κ. Thus $p \; \kappa \; q$ denotes that the events or space–time points p and q are causally connectible, which means, as Malament reminded us, that a photon or a particle of nonzero mass can travel between them (in either direction) or, equivalently, that each of the two space–time points lies within, or on the boundary of, the other's light cone.

Malament's argument for the nonconventionality of distant simultaneity did not use Reichenbach's ε-notation but was based on certain properties of the geometric structure of Minkowski's space–time R^4 which had been studied previously by Alfred A. Robb.[12]

To understand Malament's paper let us recall some definitions used in Minkowski's space–time theory. In R^4 the inner product of two points $p = (p_0, p_1, p_2, p_3)$ and $q = (q_0, q_1, q_2, q_3)$, defined by $(p, q) = p_0 q_0 - p_1 q_1 - p_2 q_2 - p_3 q_3$, induces the norm $|p| = (p,p)$ and the symmetric binary relation, the causal connectibility κ, defined by $p \; \kappa \; q = |p - q| \geq 0$. Let O denote the time-like world line of an inertial observer and Sim_O the standard simultaneity re-

[11] D. Malament, "Causal theories of time and the conventionality of simultaneity," *Nous* **11**, 293–300 (1977).

[12] A. A. Robb, *A Theory of Time and Space* (Cambridge: Cambridge University Press, 1914). In the preface of *The Absolute Relations of Time and Space* (Cambridge: Cambridge University Press, 1921), p. V, Robb admitted that, when reading Einstein's 1905 relativity paper, he was "strongly repelled by the idea that events could be simultaneous to one person and not simultaneous to another."

lation relative to O. An O causal automorphism is a bijective map $\Phi: R^4 \rightarrow R^4$ which satisfies the conditions that $\Phi(p)\ \kappa\ \Phi(q)$ if and only if $p\ \kappa\ q$ and $\Phi(p) \in O$ if and only if $p \in O$. A binary relation S on R^4 is implicitly definable from O and κ if and only if for all O causal automorphisms Φ and all p and q the following condition is satisfied: $S(p,q)$ if and only if $S[\Phi(p), \Phi(q)]$. Finally, let $\mathrm{Orth}(p, q, r, s)$ be defined by $(p - q, r - s) = 0$.

We are now in the position to understand Malament's "Proposition 1":

(i): $Sim_O(p,q)$ if and only if $(\exists\ r)\ (\exists\ s)(r \neq s\ \&\ r \in O\ \&\ s \in O\ \&\ Orth(p, q, r, s)$.

(ii): *Orth* is explicitly, first order definable in terms of κ.

Briefly expressed, proposition 1 states that the standard simultaneity relation Sim_O is definable from κ and O. That it is the *only* simultaneity relation relative to O, which is thus definable, is stated in Malament's "Proposition 2":

> *Suppose* S *is a two-place relation on* R^4 *where (i)* S *is (even just) implicitly definable from* κ *and* O; *(ii)* S *is an equivalence relation; (iii)* S *is nontrivial in the sense that there exist points* $p \in O$ *and* $q \in O$ *such that* S(p,q). *Then* S *is either* Sim_O *or the universal relation (which holds of all points).*

Propositions 1 and 2 show in what sense the relative simultaneity relation of special relativity is uniquely definable from the causal connectibility relation. That this is not the case for "absolute" simultaneity as used in pre-relativistic physics is the contents of an additional "Proposition 3":

> *Suppose* S *is a two-place relation of* R^4 *where (i)* S *is (even just) implicitly definable from* κ; *(ii)* S *is an equivalence relation; (iii)* S *is non-trivial in the sense that there exist distinct points* p *and* q *such that* S(p,q). *Then* S *is the universal relation (which holds of all points).*

Summarizing Malament's argument we can say that he showed that standard synchrony is characterized not only by the numerical choice $\varepsilon = 1/2$; but it is also uniquely characterized by the geometrical property that the simultaneity planes relative to an inertial observer are orthogonal to his world line. It is this unique geometrical property that, according to Malament, refutes the conventionality thesis according to which the relative simultaneity is not uniquely definable in terms of the relation of causal connectibility.

Without going into the mathematical details of the proof we can easily understand it if we realize that it is based on certain postulated symmetries that exist between the world-line *O* and the light-cone structure only in the case of the standard simultaneity relation. Expressed in geometrical language these postulates demand that every symmetry between the world line of *O* (the Observer) and the light-cone structure, such as translations, scale variances, or reflections about a hypersurface orthogonal to *O*, leave the world-line *O* invariant. The only simultaneity relation which satisfies these symmetries and is a nontrivial equivalence relation is the standard simultaneity relation.

In 1981 Peter Laurence Spirtes submitted his Ph.D. thesis[13] to the University of Pittsburgh. His thesis was devoted to an examination of the conventionalist doctrines and contained, in its last chapter, a critical analysis of the conventionality of simultaneity as conceived by Reichenbach, Winnie, and Malament. At first Spirtes questioned the validity of Reichenbach's notions of a "conceptual definition" and a "coordinative definition" of the concept of intrasystemic simultaneity on the ground that the "conceptual definition" of simultaneity "is not in accordance with the way scientists actually have used the world 'simultaneity.' For example, according to Newtonian physics, gravitation was a force that acted at a distance; i.e., the gravitational force causally affected distant simultaneous events. It follows that Reichenbach's definition was incorrect, and that simultaneous events can be causally connected."[14] In defense of Reichenbach it may, of course, be argued that his notion of the conceptual definition of distant simultaneity as the exclusion of causal interactions refers only to the special theory of relativity and that "simultaneity" means something different in Newtonian physics (or in the general theory of relativity). "However," said Spirtes, "it is intuitively implausible to claim that 'simultaneity' changes its meaning with every change of theory. In the absence of any compelling reason to believe that 'simultaneity' has changed its meaning, to defend Reichenbach's definition in this way is to use an ad hoc argument." Spirtes then showed that even if velocity and simultaneity form a logical circle, as Reichenbach has convincingly demonstrated, it would not imply the possibility of singling out any one statement as a coordinate definition.

Turning to Winnie's claim of the intertranslatability of the standard ($\varepsilon = \frac{1}{2}$) and the nonstandard ($\varepsilon \neq \frac{1}{2}$) versions of the Lorentz transformations, Spirtes

[13] P. L. Spirtes, *Conventionalism and the Philosophy of Henri Poincaré* (Ph.D. thesis, University of Pittsburgh, 1981).

[14] Op. cit., p. 138.

argued that this was a rather trivial result because the nonstandard versions simply use different coordinate systems[15]; and the choice of ε is certainly a matter of convention. But the intertranslatability and choice of ε are irrelevant to the question of whether the concept of simultaneity is conventional in any nontrivial sense. "In particular this intertranslatability and conventionality of ε is irrelevant to

(a) the question of whether or not the reference of the word 'simultaneity' is conventional (has a meaning which does not uniquely determine its reference);
(b) the question of whether or not the relation of simultaneity is conventional;
(c) the question of whether or not any sentences about simultaneity are true by convention.

Thus it is a misnomer to label the claim that the standard and nonstandard versions of the special theory of relativity are intertranslatable 'the thesis of the conventionality of simultaneity,'" as Winnie did at the end of his paper.

Spirtes also argued that this intertranslatability did not imply that the meaning of "simultaneity" in Reichenbach's sense (as exclusion of causal connectability) fails to establish a unique reference for "simultaneity." For in the Newtonian space–time theory this meaning of "simultaneity" does uniquely determine its reference although it is possible to set up ε-coordinates in precisely the same way as in Minkowski space–time and to express the covariant Newtonian theory in each of the standard and nonstandard coordinate systems. Because these different coordinate theories have the same underlying covariant theory they would be intertranslatable. Hence, there are intertranslatable standard and nonstandard versions of the Newtonian theory also, even though in a Newtonian space–time the meaning of "simultaneity" does uniquely determine its reference.

Turning finally to Malament's 1977 paper, Spirtes argued that *unique* definability is not a necessary condition for the nonconventionality of the simultaneity condition but that it suffices that the relation is definable in terms

[15] In a footnote Spirtes mentioned that this was also noted by Michael Friedman in 1977 in "Simultaneity in Newtonian Mechanics and Special Relativity." See chapter 13, note 28. Peter Mittelstaedt, likewise in 1977, also made this claim (see note 5).

of structures intrinsic to Minkowski space–time. Hence "Malament's first proposition by itself shows that standard simultaneity is non-conventional in the context of the special theory of relativity" whereas "Malament's second proposition shows that non-standard simultaneity relations are conventional if symmetric causal connectability is the only intrinsic relation in Minkowski space-time."[16] In addition Spirtes showed that a temporally oriented space–time admits infinitely many nonconventional simultaneity relations. He concludes his thesis with the statement that

> In a Minkowski space-time standard simultaneity is not a conventional relation. All of the non-standard simultaneity relations are conventional. However, they are conventional not because of any freedom to choose different values of ε in the equation $t_2 = t_1 + \varepsilon(t_3 - t_1)$. They are conventional because this equation does not determine a nonstandard simultaneity relation unless a conventional choice of either a temporal or spatial orientation of Minkowski space–time is made. The different possible values of ε are irrelevant to the question of whether or not any simultaneity relation is conventional in Minowski space-time.[17]

In the same year (1981) in which Spirtes presented his thesis, Russell Francis also wrote a critical analysis[18] of Malament's argument of the nonconventionality of standard simultaneity. We mention Francis' paper because it seems to be the only one that deals with an objection that the critical reader of Malament's paper may have encountered when reading Malament's postulates for his "Proposition 2." Referring to Grünbaum's proof of the intransitivity of nonstandard synchronizations[19] and to his own essay[20] of 1980, Francis pointed out that Malament's postulate (ii) in his "Proposition 2," namely that "S is an equivalence relation," already excludes nonstandard synchronizations. But Francis then mentioned that Quinn[21] has shown that non-

[16] Note 13, pp. 174–175.

[17] Ibid., pp. 187–188. The equation quoted is Reichenbach's equation (9.1) in a different notation. Whether a temporal or spatial orientability of space–time, which Spirtes regarded as a necessary condition for the conventionality of nonstandard simultaneity relations, really exists is a question that has recently been disputed. It has been argued, in particular, that particle–antiparticle annihilations furnish evidence that space–time is not time orientable. See, for example, M. J. Hadley, "The orientability of space-time," *Classical and Quantum Gravity* **19,** 4564–4571 (2002) and the literature mentioned therein.

[18] R. Francis, "Convention-free simultaneity in the causal theory of time" (unpublished, 1981).

[19] Chapter 7, note 65.

[20] Chapter 11, note 30.

[21] Chapter 11, note 26.

standard synchronisms exist which are transitive everywhere. Consequently, Francis concluded, "this objection is not fatal."

Two years later, in 1983, several opposing evaluations of Malament's 1977 essay again appeared. Roberto Torretti, in his treatise *Relativity and Geometry*,[22] called Malament's paper "short and trenchant" and spoke of the "ease and brilliance" of the proofs of its propositions. In fact, he regarded Malament's argumentation as a valid confutation of the conventionality of distant simultaneity in special relativity. He qualified his conclusion, however, by pointing out that "we have fought the alleged conventionality of simultaneity in Special Relativity invoking the structure of Minkowski space-time. Our argument is therefore seemingly powerless against those who believe that physical geometry itself is fixed by convention. This, however, is a completely different question, that must be resolved on properly philosophical—i.e. universal, not theory-dependent—grounds."[23]

In contrast, Allen I. Janis contested the importance of Malament's paper for the following reasons. Janis defined a nonstandard synchrony from standard synchrony by introducing another world-line O', which corresponds to an observer O' in uniform motion relative to the observer O. In the inertial frame F' of O', in which the hyperplanes $t' = constant$ provide standard synchrony, the original world-line O represents an observer in uniform motion and Sim_O a nonstandard simultaneity relation. "It thus appears to me," Janis contended, "that if one wishes to use Malament's propositions to argue that the unique choice of standard synchrony is not conventional, it must be on the basis that it is an artificial complication to introduce the parameters that specify a particular O', for it is only their introduction that is needed to specify a particular nonstandard synchrony."[24]

Janis continued: "What I like to suggest, then, is that requiring implicit definability from κ and O . . . is not very different from requiring equal speeds of light in opposite directions." Janis concluded therefore that Malament's argumentation carried no more weight than the simplicity argument, according to which the choice $\varepsilon = 1/2$ should be made on the basis of simplicity, for only that choice makes the velocity of light isotropic. "Those that

[22] Chapter 2, note 75, p. 229.

[23] Ibid., pp. 230–231.

[24] A. I. Janis, "Simultaneity and Conventionality" in R. S. Cohen and L. Laudan (eds.), *Physics, Philosophy and Psychoanalysis* (Dordrecht: Reidel, 1983), pp. 101–110.

found the simplicity argument unconvincing should be similarly unconvinced by an argument on Malament's results; those that are persuaded by Malaments's results should have been similarly persuaded by the simplicity argument. If not, wherein lies the essential difference?"[25]

An even stronger opponent of the conventionality thesis was Michael Friedman, who in his influential study *Foundations of Space-Time Theories*, published in 1983, claimed: "there is a fundamental fact about the standard simultaneity relation . . . that both Reichenbach and Grünbaum overlook: namely, in Minkowski space-time the standard relation is *explicitly definable* from the space-time metric g (in fact, from the conformal structure of g), whereas the nonstandard ($\varepsilon \neq \frac{1}{2}$) relations are not so definable (an observation first made by Malament)."[26] That the introduction of time-orthogonal coordinates implies $\varepsilon = \frac{1}{2}$ is of course quite right. But, as Havas in his previously mentioned essay[27] emphasized, Friedman's conclusion that "we cannot dispense with standard simultaneity without dispensing with the entire conformal structure of Minkowski space-time"[28] "not only does not follow from it, but is patently wrong."

In an essay written in 1989 and designed to disprove the sometimes contended incompatibility of the special theory of relativity with the assumption of the indeterminateness of the future, Howard Stein[29] referred also to Malament's 1977 paper. Stein credited Malament with having shown that what Stein calls "the Einstein-Minkowski conception of relative simultaneity" is not only definable "in a direct geometrical way within the framework of Minkowski's geometry . . . but is the only possible such conception that satisfies certain very weak 'natural' constraints." Although Stein accepted Malament's thesis he qualified his approval in an interesting footnote that states: "There is a slightly delicate point to be noted: Malament's discussion, which is concerned with certain views of Grünbaum, follows the latter in treating space-time without a distinguished time-orientation. To obtain Malament's conclusion for the (stronger) structure of space-time within a time-orientation, one has to strengthen somewhat the constraints he imposes on

[25] Ibid., p. 109.

[26] M. Friedman, *Foundations of Space-Time Theories* (Princeton, New Jersey: Princeton University Press, 1983), p. 310.

[27] Chapter 12, note 65, p. 144.

[28] Note 26, p. 312.

[29] H. Stein, "On relativity theory and openness of the future," *Philosophy of Science* **58,** 147–167 (1991).

the relation of simultaneity; it suffices, for instance, to make that relation . . . relative to a *state of motion* (i.e. a time-like *direction*) rather than—as in Malament's paper—to an *inertial observer* (i.e., a time-like *line*)."[30] Stein seemed not to have been aware that Spirtes had already noted in his Ph.D. dissertation[31] this minor defect in Malament's argumentation although only for the purpose of showing that the conventionality concept involved differs from that espoused by Reichenbach or Grünbaum.

In a paper entitled "The Conventionality of Simultaneity,"[32] Michael Redhead accepted Malament's argument of the unique definability of standard synchrony from the relation of causal connectibility as logically faultless. Nevertheless he claims that the Reichenbach–Grünbaum conventionality thesis can still be defended on grounds similar to those mentioned by Janis, namely that any procedure of establishing standard simultaneity in a moving reference system automatically defines nonstandard simultaneity in a stationary system, a fact which restores conventionality in the latter system insofar as one has to choose which moving system to use for this purpose.

In addition Redhead also referred to the so-called neo-Lorentzian interpretation of the theory of relativity which, as propounded for example by S. J. Prokhovnik,[33] asserts the existence of a privileged ether-frame and reinstates thereby an absolute simultaneity and hence a nonconventional simultaneity relation. That Redhead did not attach much importance to this reintroduction of absolute simultaneity is shown by the fact that in a paper, written in cooperation with Talal A. Debs,[34] he based the understanding of the famous "twin paradox" on the conventionality of simultaneity.

Debs and Redhead accepted Malament's contention of the equivalence between standard synchrony and the Minkowski orthogonality to the time axis of the reference system; they also accepted his demonstration of the definability of this orthogonality in terms of the causal structure of Minkowski space–time. "Nevertheless," they continued, "the conventionality thesis can

[30] Ibid., p. 153.

[31] Note 13.

[32] M. Redhead, "The conventionality of simultaneity," in J. Earman, A. I. Janis, G. J. Massey, and N. Rescher (eds.), *Philosophical Problems of the Internal and External Worlds: Essays on the Philosophy of Adolf Grünbaum* (Pittsburgh: University of Pittsburgh Press, 1993), pp. 103–128.

[33] S. J. Prokhovnik, *The Logic of Special Relativity* (Cambridge: Cambridge University Press, 1967).

[34] T. A. Debs and M. L. G. Redhead, "The twin 'paradox' and the conventionality of simultaneity," *American Journal of Physics* **64,** 384–392 (1996).

still be defended on the grounds that any method that establishes standard synchrony in a moving frame will automatically define nonstandard synchrony in a stationary frame, so the conventional element is restored in specifying simultaneity in the stationary frame, viz., the choice of whether to import into *that* frame the standard synchrony defined in any of the moving frames."[35]

A detailed critical analysis of Malament's essay and of the papers of some of its commentators may be found in the comprehensive review article by Ronald Anderson, Indrakumar Vetharaniam, and Geoffrey E. Stedman.[36]

Although these authors expressed[37] some dissatisfaction concerning the logical rigor of Malament's argumentation, they did not specify any particular deficiency. That, in fact, Malament's proof contains an unwarranted physical assumption was claimed by Sahotra Sarkar and John Stachel in their 1999 article[38] on Malament's paper. Because their article involved many mathematical details we shall content ourselves with only a brief outline of its contentions. First, we must note that for the causal relation under discussion they employ instead of κ the lightlike relatedness λ between two events p and q. This relation, defined by $p\lambda q = |p - q| = 0$, is definable in terms of κ.[39] Use of the lightlike relation enables the authors to define the concepts of a backward (or forward) null cone and to show that Malament's argumentation is logically untenable when he assumes that "the class of O causal automorphisms includes all rotations, translations . . . and reflections of R^4 with respect to hypersurfaces orthogonal to O which map O onto itself."[40] As Sarkar and Stachel showed, Malament's contention that any simultaneity relation has to be invariant under temporal reflections is an unwarranted physical assumption.

[35] Ibid., p. 387.

[36] R. Anderson, I. Vetharaniam, and G. E. Stedman, "Conventionality of synchronisation, gauge dependence and test theories of relativity," *Physics Reports* **295,** 93–180 (1998).

[37] Note 36, § 2.2.2.

[38] S. Sarkar and J. Stachel, "Did Malament prove the non-conventionality of simultaneity in the special theory of relativity?" *Philosophy of Science* **66,** 208–220 (1999).

[39] This (explicitly first-order) definability had already been emphasized by Malament by pointing out that $p \lambda q$ is logically equivalent to $p\kappa q$ & $(p = q \vee (\exists r)(r \neq p$ & $r \neq q$ & $[s] [s \kappa p$ & $s \kappa q \rightarrow s \kappa r]$) See note 11, p. 300. Or: event p is lightlike related to event q if and only if the following condition is satisfied: p is causally connectible with q and in addition p is either identical with q or there exists an event r different from p and from q such that for every event s its causal connectibility with p and with q implies also its causal connectibility with r.

[40] Note 11, p. 298.

If this assumption is ignored, Sarkar and Stachel continued, Malament's "other criteria for defining simultaneity are also satisfied by membership in the same backward (forward) null cone of the family of such cones with vertices on an inertial path. What is then unique about the standard convention is its independence of the choice of inertial path in a given inertial frame, confirming a remark made by Einstein in 1905. Similarly, what is unique about the backward (forward) null cone definition is that it is independent of the state of motion of an observer at a point on the inertial path."[41]

Sarkar and Stachel referred here to a remark in Einstein's seminal paper on special relativity in which Einstein illustrated the conventionality of the simultaneity relation, prior to his definition of standard simultaneity, by pointing out that simultaneity could be defined by using the family of backward null cones of an inertial world line with these words: "We might, of course, content ourselves with time values determined by an observer stationed together with his watch at the origin of the coordinates, and co-ordinating the corresponding positions of the hands [of the clock] with light signals, given out by every event to be timed, and reaching him through empty space. But this co-ordination has the disadvantage that it is not independent of the standpoint of the observer with the watch or clock, as we know from experience."[42] Einstein's proposal of using the backward null cone for the purpose of defining simultaneity can, as Sarkar and Stachel show, "be incorporated into a causal definition of non-standard simultaneity within the standard structure of the special theory of relativity."

Grünbaum pointed out that Malament postulating, in assumption (iii) of his "Proposition 3," that *S* is an equivalence relation, rather than deriving it, is a serious defect in Malament's argumentation against the conventionality thesis.[43] For, as Norton commented, it "eliminates by decree any simultaneity relation that does not partition the spacetime into disjoint sets of mutually simultaneous events." The above-mentioned "topologically simultaneity" (the temporal relation between events that are causally inconnectible by genidentical causal chains) is an example. For two events, e_1 and e_2, may each be causally inconnectible with a third event e_3, but they may be causally con-

[41] Note 38, p. 219.

[42] Chapter 7, note 6, pp. 892–893.

[43] See John D. Norton's essay "Philosophy of Space and Time" in M. H. Salmon et al. (eds.), *Introduction to the Philosophy of Science* (Englewood Cliffs, New Jersey: Prentice Hall, 1992), chapter 5.

nectible among themselves when belonging, for example, to the same world line. Clearly, such a relation cannot partition space–time into disjoint sets of mutually simultaneous events. Because Malament began his paper[44] with the remark that it was primarily directed against Grünbaum's specific conventionalism of simultaneity, Grünbaum wrote in December 2000 an essay entitled "David Malament and the Conventionality of Simultaneity: A Reply." Grünbaum referred to the above-mentioned paper by Janis[45] and explained "how Janis's idea in that paper can be directly harnessed to show that Malament's proof does NOT undermine what I [Grünbaum] actually said about the conventionality of simultaneity relative to an inertial observer. And I show clearly that IF Malament *had* refuted what I said, then he would also have refuted what Einstein had said unambiguously."[46]

The statement in Malament's causal theory of time, that only for standard simultaneity the simultaneity planes relative to an inertial observer are orthogonal to his world line, which defines his inertial frame, greatly influenced subsequent discussions on the conventionality thesis both among physicists and philosophers. Thus Dennis Dieks who claimed that "relativistic time is *more strongly* connected with physical phenomena than its classical counterpart," stated, though without reference to Malament, that "in geometrical terms, standard simultaneity has unique significance as the orthogonality relation in Minkowski spacetime with respect to time-like geodesics," and concluded that "the difference between classical and relativistic simultaneity is not in an interesting way connected with questions of 'conventionality.' "[47] In an earlier essay Dieks contended that simultaneity "is a *physical relation*, reflecting a particular physical condition, rather than a merely conventionally chosen equality of time coordinates."[48] Dieks's argument that relativistic time and, in particular, relativistic simultaneity are no more conventional than their correlates in classical physics is directed against the view that "relativistic time is inherently 'conventional' and therefore not of importance for questions about the nature of reality." Dieks emphasized, however, that

[44] Note 11.

[45] Note 24.

[46] Letter from Prof. Grünbaum to the author, dated 25 December 2000.

[47] D. Dieks, "Time in special relativity and its philosophical significance," *European Journal of Physics* **12,** 253–259 (1991).

[48] D. Dieks, "Newton's conception of time in modern physics and philosophy," in P. B. Scheurer and G. Debrock (eds.), *Newton's Scientific and Philosophical Legacy* (Dordrecht: Kluwer, 1988), pp. 151–159.

his anticonventionalism does not imply an approval of the other view of relativistic time according to which we live in a four-dimensional "block-universe" in which past and future are mere illusions. For according to Dieks the relativistic structure of time and the relativistic notion of simultaneity do not imply that "past and future, and a shifting 'now,' are unreal."

Prompted by the somewhat antithetical essays by Winnie and Malament, the debate about the conventionality thesis of distant simultaneity became the subject of so many discussions during the past few decades that it would be impossible to report on all of them in a book of reasonable size. We will therefore confine ourselves mainly to the philosophically more important arguments and refer the reader, interested in technical details, to the previously mentioned comprehensive 1998 review article by Anderson, Vetharaniam, and Stedman.[49] We will also discuss more recent publications and some earlier studies, not mentioned in this review, provided they seem to be of philosophical importance.

In 1983 Felix Mühlhölzer,[50] a research student of Wolfgang Stegmüller at the University of Munich, published his Ph.D. thesis on the concept of time in the special theory of relativity. The major part of it contained a severe criticism of the conventionality thesis of distant simultaneity. Mühlhölzer challenged the claim of Reichenbach and Grünbaum that causal connectibility determines the temporal relations in space–time and that its absence defines simultaneity for three reasons. First, even if the contention that the whole metrical space–time structure is a logical consequence of its causal structure would be valid for the special theory of relativity, it certainly loses its validity in the general theory of relativity, which contains the special theory as a special case. But a contention of such fundamental import for a given theory should not be vulnerable if that theory becomes part of a more comprehensive theory. Second, because the quantum theory, even in its relativistic formulation, questions the universal validity of causality, causality should not be given such a central role as Reichenbach or Grünbaum contend. Third and most important, it is possible to construct axiomatically the special theory of relativity by replacing the notion of causal connectibility by the concept of symmetry.

In § 12 of his book Mühlhölzer presented his deductive construction of the special theory of relativity "without causality" on the basis of five ax-

[49] Note 36.

[50] F. Mühlhölzer, *Der Zeitbegriff in der Speziellen Relativitätstheorie* (Frankfurt am Main: Peter Lang, 1983).

ioms, A_1 to A_5, which, as he puts it, "describe very simple and fundamental facts of nature." A_1 expresses the principle of relativity, A_2 is the principle of homogeneity, A_3 is an axiom of continuity, A_4 "postulates Euclidicity and thus, in particular, the isotropy of space and implies consequently a unique synchronization of clocks in an inertial system," and A_5 implies the existence of an upper limit of the velocities of inertial systems. The concept of causality is nowhere referred to in the axioms, nor is that of "light," which serves as the transmitter of causality in the usual presentations of the theory. Of importance for us is axiom A_4, for it implies the isotropy of space, which, if applied to the velocity of light, as Einstein phrased it in his definition of simultaneity, says that "the time needed for the light to travel from *A* to *B* is equal to the time it needs to travel from *B* to *A*."[51] In other words, Mühlhölzer's axiomatization of the special theory of relativity is compatible only with standard simultaneity ($\varepsilon = 1/2$) and has been claimed therefore to have refuted the conventionality thesis of distant simultaneity.

Historically viewed, Mühlhölzer's proposal follows a long tradition of axiomatizing the special theory without using the light postulate and obtaining the Lorentz transformation with a universal constant that replaces the velocity of light *c*. In 1910 W. v. Ignatowsky had already published such an approach[52] which was followed by P. Frank and H. Rothe[53] and later by Y. P. Terletzki (1968), G. Süssmann (1969), H. Brennich (1969), V. Gorini (1973), A. R. Lee and T. M. Kalotas (1975), P. Mittelstaedt (1976), J. M. Levy-Leblond (1976), among others.

But did Mühlhölzer or any of his precursors really dispense with the synchronization problem? All these approaches started with the assumption of some space–time coordinate system $S(x, y, z, t)$,[54] where *t* denotes of course the time coordinate of the inertial system *S*. But the introduction of such a temporal coordinatization cannot be carried out without some prior clock synchronization. Einstein was aware of this condition when he declared in his 1905 paper that "we have not defined a common 'time' for [the points] *A* and *B* for the latter cannot be defined at all unless we establish *by defini-*

[51] Chapter 7, note 6.

[52] W. v. Ignatowsky, "Einige Bemerkungen zum Relativitätsprinzip," *Physikalische Zeitschrift* **11,** 972–976 (1910).

[53] P. Frank and H. Rothe, "Über die Transformation der Raumkoordinaten von ruhenden auf bewegte Systeme," *Annalen der Physik* **34,** 825–855 (1911).

[54] In Mühlhölzer's axiomatization the disjunction into three spatial and one temporal coordinate occurs only after the introduction of axiom A_4.

tion that the 'time' required by light to travel from A to B equals the 'time' it requires to travel from B to A."[55] Einstein's admonition that the very use of t as a time coordinate in an inertial system presupposes the choice of a synchronization procedure does not seem to have been considered by those who, like Mühlhölzer, claimed to be able to construct the special theory without any conventional definition of distant simultaneity.

In recent years the role of the notion of simultaneity in the *geometry* of relativistic space–time has become an issue that conventionalists and their opponents use to support their respective assertions. As mentioned earlier, Reichenbach had already recognized that the notion of simultaneity plays an important role in the metrical geometry of special relativity when he defined the length of a moving line segment as the distance between *simultaneous* positions of its end points or when he argued that "space measurements are reducible to time measurements" and that "time is therefore logically prior to space."[56] It is important to distinguish between the "rest length," say, of a moving rod, and its "proper length," although most authors of texts on relativity explicitly identify these two terms.[57] The "rest length" of a moving rod is defined by the repeated superposition of the standard of length moving together with the rod and is therefore independent of the velocity of the rod and of the synchronization parameter ε. On the other hand, the "proper length" of a moving rod, being the distance between the simultaneous positions of its endpoints, depends on the choice of the Reichenbach parameter ε. Reichenbach, therefore, defined what he called "the characteristic length" (Eigenlänge) of a measuring rod as the length that is obtained "if two simultaneous events happen at its ends where simultaneity is defined in such a way that the velocity of light along the rod in both directions is equal"[58] (i.e., by standard synchronization).

In the modern literature the term "proper time" refers to what Erwin Schrödinger called "the distance of simultaneity or, shorter, the simultaneous distance."[59] To understand the meaning of this term note that in the modern approach the space–time of special relativity is regarded as a four-

[55] Chapter 7, note 6.

[56] Chapter 9, note 14 (1958), p. 155.

[57] See, for example, W. Rindler, *Introduction to Special Relativity* (Oxford: Clarendon Press, 1982), p. 28; H. M. Schwartz, *Introduction to Special Relativity* (New York: McGraw-Hill, 1968), p. 62.

[58] Chapter 9, note 12, definition 26.

[59] E. Schrödinger, *Space-Time Structure* (Cambridge: Cambridge University Press, 1950), p. 78.

dimensional real affine differential manifold, the coordinated points of which are called "events." To obtain a temporal and spatial ordering of these events the manifold is endowed (1) with an affine structure or "rigging" of world lines that represent the positions of inertial observers at rest relative to each other and (2) with nonintersecting hypersurfaces of simultaneity that intersect these worldlines.[60] "The temporal metrical properties of events are determined by the spacing along a world line between its points of intersection with the hypersurfaces of simultaneity on which the events lie, and the spatial metrical properties are given by the 4-dimensional spacetime intervals between simultaneous events on the worldlines on which the events lie. The latter stipulation means that such spatial distances are the 'proper lengths' as *given along the hypersurfaces of simultaneity*. Clearly they are dependent on the hypersurface of simultaneity and thus on the choice of synchronization."[61] As Anderson and Stedman proved in their paper, the proper length (or "distance of simultaneity") has a maximum value in the case of standard synchronization.

In a study of the operational procedures for the determination of proper lengths Graham Nerlich concluded that standard synchronization and hence standard simultaneity are not a matter of convention from the fact that the maximum of the proper length of a rod, given by standard synchrony, equals its rest length. For "there clearly is an essential metrical construction which picks out a class of simultaneous events, in a non-arbitrary way, to a given class of rest points. . . . No assumption or connection about the speed of light is made at any point."[62]

Anderson and Stedman also criticized the claim, made by Robert Alan Coleman and Herbert Korte, that proper lengths, or what they call the "spatial metric induced on a hyperplane of simultaneity," can be empirically determined.[63] Furthermore, by defining a space-dependent three-dimensional vector $\boldsymbol{\kappa}$ as the synchronization parameter and the vector $\boldsymbol{h}$ as the gradient of $\boldsymbol{\kappa x}$ [i.e., $\boldsymbol{h} = \nabla (\boldsymbol{\kappa} \cdot \boldsymbol{x})$], where $\boldsymbol{x}$ is the three-dimensional spatial vector, the authors suggest to view $\boldsymbol{h}$ as a "gauge field" in a manner analogous to the well-

[60] For details see R. Anderson and G. E. Stedman, "Distance and the conventionality of simultaneity in special relativity," *Foundations of Physics Letters* **5,** 199–220 (1992).

[61] Ibid., pp. 201–202.

[62] G. Nerlich, "Special relativity is not based on causality," *The British Journal of the Philosophy of Science* **33,** 361–388 (1982).

[63] R. A. Coleman and H. Korte, "An empirical, purely spatial criterion for the planes of simultaneity," *Foundations of Physics* **21,** 417–437 (1991).

known vector potential of electromagnetism. They thus regarded the resynchronization at a given point as a local gauge transformation of $\boldsymbol{h} \rightarrow \boldsymbol{h} + \nabla\chi(\boldsymbol{x})$. Anderson and Stedman thus concluded that "such a perspective towards the conventionality of the one-way speed of light as a local 'gauge' freedom makes perspicuous two insights at the heart of the conventionalist position. The first is that our knowledge of the world is essentially localized in space and time, and the second is that our knowledge of global features of the world is only obtained operationally by measurements using round-trip phenomena. Such features tend to be hidden in the usual formulations of the conventionality of the one-way speed of light where constant choices of $\boldsymbol{\kappa}$ are considered."[64]

In his well-known textbook on relativity Christian Møller[65] presented a mathematical exposition of the use of gauge transformations in general relativity. That the afore-mentioned special relativistic synchronization transformations by Edwards, Winnie, and others can be considered as perhaps the simplest nontrivial examples of such gauge transformations was shown by Anderson and Stedman[66] in 1977. In cooperation with Vetharaniam, these authors offered an excellent survey of the use of the formalism of gauge theories in relativity.[67]

Recently, Ettore Minguzzi, a Ph.D. student at the University of Milano-Bicocca, published a paper initially conceived as an analysis of the conventional nature of simultaneity, using in part the gauge transformation approach.[68] As he later recognized, however, the proof offered in this paper, that the structure of Minkowski space–time is a logical consequence solely of the constancy of the two-way light velocity, "should be considered as more interesting because it relates two convention-free quantities (thus avoiding a conventional choice of simultaneity)."[69]

[64] Note 60, pp. 213–214.

[65] C. Møller, *The Theory of Relativity*, 2nd edition (Oxford: Clarendon Press, 1972).

[66] R. Anderson and G. E. Stedman, "Dual Observers in Operational Relativity," *Foundations of Physics* **7,** 29–33 (1977).

[67] Note 36.

[68] E. Minguzzi, "On the conventionality of simultaneity," *Foundations of Physics Letters* **15,** 153–169 (2002).

[69] Letter from E. Minguzzi to the author of 20 December 2002.

CHAPTER FIFTEEN

Simultaneity in General Relativity and in Quantum Mechanics

In contrast to the voluminous literature on the notion of distant simultaneity in the special theory of relativity, the study of this concept in the general theory of relativity, or only in noninertial space–time systems, has been given rather limited attention. One reason is that a serious study of this subject requires some knowledge of the tensor calculus and differential geometry that philosophers seldom possess. Another reason is the restricted applicability of this notion in the general theory. In fact, most treatises on general relativity ignore this topic completely.[1]

In 1907 Einstein himself recognized that the standard operational definition of this concept, as he proposed in his 1905 relativity paper, and hence also his definition of the concept of time (*Edt* in chapter 11), lose their ap-

[1] Exceptions, which will be used in our presentation, are L. Landau and E. Lifshitz, *The Classical Theory of Fields* (Cambridge, Massachusetts: Addison-Wesley, 1951, 1962); H. Arzeliès, *Relativité Généralisé–Gravitation* (Paris: Gauthier-Villars, 1961); S. A. Basri, "Operational Foundations of Einstein's General Theory of Relativity," *Reviews of Modern Physics* **37,** 288–315 (1965); J. W. Kummer and S. A. Basri, "Time in general relativity," *International Journal of Theoretical Physics* **2,** 255–265 (1969); I. Ciufolini and J. A. Wheeler, *Gravitation and Inertia* (Princeton, New Jersey: Princeton University Press, 1995).

plicability in noninertial systems and, in particular, accelerated systems. For in part V of his comprehensive 1907 essay on relativity (listed in note 40 of chapter 7), from what later became known as the equivalence principle, Einstein derived the result that the velocity of light c' ceases to be a universal constant and depends on the gravitational potential Φ in accordance with the equation

$$c' = c(1 + \Phi/c^2). \tag{15.1}$$

The mathematical apparatus Einstein used in these considerations did not yet include the possibility that the metric of the space–time under consideration can itself be a function of time. The possibility of a time-dependent metric seems to have occurred to Einstein not before 1914 or 1915 when he recognized the importance of formulating generally covariant equations for physical processes by means of the tensor calculus and Riemannian space–time geometry.

At the end of chapter 7, it was pointed out that Einstein proposed that the coordinate system to be dealt with should be a reference system "in which the Newtonian equations hold." It was also shown that within the context of his 1905 relativity paper such an assumption involved a vicious circle. Further, it was claimed that not every kind of reference system admits the standard definition of distant simultaneity. As we will now see, by using the Riemannian metric, it is possible to determine exactly what kind of coordinate system admits Einstein's standard definition of distant simultaneity.

To make the following considerations accessible to more than the mathematical expert we will simplify their treatment as much as possible. We must assume, however, that the reader knows that in the general theory the invariant separation ds between two infinitesimally close events x^a and $x^a + dx^a$ is given by

$$ds^2 = g_{\alpha\beta}\, dx^\alpha\, dx^\beta \quad (\alpha, \beta = 0, 1, 2, 3), \tag{15.2}$$

where the coefficients $g_{\alpha\beta}$ are functions of the coordinates x^a etc., x^0 is the time coordinate, and the Einstein summation rule over repeated Greek indices is applied. The metric (or gravitational field) is called *stationary* if the $g_{\alpha\beta}$ values are time independent, that is, when $g_{\alpha\beta,0} = 0$ (which is the case if it admits a timelike Killing vector field). It is called *static* if it is stationary and, in addition, $g_{\alpha 0} = 0$ for $\alpha = 1, 2$, and 3 (which is the case if the timelike Killing vector field is orthogonal to a foliation of spacelike hypersurfaces).

To study the problem of whether it is possible to define simultaneity in general relativity analogously with its standard definition in special relativity we consider a clock U_A at a location A and a clock U_B at a location B infinitesimally near to A. Let dx^0_{AB} denote the coordinate time interval required by a light signal to travel from A to B and dx^0_{BA} the coordinate time interval for the signal to travel from B to A. For a light signal $ds^2 = 0$ so that equation (15.1) can be written in the form

$$0 = g_{\alpha\beta}\, dx^\alpha\, dx^\beta - 2\, g_{0\alpha}\, dx^0\, dx^\alpha + g_{00}\, dx^0\, dx^0, \tag{15.3}$$

where every Greek index is now to be summed over from 1 to 3. This quadratic equation in dx^0 has two roots, one corresponding to the interval required for the motion of the signal from A to B.

$$dx^0_{AB} = g_{00}{}^{-1}\, [-g_{0\alpha}\, dx^\alpha - \{(g_{0\alpha}\, g_{0\beta} - g_{\alpha\beta}\, g_{00})\, dx^\alpha\, dx^\beta\}^{1/2}] \tag{15.4}$$

and the other to the coordinate time interval required for the motion of the signal from B to A.

$$dx^0_{BA} = g_{00}{}^{-1}[-g_{0\alpha}\, dx^\alpha + \{(g_{0\alpha}\, g_{0\beta} - g_{\alpha\beta}\, g_{00})\, dx^\alpha\, dx^\beta\}^{1/2}]. \tag{15.5}$$

To synchronize the clock U_A at A with the clock U_B at B a light signal is sent from B to A and immediately reflected to B, where the coordinate time intervals are dx^0_{BA} and dx^0_{AB}, respectively, as given in (15.3) and (15.4) and distances are measured from A to B (see fig. 15.1). If the signal reaches A at the instant x^0 it must have left B at the instant $x^0 + dx^0_{BA}$ and is due to return to B at $x^0 + dx^0_{AB}$. In accordance with the definition of standard simultaneity the reading of U_B which is to be regarded as simultaneous with x^0 on U_A is that reading of U_B that lies midway between the instants of the signal's departure and arrival at B, that is, the instant $x^0 + \frac{1}{2}\,(dx^0_{AB} + dx^0_{BA})$. Substituting (15.4) and (15.5) we conclude that the coordinate time difference Δx^0 between two simultaneous infinitesimally separated events is

$$\Delta x^0 = g_\alpha\, dx^\alpha \quad \text{where } g_\alpha = -g_{0\alpha}/g_{00}. \tag{15.6}$$

Equation (15.6) enables us to synchronize step by step infinitesimally near clocks along any open curve but, in general, not along any closed curve because the difference between the initial Δx^0 and the final Δx^0 differs from zero along a closed curve. Thus, as Landau and Lifshitz phrased it, "in the general theory of relativity, simultaneity of events not only has a different meaning in different systems of reference, as in the special theory, but gen-

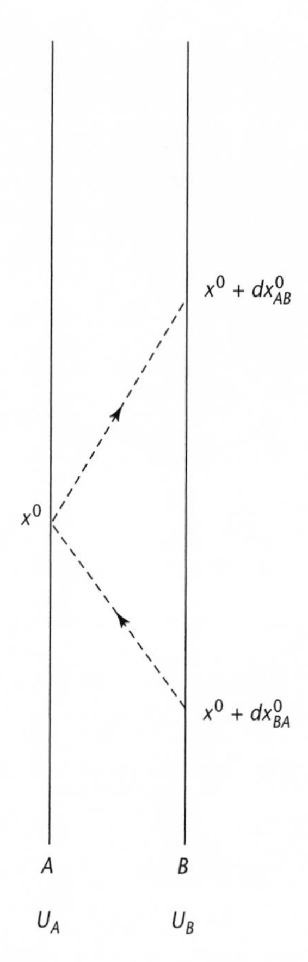

Figure 15.1 The dashed lines represent the world lines of the light signals. It is assumed the $dx^0_{BA} < 0$ and $dx^0_{AB} > 0$.

erally speaking cannot be achieved even within a single reference frame. The only case in which synchronization of clocks is possible is that of a reference system in which all the quantities $g_{0\alpha}$ are zero (or can be made to vanish by a suitable choice of the coordinates x^0)."[2] In other words, in the general theory of relativity only a space–time with a static metric admits standard synchronization. In the special theory of relativity, regarded as a special case of the general theory, the metric is usually expressed in the form $ds^2 = dx^2 + dy^2 + dz^2 - c^2\, dt^2$ so that $x^0 = c\,t$, $x^1 = x$, $x^2 = y$, and $x^3 = z$, $g_{00} = -1$, $g_{11} = g_{22} = g_{33} = 1$, and all other $g_{\alpha\beta} = 0$. Hence, because $g_{0\alpha} = 0$ for $\alpha = 1, 2, 3$ the

[2]Note 1 (1951), pp. 259–260; (1962), p. 236.

metric is static. This conclusion legitimizes Einstein's 1905 definition of standard simultaneity in the special theory from the viewpoint of the general theory of relativity.

In a stationary field, where $g_{0\alpha} \neq 0$, the coordinate time difference between two simultaneous events, which occur at separated points, is given, according to (15.6), by

$$\Delta x^0 = -\int g_{00}{}^{-1} g_{0\alpha}\, dx^\alpha, \tag{15.7}$$

where the integral is taken along the curve joining the two points. Finally, in the case of a closed contour the coordinate time difference between the initial event and the final event, reached after returning to the location of the starting event, is given by

$$\Delta x^0 = -\int g_{00}{}^{-1} g_{0\alpha}\, dx^\alpha \tag{15.8}$$

where the integration is carried out along the closed contour.

To obtain the relation between the coordinate time variable x_0, defined by an arbitrarily running clock, and the corresponding proper time denoted by τ we consider the interval ds between two infinitesimally separated events that occur at one and the same point in space. Because in this case $dx^i = 0$ ($i = 1, 2, 3$) and $ds = c\, d\tau$, equation (15.1) reduces to

$$c^2\, d\tau^2 = g_{00}\, (dx^0)^2, \tag{15.9}$$

which shows that $g_{00} > 0$ and $g_{00}{}^{1/2}$ is a real number. This enables us to conclude that the proper time interval between two events e_1 and e_2 occurring at the same point in space is given by

$$\tau = c^{-1} \int g_{00}{}^{1/2}\, dx^0, \tag{15.10}$$

where the integration extends over the interval from the x^0 coordinate of the event e_1 to that of e_2.

To find out whether the above-described synchronization in general relativity is an equivalence relation, and therefore by implication at least a transitive relation, and enables us to define equivalence classes, as we have done in chapter 11 with respect to the synchronization relation in the special theory, it is useful to define

$$\gamma_\alpha = g_{0\alpha}/g_{00}{}^{1/2} \quad \text{and} \quad \tau_{\alpha\beta} = \gamma_{\alpha,\beta} - \gamma_{\beta,\alpha}. \tag{15.11}$$

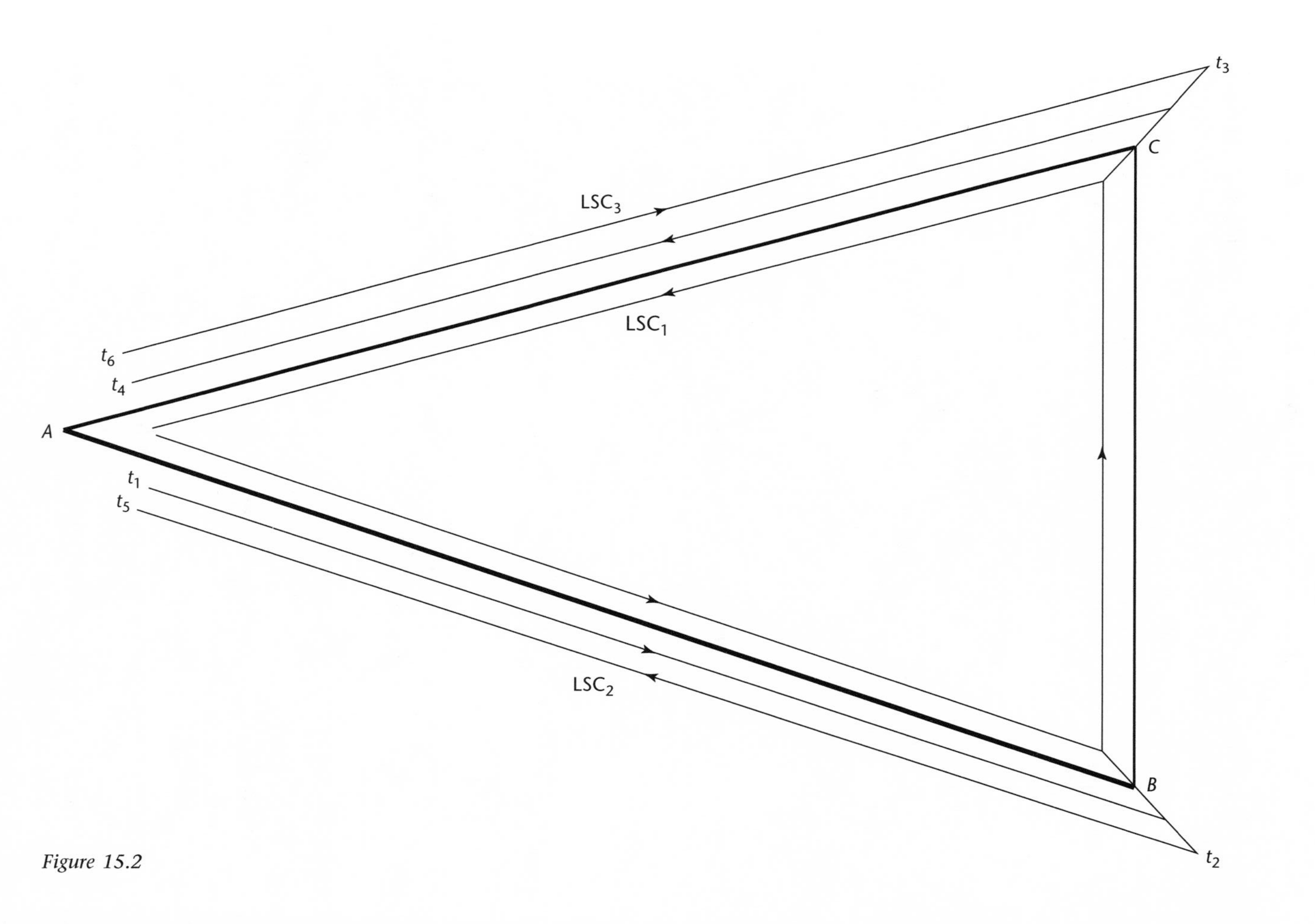

Figure 15.2

It is then not difficult to prove[3] that if clocks are synchronized along a closed curve the reading of the last clock will differ from that of the first clock by the amount $c\int\delta\tau = \int\tau_{\alpha\beta}\, df^{\alpha\beta}$, where $df^{\alpha\beta}$ is an area element of the $x^{\alpha}x^{\beta}$ surface ($\alpha \neq \beta$). It then follows that in the general theory synchronization as defined above is a transitive relation if and only if

$$\tau_{\alpha\beta} = 0. \tag{15.12}$$

To prove that in the general theory of relativity the clock synchronization, despite its nontransitivity, does not lead to such violations of causality we use, following Basri but with slight variations, the following abbreviations: Let U_A, U_B, and U_C be three synchronized clocks located, respectively, at the vertices A, B, and C of an arbitrary triangle (see fig. 15.2). The fact that a light signal leaves U_A at time t_1, arrives at U_B at t_2, thereafter at U_C at t_3 and so forth will be denoted by $[A(t_1) \rightarrow B(t_2) \rightarrow C(t_3) \rightarrow \ \ldots\ ]$, and called "a light signal chain" or, in brief, LSC.

Consider the three light signal chains: LSC_1 defined by $[A(t_1) \rightarrow B(t_2) \rightarrow C(t_3) \rightarrow A(t_4)]$, LSC_2 defined by $[A(t_1) \rightarrow B(t_2) \rightarrow A(t_5)]$, and LSC_3 defined by $[A(t_6) \rightarrow C(t_3) \rightarrow A(t_4)]$. Obviously, because LSC_2 and LSC_3 are more direct than LSC_1,

$$t_5 - t_1 \leq t_4 - t_1 \quad \text{and} \quad t_4 - t_6 \leq t_4 - t_1 \tag{15.13}$$

and therefore

$$t_5 \leq t_4 \quad \text{and} \quad t_1 \leq t_6 \tag{15.14}$$

Note that all four time intervals in (15.13) are recorded by only one clock, namely U_A. Let us now assume that U_B and U_C are synchronized with U_A. Then by definition

$$t_2 = t_1 + \varepsilon(t_5 - t_1) \quad \text{and} \quad t_3 = t_6 + \varepsilon(t_4 - t_6) \tag{15.15}$$

where $0 < \varepsilon < 1$. Because, in accordance with LSC_2, $t_1 < t_5$ or $t_5 - t_1 > 0$ it follows from (15.15) that

$$t_2 > t_1, \tag{15.16}$$

which proves that in LSC_2 the signal left A *before* it arrived at B. Furthermore, (15.15) implies that $t_5 - t_2 = t_5 - [t_1 + \varepsilon(t_5 - t_1)] = (1 - \varepsilon)(t_5 - t_1)$. But since

[3] See, for example, S. A. Basri, note 1 (1965).

in accordance with LSC_2 $t_5 > t_1$ it follows that $t_5 > t_2$ which proves that in LSC_2 the signal left *B* before it arrived at *A*.

Finally, (15.15) implies that $t_3 - t_2 = t_6 + \varepsilon(t_4 - t_6) - t_1 - \varepsilon(t_5 - t_1) = (1 - \varepsilon)(t_6 - t_1) + \varepsilon(t_4 - t_5)$ and hence from (15.14) and $0 < \varepsilon < 1$ we obtain $t_2 < t_3$ which shows that the signal from *B* to *C* cannot arrive at *C* *before* it has left *B*.

Although general relativity agrees with special relativity insofar as it does not violate causality, it differs from the special theory in another simultaneity-related aspect. Whereas in the special theory clocks once synchronized continue to be synchronized, in the general theory, at least in the case of time-dependent gravitational fields, clocks, once synchronized, cease to be so unless the process of synchronization is immediately renewed and incessantly repeated.

The most widely debated problem concerning simultaneity in noninertial reference systems deals undoubtedly with the relativistic rigidly rotating disk. Although strictly speaking this topic belongs to the special theory rather than to the general theory of relativity the mathematics used in it relies heavily on the mathematics used in the latter. Moreover, as John Stachel[4] rightly pointed out, it was the study of this problem that led Einstein to realize that the metric required for a relativistic treatment of gravitational fields cannot be flat.

The early discussions concerning the rotating disk, prompted by the famous "Ehrenfest's paradox,"[5] were almost exclusively confined to the problem of the spatial geometry of the disk. One of the early studies of temporal relations on the rotating disk can be found in Einstein's 1916 review paper.[6]

Considering

> a space which is free of gravitational fields we introduce a Galilean system of reference $S(x,y,z,t)$ and also a system of co-ordinates $S'(x',y',z',t')$ in uniform rotation relatively to S. . . . We are unable to introduce a time corresponding to physical requirements in S', indicated by clocks at rest relative to S'. To convince ourselves of this impossibility, let us imagine two clocks of identical constitution placed, one at the origin of co-ordinates, and the other at the circumference of the

[4] J. Stachel, "Einstein and the rigidly rotating disk," in A. Held (ed.), *General Relativity and Gravitation* (New York: Plenum Press, 1980), vol. 1, pp. 1–15.

[5] If r denotes the radius of the disk when at rest and r' is its radius as measured by a comoving observer, then because of the Lorentz contraction the circumference shrinks and $2\pi r' < 2\pi r$ or $r' < r$. But because the radius is moving normally to the direction of rotation $r = r'$. The contradiction $r' < r$ and $r = r'$ is called the Ehrenfest paradox. See P. Ehrenfest, "Gleichförmige Rotation starrer Körper und Relativitätstheorie," *Physikalische Zeitschrift* **10,** 918 (1909).

[6] A. Einstein, "Grundlage der allgemeinen Relativitätstheorie," *Annalen der Physik* **49,** 769–822 (1916).

> circle, and both envisaged from the "stationary" system S. By a familiar result of the special theory of relativity, the clock at the circumference—judged from *S*—goes more slowly than the other, because the former is in motion and the latter at rest. An observer at the common origin of co-ordinates, capable of observing the clock at the circumference by means of light, would therefore see it lagging behind the clock beside him. As he will not make up his mind to let the velocity of light along the path in question depend explicitly on the time, he will interpret his observations as showing that the clock at the circumference "really" goes more slowly than the clock at the origin. So he will be obliged to define time in such a way that the rate of a clock depends upon where the clock may be. . . . We therefore reach this result: In the general theory of relativity, space and time cannot be defined in such a way that the differences of the spatial co-ordinates be directly measured by the unit measuring-rod, or differences in the time co-ordinate by a standard clock.[7]

Although it is implicit in these statements, Einstein did not yet realize that the impossibility of measuring time in a rotating co-ordinate system or rotating disk results from the nontransitivity of clock synchronization in such a system. In fact, if two clocks, say U_A and U_B, on the rim of the rotating disk have been individually synchronized with a clock U_C at the center of the disk, U_A and U_B will not be synchronized among themselves.

Apparently, Reichenbach was the first to study systematically the possibility of using the concept of simultaneity in the general theory of relativity. In § 37 and § 40 of *Axiomatization*[8] and in § 41 and § 42 of *The Philosophy of Space and Time*[9] he arrived at the following conclusions. In a space–time with a static metric it is possible to introduce a standard synchronization that is both symmetric and transitive. The only difference between a static system and an inertial reference frame is that in the static system the unit of time depends on the location and is therefore not transportable by natural clocks. In stationary space–time, synchronization and simultaneity are symmetric but not transitive temporal relations. Reichenbach illustrates his conclusions by the example of a rotating disk and shows that in this case a transitive synchronization is impossible, although every clock on the disk momentarily

[7] Ibid., English translation: "The foundation of the general theory of relativity," chapter 7, note 15, p. 116.

[8] Chapter 9, note 12.

[9] Chapter 9, note 14.

shows the same time as the clock at rest on the stationary plane under the disk with which it just happens to coincide.

More recently, Peter Mittelstaedt[10] carried out a profound study of the concept of simultaneity in general relativity. An interesting facet of his work was his presentation of the relation between the conditions of distant simultaneity in the general theory and the conventionality problem. Thus he showed that under certain conditions the velocity of light in a direction defined by the unit vector n^i is given by the expression

$$c(n^i) = c\, g_{00}{}^{1/2}/(1 - g_{0i}\, n^i\, g_{00}{}^{-1/2}), \tag{15.17}$$

from which it follows that light propagates isotropically, that is, independently of n_i, and satisfies thereby the conditions for standard synchronization, if and only if $g_{0i} = 0$. These considerations led Mittelstaedt to the conclusion that "the Einstein-synchronization corresponds to a time-orthogonal or synchronous coordinate system and loses thereby much of its generally ascribed conventionality. This suggests, retrospectively, that also the choice $\varepsilon = 1/2$ is more than an arbitrary convention."[11] Mittelstaedt's conclusion agrees, of course, with the previously demonstrated theorem, that only a static metric, in which the coordinate time difference $\Delta x^0 = 0$, admits a standard synchronization.

This important result raises the critical question of whether the metric of the space–time of the cosmos in which we live is really static. For if it is not static, no standard synchronization can be established and Einstein's conception of time, as the equivalence class of all events induced by the standard simultaneity relation, that is, S/σ, is not merely a convention but an empty illusion. In other words, the reality of time, as conceived by Einstein, depends on the validity of (15.7). It therefore poses a question that can be resolved only by experimental evidence, that is, by observational measurements that confirm or disprove that the coordinate time difference $\Delta x^0 = 0$.

A geophysical experiment of this kind was suggested by Jeffrey M. Cohen and Harry E. Moses[12] in 1977 by using a closely spaced set of synchronous satellites, ringing the equator at a distance of 42,000 km from the center of the Earth, or by using simply earthbound clocks rotating with the Earth. In

[10] P. Mittelstaedt, *Der Zeitbegriff in der Physik* (Mannheim: Bibliographisches Institut, 1980), pp. 156–165.

[11] Op. cit., p. 165. See also chapter 14, note 5.

[12] J. M. Cohen and H. E. Moses, "New test of the synchronization procedure in noninertial systems," *Physical Review Letters* **39**, 1641–1643 (1977).

this case the line element ds in the rotating frame, expressed in polar coordinates r, φ, and θ, is given by the formula

$$ds^2 = -\gamma^{-2} c^2(dt - r^2 c^{-2} \omega\gamma^2 \sin^2 \theta \, d\varphi)^2 + dr^2 + r^2 d\theta^2 + r^2 \gamma^2 \sin^2 \theta \, d\varphi^2 \quad (15.18)$$

where

$$\gamma^2 = 1 - r^2 \omega^2 c^{-2} \sin^2 \theta \quad (15.19)$$

and $\omega = 7 \times 10^{-5}$ sec^{-1} is the angular velocity of the Earth.

They assume that one of the satellites carries two clocks, U_1 and U_2. One of these clocks, say U_1, has to be synchronized with the clock in the adjacent satellite and this procedure has to be continued consecutively with each of the satellites along the circle formed by them. When this procedure finally reaches the first satellite, U_2 has to be synchronized with the clock in the preceding satellite. Then the difference in time Δt, indicated by U_1 and U_2, is given (to first order in ω) by

$$\Delta t = 2 \pi r^2 \omega c^{-2} \quad \text{or} \quad \Delta x^0 = 2 \pi r^2 \omega c^{-1} \quad (15.20)$$

In the case of the satellites (r = 42,000 km) this difference is approximately 9 μsec and in the case of the earthbound clocks (r = 6,400 km) it is about 0.2 μsec, time intervals that are easily measurable by atomic or even first-rate quartz clocks.

Commenting on this paper D. B. Lichtenberg and R. G. Newton[13] claimed that equation (15.20), to the accuracy obtained by Cohen and Moses, can be derived by using merely the special theory of relativity and, moreover, can be generalized to the case of large separations between the clocks. Using the Lorentz transformations of the special theory, Lichtenberg and Newton showed that, if the synchronization of two clocks A' and B' is performed by synchronizing in steps closely spaced intermediary clocks, their time difference amounts to

$$\Delta t = 2 \pi r^2 \omega \gamma c^{-2} \quad (15.21)$$

where γ has, as usual in the special theory, the value

$$\gamma = (1 - v^2/c^2)^{-1/2}. \quad (15.22)$$

[13] D. B. Lichtenberg and R. G. Newton, "Comment on a 'New test of the synchronization procedure in noninertial systems,' " *Physical Review D* **19,** 1268–1270 (1979).

According to these authors this result "must hold even if A' and B' are at rest on the rim of a rotating disk of zero mass. This follows because two clocks which are spaced closely enough on the rim can be considered to be in the same co-moving inertial frame with negligible error." Clearly, equation (15.10) agrees with (15.21) to the lowest order in v/c.

If there are n such clocks, equally spaced and pairwise consecutively synchronized by light signals, then the synchrony discrepancy, as measured in the reference frame relative to which the disk is rotating, amounts to

$$\Delta t = (n\,\omega\, r^2/c^2)\sin(2\pi/n) \qquad (15.23)$$

which in the limit $n \rightarrow \infty$ agrees with (15.21).

Independently of Lichtenberg and Newton F. Curtis Michel also contended that the result obtained by Cohen and Moses to the first order, that is, $\Delta t = 2\pi r^2 \omega c^{-2}$, is a purely special relativistic effect and is simply the usual relativistic time dilation. If T denotes the synchronous time as measured in a moving frame and t and x denote the time and position coordinates in the stationary frame, then according to the well-known Lorentz time transformation,

$$\Delta T = \gamma(\Delta t - v\Delta x/c^2) \qquad (15.24)$$

where the last term expresses the fact that clocks synchronized in one frame, in general, cease to be synchronized in another frame. Thus, because around a ring $\Delta x = 2\pi r$ and $v = r\omega$, for a given moment ($\Delta t = 0$)

$$\Delta T = 2\pi r^2\, \omega c^{-2} \qquad (15.25)$$

"which is the Cohen and Moses relation without appeal to metrics or general relativity. . . . The reason that we can 'get away' with applying the rectilinear relationship (13) for circular motion is that, *to first order*, the circular motion is rectilinear."[14]

In a rejoinder to the comments made by Lichtenberg, Newton, and Michel, Cohen, his colleague Christ Ftaclas, and Moses pointed out that because a global inertial frame was assumed,

> it is clear that the results for this example can be obtained (to all orders in v/c) using special relativity. . . . In the literature synchronization on a rotating disk has been treated repeatedly using special relativity. A possibly misleading aspect

[14] F. C. Michel, " 'New' synchronization procedure in noninertial systems," *Physical Review D* **19**, 1271–1272 (1979).

of special-relativistic derivations is that the δt which normally arises in special relativity can be made to vanish with a Lorentz transformation, but the synchronization time difference discussed by Cohen and Moses is measured between two clocks at the same point in spacetime. *It is agreed upon by all observers and cannot be transformed away.*[15]

Cohen and Moses, in collaboration with Arnold Rosenblum, also studied the problem whether clock-transport synchronization, with infinitesimally slow transport velocity, and light-signal synchronization yield the same results for noninertial frames or gravitational fields as they do for inertial systems. They show that both methods give path-dependent, but not necessarily identical results even for the same path. Thus for the same path about 100 km above the Earth the daily "error," defined by Δt of equation (15.10), amounts to about 89 μsec for clock-transport synchronization and to about 60 μsec for light signal synchronization.[16]

In their paper Lichtenberg and Newton[17] rightly pointed out that the method of testing synchronization by means of closely spaced clocks on satellites moving in eastward and westward directions around the Earth, as proposed by Cohen and Moses in 1977, is in principle identical with the experimental procedure applied in measuring the so-called "Georges Sagnac effect."[18]

What is the relation between the Sagnac effect, conceived by Oliver J. Lodge[19] twenty years before Sagnac performed it to verify the existence of the

[15] J. M. Cohen, C. Ftaclas, H. E. Moses, "Clock synchronization in a rotating frame," *Physical Review D* **19,** 1273–1274 (1979).

[16] J. M. Cohen, H. E. Moses, and A. Rosenblum, "Clock-transport synchronization in noninertial frames and gravitational fields," *Physical Review Letters* **51,** 1501–1502 (1983). See also A. Rosenblum, "Clock transport synchronization and the dragging of inertial frames," *Classical and Quantum Gravitation* **4,** L215–L216 (1987); A. Rosenblum and M. Treber, "Clock transport synchronization and the dragging of inertial frames for elliptical orbits," *International Journal of Theoretical Physics* **27,** 921–924 (1988).

[17] Note 13, p. 1269.

[18] G. Sagnac, "L'éther lumineaux démontré par l'effet du vent relatif d'éther dans un interféromètre en rotation uniforme," *Comptes Rendue* **157,** 708–710 (1913); "Sur la preuve de la réalité de l'éther lumineux par l'experience de l'interférographe tournant," ibid., **157,** 1410–1413 (1913). For historical details see E. J. Post, "Sagnac effect," *Reviews of Modern Physics* **39,** 475–493 (1967); R. Anderson, H. R. Bilger, and G. E. Stedman, " 'Sagnac' effect: A century of Earth-rotated interferometers," *American Journal of Physics* **62,** 975–985 (1994).

[19] O. J. Lodge, "Aberration problems—a discussion concerning the motion of the ether near the earth, and concerning the connection between ether and gross matter, with some new experiments," *Philosophical Transactions of the Royal Society, London,* A **184,** 727–807 (1893); "Experiments on the absence of mechanical connection between ether and matter," ibid., **189,** 149–165 (1897).

ether, and the problem of the possibility of standard simultaneity in noninertial reference systems? The answer was given by A. Tartaglia in his statement that "basically the Sagnac effect is a consequence of the break of the univocity of simultaneity in rotating systems."[20] He explained this statement in an essay written with G. Rizzi, which studied the possibility of globally splitting four-dimensionally space–time into a three-dimensional space and a one-dimensional time. It is shown that according to the general theory of relativity this is possible only for extended reference frames that are defined by a congruence of timelike worldlines in general Riemann space–times, for which the "vortex tensor"[21] vanishes. Recall that the space–time vortex tensor Ω_{ij} is defined by $\delta_i\gamma_j - \delta_j\gamma_i$, where γ_i values are the components of a unitary vector defined by $\gamma_i = g_{i0}/(-g_{00})$. "As a consequence, in any reference frame for which the vortex tensor differs from zero, the concept of 'the whole physical space at a given instant' turns out to be conventional, in the sense that it is lacking an operational meaning because of the impossibility of a symmetrical and *transitive* synchronization at large."[22] Hence standard synchronization or simultaneity, a symmetric and transitive relation, loses its general validity.

Tartaglia and Rizzi were not the first who proved that there are space–time reference systems in general relativity that do not admit a global standard simultaneity. This limitation had already been demonstrated more than twenty-five years earlier by Christian Møller and published in the second edition of his comprehensive textbook on relativity.[23] Møller defined the gauge-invariant unit vector Γ_i, just as Tartaglia later defined γ_i, by $\Gamma_i = g_{i0}\ (g_{00})^{-1/2}$ and showed that the standard simultaneity between two events P and P', whose coordinates are (x^i) and $(x^i + dx^i)$, respectively, is defined by the condition that the corresponding standard time differential $-c^{-1}\,\Gamma_i\,dx^i$ vanishes, that is, that

$$c\,dt = -\Gamma_i\,dx^i = 0. \qquad (15.26)$$

Because dt is gauge invariant, simultaneity defined in this way depends on the system of reference alone. However, if one extends this definition to spatially separated events connected by a curve and uses rule (15.15) for each

[20] A. Tartaglia, "General relativistic corrections to the Sagnac effect," *Physical Review* D **58,** 064009/1–9 (1998).

[21] For details see C. Cattaneo, "Conservation laws in general relativity," *Nuovo Cimento* **13,** 237–240 (1959).

[22] G. Rizzi and A. Tartaglia, "Speed of light on rotating platforms," *Foundations of Physics* **28,** 1663–1683 (1998).

[23] C. Møller, *The Theory of Relativity* (Oxford: Clarendon Press, 1952, 1972), p. 34.

infinitesimal part of the curve, one finds that the simultaneity obtained in this way depends on the connecting curve. "Thus, in a general system of reference it is impossible to define globally standard simultaneity between any two events."[24] It thus follows that, as mentioned in the Preface, distant simultaneity, the very same concept that in 1905 was instrumental for the creation of the theory of relativity was finally disqualified by the generalized version of the same theory as having lost its general validity.

Let us begin our study of the status of the concept of simultaneity in the quantum theory with a brief discussion of a paper by Mark Zangari,[25] in which he used the mathematics[26] but not the physics of quantum mechanics to disprove the conventionality thesis of distant simultaneity. In contrast to all previous attempts to refute the thesis on physical grounds Zangari contended to have found a purely mathematical proof of the untenability of nonstandard simultaneity by demonstrating the impossibility of a mathematical transformation from standard to nonstandard coordinates within the special theory of relativity. In other words, Zangari claimed, by using Reichenbach's synchronization parameter ε, to have shown that "transformations of the value of ε to values other than $1/2$ *cannot be defined* in a way that is consistent with the special theory of relativity."

A space–time point x, which in standard synchronization ($\varepsilon = 1/2$) is given by $x = (ct, x^1, x^2, x^3)$, is represented by him in nonstandard coordinates x_ε ($\varepsilon \neq 1/2$) by

$$x_\varepsilon = (c\ t_\varepsilon,\ x^1_\varepsilon, x^2_\varepsilon, x^3_\varepsilon) = (ct + (2\varepsilon - \mathbf{1})\ x,\ x^1,\ x^2,\ x^3) = (ct + \boldsymbol{n}\ \mathbf{x},\ x^1, x^2, x^3) \quad (15.27)$$

where the parameter

$$\boldsymbol{n} = 2\ \varepsilon - 1 \quad (15.28)$$

has been used so that for standard synchronization $\boldsymbol{n} = (0,0,0)$ and for one-dimensional transformations

$$\boldsymbol{n} = (n,0,0) \quad (15.29)$$

[24] Ibid., p. 378.

[25] M. Zangari, "A new twist in the conventionality of simultaneity debate," *Philosophy of Science* **61,** 267–275 (1994).

[26] See, for example, C. W. Misner, K. S. Thorne, and J. A. Wheeler, *Gravitation* (San Francisco: W. H. Freeman, 1973), chapter 41, or R. M. Wald, *General Relativity* (Chicago: The University Press, 1984), chapter 13.

Rather than using a real four-vector Zangari represents a space–time point by a 2×2 complex matrix X, where $X = \sigma_a x^\alpha$ and

$$\sigma_1 = \begin{pmatrix} 0 & 1 \\ 1 & 0 \end{pmatrix} \quad \sigma_2 = \begin{pmatrix} 0 & -1 \\ 1 & 0 \end{pmatrix} \quad \sigma_3 = \begin{pmatrix} 1 & 0 \\ 0 & -1 \end{pmatrix} \quad \sigma_4 = \begin{pmatrix} 1 & 0 \\ 0 & 1 \end{pmatrix} \tag{15.30}$$

are the well-known Pauli matrices. In the case with $\boldsymbol{n}$ as given by (15.28) the transformation from standard to nonstandard synchronization would therefore require the applicability of a matrix $\boldsymbol{P(n)}$, assumed to have the form

$$\begin{pmatrix} a & b \\ c & d \end{pmatrix}, \tag{15.31}$$

a, b, c, and d are complex numbers, which maps X into $X\varepsilon$, that is, which satisfies the equation

$$\begin{pmatrix} ct + n\,x^1 + x^3 & x^1 - i\,x^2 \\ x^1 + i\,x^2 & ct + n\,x^1 - x^3 \end{pmatrix} = \begin{pmatrix} a & b \\ c & d \end{pmatrix} \begin{pmatrix} c\,t + x^3 & x^1 - i\,x^2 \\ x^1 + i\,x^2 & ct - x^3 \end{pmatrix} \begin{pmatrix} a^* & c^* \\ b^* & d^* \end{pmatrix} \tag{15.32}$$

where * denotes complex conjugation. A simple algebraic calculation shows that equation (15.32) has a solution only for $n = 0$, which means that $\boldsymbol{P}$ is the identity (or unit) matrix and $\varepsilon = 1/2$.

Zangari thus arrived at the conclusion that "it is not always possible to define coordinates with $\varepsilon \neq 1/2$ on which representations of the Lorentz group can act." The fact that spinor transformations have Lorentz transformations only in space–time coordinate systems with standard synchrony, Zangari concluded, disproves the conventionality thesis of distant simultaneity.

Although Zangari emphasized that his argument is not based on any particular *physical* phenomenon because it proves that any value of ε between 0 and 1 is not *mathematically* acceptable, it can easily be turned into an argument based on empirical evidence. It suffices to point out that experimental physics reveals the existence of spin-1/2 particles, which according to Zangari's argument constrain in the theory of relativity the synchrony choice to $\varepsilon = 1/2$. As David Gunn and Indrakumar Vetharaniam in their critical review[27] of Zangari's paper emphasized, (a) Zangari's argument assumes the existence

[27] D. Gunn and I. Vetharaniam, "Relativistic quantum mechanics and the conventionality of similarity," *Philosophy of Science* **62**, 599–608 (1995).

of a representation of space–time points in the special theory of relativity, called the complex representation, in which rotations and boost transformations are elements of the special linear group in two dimensions consisting of all 2×2 Hermitian matrices with complex elements and determinant unity, usually[28] denoted by SL(2,C). (b) Only SL(2,C) admits the spinorial description of particles. (c) SL(2,C) transformations cannot consistently be generalized to include nonstandard synchrony coordinate systems. (d) Hence in the special theory of relativity spin-½ particles can be described only in standard synchronization. (e) Since space–time simultaneity relations are to be valid independently of the characteristics of the particles under considerations, only standard simultaneity can have general validity.

As Gunn and Vetharaniam explained in detail, Zangari's conclusion (d) is wrong not only because it would be incompatible with Erich Kretschmann's famous proof[29] that *any* physical theory can be formulated in a generally covariant way, but because point (b) is invalid. "It is just not the case that the complex representation of spacetime points is necessary for a description of spin-½ particles." For, as Gunn and Vetharaniam showed in full detail, Dirac's famous equation of the electron can be derived not only, as usual, in the case of standard synchronization, but also for inertial systems in which $\varepsilon \neq 1/2$.

Gunn and Vetharaniam's refutation of Zangari's argument[30] against the validity of the conventionality thesis was in turn refuted by Vassilios Karakostas,[31] who agreed with Zangari's conclusion that "ε-transformations are not possible on spinors" but not with Zangari's proof of this conclusion. Karakostas also reminded us that this conclusion, though not expressed with specific reference to ε-synchronization, had already been obtained in 1938 by Élie Cartan[32] in his presentation of the relativistic covariance of spinor field equations and later by Gelfand, Milnos, and Shapiro[33] in their analysis

[28] See, for example, É. Cartan, *Leçons sur la Théorie des Spineurs* (Paris: Hermann, 1938); *The Theory of Spinors* (Cambridge, Massachusetts: MIT Press, 1966; New York: Dover, 1981).

[29] E. Kretschmann, "Über den physikalischen Sinn der Relativitätspostulate," *Annalen der Physik* **53,** 575–613 (1917).

[30] Note 27.

[31] V. Karakostas, "The conventionality of simultaneity in the light of the spinor representation of the Lorentz group," *Studies in the History and Philosophy of Modern Physics* **28,** 249–276 (1997).

[32] É. Cartan, *Leçons sur la Théorie des Spineurs* (Paris: Hermann, 1983); *The Theory of Spinors* (New York: Pergamon, 1963; Dover, 1983).

[33] I. M. Gelfand, R. A. Milnos, and Z. Y. Shapiro, *Representations of the Rotation and Lorentz Groups and their Applications* (New York: Pergamon, 1963).

of the difference between the spinor and the tensor coordinate transformations of the Lorentz group.

For a detailed presentation of Karakosta's exposition, which is based on several group-theoretical theorems, the interested reader should refer to the original literature.

The most important reason why Zangari's proof that the equation

$$X = \sigma_\alpha \, x_\alpha = \begin{pmatrix} c\,t + x^3 & x^1 - i\,x^2 \\ x^1 + i\,x^2 & c\,t - x^3 \end{pmatrix}, \tag{15.33}$$

which, as Zangari explicitly admitted, is taken from the classical book on spinors by Roger Penrose and Wolfgang Rindler,[34] cannot be solved for nonzero n or $\varepsilon \neq 1/2$ is that this matrix corresponds exclusively to *restricted* Lorentz transformation. As such, it cannot even describe reflections or inversions and hence also no change from a left-handed to a right-handed coordinate system or vice versa. This restriction had been emphasized explicitly by Penrose and Rindler when they wrote: "for a Lorentz transformation continuous with the identity must be restricted, since no continuous Lorentz motion can transfer the positive time axis from inside the future null cone to inside the past null cone, or achieve a space reflection." It should not have come as a surprise to Zangari, therefore, that equation (15.32) has no solution for nonzero $\boldsymbol{n}$, that is, for an $\varepsilon \neq 1/2$.

Chapter 13 showed that in an inertial system standard synchronization, as proposed by Einstein in 1905, for example, and slow clock-transport synchronization, as proposed by Ellis and Bowman in 1967, for example, are equivalent synchronization procedures provided the velocity of the third synchronizing clock used in the clock-transport synchronization tends toward zero. The question of whether such an equivalence also holds if the clocks to be synchronized move along noninertial trajectories or in a gravitational field, as far as these conditions admit such synchronizations, was studied by Helmut Rumpf in 1985. Rumpf showed that such an equivalence holds in an arbitrary stationary metric provided the following three conditions are satisfied: (1) the trajectories of the two clocks to be synchronized belong to a timelike Killing vector field; (2) proper time is replaced by the Killing parameter; (3) the light rays in the standard synchronization move along the path of the

[34] R. Penrose and W. Rindler, *Spinors and Space-Time* (Cambridge: Cambridge University Press, 1984), vol. 1.

synchronizing clock. Rumpf calls the surface specified by the curve of the synchronization path and the Killing trajectories a two-dimensional "synchronization sheet" and deduces from condition (3) that the null lines of the synchronizing light rays and the world line of the synchronizing clock lie within such a synchronization sheet. Because the metric induced on the two-dimensional synchronization sheet by the full space–time metric is itself stationary there exists a coordinate system in which this metric can be written in the form

$$ds^2 = g_{00}\, dt^2 + 2\, g_{01}\, dt\, dx + g_{11}\, dx^2$$

and in which the curves x = constant are Killing trajectories. But because every two-dimensional stationary metric is also static the conditions for standard synchronization, as explained at the beginning of this chapter, are fulfilled. We confine ourselves with this conclusion and refer the reader for the rest of the proof to Rumpf's paper.[35]

As to the physical status of the concept of simultaneity in quantum mechanics we find ourselves in an extremely difficult situation. As the authors of a recent treatise *Time in Quantum Mechanics* declared, "the treatment of time [let alone simultaneity] is one of the important and challenging open questions in the foundations of quantum theory."[36] One reason for these difficulties is that, as Wolfgang Pauli had shown in 1926, in contrast to energy E, with which it satisfies the uncertainty relation $\Delta E\ \Delta t = h$, the time variable t is not an observable of the system but merely an extraneous topological ordering parameter, because the Hamiltonian with a semibounded spectrum is incompatible with a group of shifts that are generated by some self-adjoint operator, which would then represent the canonically conjugate time observable.

For the widely discussed problem of whether one could nevertheless define an operator T that satisfies with the Hamiltonian H the commutation relation $[T,H] = ih/2\pi$, so that the time-energy and the position-momentum relations would have the same logical status, see section 5.4 in *The Philosophy of Quantum Mechanics*.[37]

[35] H. Rumpf, "On the equivalence of electromagnetic and clock-transport synchronization in noninertial frames and gravitational fields," *Zeitschrift für Naturforschung* **40a,** 92–93 (1985).

[36] J. G. Muga, R. S. Mayato, and I. L. Egusquiza (eds.), *Time in Quantum Mechanics* (Berlin: Springer, 2002), p. V.

[37] M. Jammer. *The Philosophy of Quantum Mechanics* (New York: John Wiley, 1974), pp. 136–156.

The energy-time uncertainty relation might jeopardize the rigorousness of Einstein's 1905 simultaneity definition insofar as the emission of a light signal, that is, of at least one photon, involves an uncertainty ΔE in the energy of the emitter. Consequently, it also causes an uncertainty Δt in the time of the emission process related to the former by the equation $\Delta E\, \Delta t \geq h/2$. Therefore, the t_A, t_B, and t'_A in the equation $t_B - t_A = t'_A - t_B$, which defines synchrony of clocks and hence simultaneity, cannot be sharply determined.[38] Yakir Aharonov and David Bohm, who claimed to have shown that an accurate measurement of the energy can be carried out in an arbitrarily short time interval rejected this conclusion in 1961.[39]

This leads us to the general question of whether quantum mechanics implies the existence of an absolute lower limit to the operational measurability of time intervals. A positive answer would clearly confound a precise definability of distant simultaneity. In 1958 H. Salecker and E. P. Wigner[40] showed that, apart from experimental errors, quantum limitations indeed exist that affect the accuracy of space–time measurements.

Note that the absolute lower limit to the measurability of time intervals, which one obtains if the Salecker–Wigner restrictions are taken into account, amounts to a time interval on the order of 10^{-24} sec, which coincides with the time interval ascribed for completely different reasons to the so-called *chronon*, the smallest possible time interval in the modern theory of temporal atomicity. If we divide the smallest existing space interval, namely the effective diameter of the electron, that is, about 10^{-13} cm, by the fasted possible velocity, that of the propagation of light, 3×10^{10} cm/sec, then clearly the result is a time interval of the order of the chronon.[41]

An even greater challenge to the validity of the standard definition of simultaneity than this atomicity of time in the theory of quantum phenom-

[38] See chapter 7, note 6.

[39] Y. Aharonov and D. Bohm, "Time in the quantum theory and the uncertainty relation for time and energy," *Physical Review* **122,** 1649–1658 (1961). Reprinted in J. A. Wheeler and W. H. Zurek, *Quantum Theory and Measurement* (Princeton, New Jersey: Princeton University Press, 1983), pp. 715–724. For details of the Aharonov-Bohm argument and criticisms of it by V. A. Fock, G. R. Allcock, F. Engelmann, and E. Fick, see M. Jammer, *The Philosophy of Quantum Mechanics* (New York: John Wiley, 1974), pp. 148–152.

[40] H. Salecker and E. P. Wigner, "Quantum limitations of the measurement of space-time distances," *The Physical Review* **109,** 571–577 (1958).

[41] For a brief history of the chronon and additional information of its role in the modern history of the concept of time see G. J. Whitrow, *The Natural Philosophy of Time* (London: Thomas Nelson, 1961), pp. 153–157, 234–237.

ena is the theoretical possibility of the action-at-a-distance effects associated with so-called quantum-mechanical "entangled states." One of the earliest, though not the first, of such presentations was the famous Einstein–Podolsky–Rosen thought experiment that seemed to prove the existence of "faster than light" influences between locally separated parts of a quantum system.[42] Although originally intended to disprove the completeness of quantum mechanics, that is, the contention that "every element of the physical reality has a counterpart in the physical theory," it is important in our present context because it seems to prove the existence of what Einstein once referred to as "spooky, action-at-a-distance" effects. These effects seem to be instantaneously transmitted over arbitrarily great distances and could therefore reestablish the absolute simultaneity of Newtonian physics.

The published version of the Einstein–Podolsky–Rosen (EPR) argument refers to the concept of simultaneity when it declares that:

> either (1) the quantum-mechanical description of reality given by the wave function is not complete or (2) when the operators corresponding to two physical quantities do not commute the two quantities cannot have *simultaneous* reality. Starting then with the assumption that the wave function does give a complete description of the physical reality, we arrived at the conclusion that two physical quantities, with noncommuting operators, can have *simultaneous* reality. Thus the negation of (1) leads to the negation of the only other alternative (2). We are thus forced to conclude that the quantum-mechanical description of physical reality given by wave functions is not complete.[43]

The two physical quantities discussed in this EPR thought experiment are position and momentum and their "simultaneous reality" contradicts Heisenberg's indeterminacy relation, which excludes the possibility of simultaneously specifying the values of two canonically conjugate observables.

The logical cogency of this conclusion has been called into question by the argument that the *simultaneous* reality of the two conjugate observables ascribed to one of the systems has been inferred merely from the *possibility* of measuring *either* of the corresponding observables on the other system and

[42] A. Einstein, B. Podolsky, and N. Rosen, "Can quantum-mechanical description of physical reality be considered complete?," *Physical Review* **47,** 777–780 (1935); reprinted in J. A. Wheeler and W. H. Zurek, *Quantum Theory and Measurement* (see note 39), pp. 138–141.

[43] Op. cit., p. 780 (italics added).

not from the possibility of measuring *both* of these observables, a procedure that is inapplicable for conjugate observables.

Einstein's own argumentation against the completeness of quantum mechanics, as we know from his correspondence with Schrödinger and from other sources, differed considerably from Podolsky's version published in the *Physical Review*. In brief, Einstein based his argument for the incompleteness of quantum mechanics on the conjugation of two principles that, following Don Howard,[44] are usually called the "separability principle" and the "locality principle." According to the separability principle two spatially separated systems possess their own separate real states, and according to the locality principle, no physical effect can be propagated in space with superluminal velocity. That the locality principle, which clearly denies the physical possibility of an instantaneous distant simultaneity, was for Einstein the main reason for his denial of the completeness of the quantum mechanical description of physical reality can be confirmed by his repeated insistence on this principle. Thus, for example, in a letter to Max Born, dated 18 March 1948, Einstein wrote:

> If a physical system is spread out over the parts of space *A and B*, then what exists in *B* should somehow possess an existence which is independent of what exists in *A*. What actually exists in *B* should therefore not depend on what kind of measurement is being performed at the part of space in *A*; it should even not depend on whether at *A* a measurement is being performed or not. One who accepts this program can hardly regard the quantum-theoretical description as a *complete* presentation of physical reality. One who never-the-less tries to regard it as such must assume that the physically real in *B* incurs an instantaneous change because of a measurement in *A*. This would be incompatible with my physical instinct.[45]

A mathematically simpler reformulation of the EPR argument by David Bohm[46] in terms of the states of a pair of spin-½ particles in the singlet state became the standard presentation of this argument in the philosophical lit-

[44] D. Howard, "Einstein on locality and separability," *Studies in History and Philosophy of Science* **16,** 171–201 (1985).

[45] "Dagegen sträubt sich mein physikalischer Instinkt." *Albert Einstein–Max Born, Briefwechsel 1916–1955* (Munich: Nymphenburger Verlagshadlung, 1969), pp. 223–224; *The Born–Einstein Letters* (London: Macmillan, 1971; New York: Walker and Company, 1971), pp. 169–170.

[46] D. Bohm, *Quantum Theory* (Englewood Cliffs, New Jersey: Prentice Hall, 1951), pp. 614–619.

erature on this issue. Its claim for the incompleteness of quantum mechanics and its emphasis on "elements of physical reality" were soon regarded as an incentive for the construction of hidden-variable theories. In fact, in 1952 Bohm[47] himself presented a logically consistent hidden-variable theory that ascribes to every particle definite values of position and momentum. In 1966 John Stewart Bell[48] showed that this theory does not contradict John von Neumann's 1932 "proof" of the impossibility of hidden variables, because von Neumann's proof rests on a certain premise, the so-called additivity postulate, which needs not necessarily be satisfied. Moreover, using Bohm's EPR version, Bell showed that every local deterministic hidden-variable theory involves a certain inequality, called "Bell's inequality," which is *not* satisfied by quantum physics. This result, called "Bell's theorem," namely that no *local* deterministic hidden-variable theory can reproduce all the predictions of quantum mechanics, had been formulated by Bell already in 1964 as follows: "In a theory in which parameters are added to quantum mechanics to determine the results of individual measurements, without changing the statistical predictions, there must be a mechanism whereby the setting of one measuring device can influence the reading of another instrument, however remote. Moreover, the signal involved must propagate instantaneously, so that such a theory could not be Lorentz invariant."[49]

It should not be surprising, therefore, that in an article entitled "How to teach special relativity?" Bell recommended the reintroduction of the idea of an ether in the sense of a preferred inertial reference frame, but one that is intrinsically unobservable.[50] When asked whether he "would prefer the notion of objective reality and throw away one of the tenets of relativity, that signals cannot travel faster than the speed of light," he answered: "Yes. The idea that there is an aether, and those Fitzgerald contraction and Larmor dilations occur, and that as a result the instruments do not detect motion through the aether—that is a perfect coherent point of view. . . . The reason I want to go back to the idea of an aether here is because in the EPR exper-

[47] D. Bohm, "A suggested interpretation of the quantum theory in terms of 'hidden variables,' " *Physical Review* **85,** 166–179, 180–193 (1952).

[48] J. S. Bell, "On the problem of hidden variables in quantum mechanics," Preprint (SLAC-PUB-44), 1964; *Reviews of Modern Physics* **38,** 447–452 (1966).

[49] J. S. Bell, "On the Einstein Podolsky Rosen paradox," *Physics* **1,** 195–200 (1964).

[50] "How to teach special relativity," in J. S. Bell, *Speakable and Unspeakable in Quantum Mechanics* (Cambridge: Cambridge University Press, 1987), pp. 67–80.

iments there is the suggestion that behind the scenes something is going faster than light."[51]

That indeed Bell's inequality is violated and that quantum theory therefore confirms the existence of superluminal velocities and nonlocality, those spooky actions that Einstein hated so much, has been most convincingly confirmed by experiments performed by Alain Aspect and his team.[52] But can their conclusion that a measurement of certain properties of one particle apparently instantaneously affect the outcome of a measurement on another particle be capitalized to establish absolute simultaneity as conceived in prerelativistic physics?

The question of whether quantum mechanics, because of its instantaneous interactions, contradicts relativity has been the subject of numerous investigations.[53] In any case, as Ballentine and Jarrett[54] convincingly showed, Bell's theorem requires a locality principle that is stronger than the locality principle needed to meet the requirements of the theory of relativity. But the question in its generality is still a matter of dispute.[55]

[51] Interview with J. S. Bell in P. C. W. Davies and J. R. Brown, *The Ghost in the Atom* (Cambridge: Cambridge University Press, 1986), p. 57.

[52] A. Aspect, "Proposed experiment to test the nonseparability of quantum mechanics," *Physical Review D* **14,** 1944–1951; reprinted in J. A. Wheeler and W. H. Zurek (see note 14), pp. 435–442. A. Aspect, J. Dalibard, and G. Roger, "Experimental tests of Bell's inequalities," *Physical Review Letters* **49,** 1804–1807 (1982).

[53] P. H. Eberhard, "Bell's Theorem and the Different Concepts of Locality," *Nuovo Cimento* **46B,** 392–419 (1978); G. C. Ghirardi, A. Rimini, and T. Weber, "A General Argument against Superluminal Transmission through the Quantum Mechanical Measurement Process," *Lettere a Nuovo Cimento* **27,** 293–298 (1980); P. J. Bussey, "Superluminal communication in EPR experiments," *Physics Letters* **90A,** 9–12 (1982); J. P. Jarrett, "An Analysis of the Locality Assumption in the Bell Argument," in L. M. Roth and A. Inomata, *Fundamental Questions in Quantum Mechanics* (New York: Gordon and Breach, 1982), pp. 21–28.

[54] L. E. Ballentine and J. P. Jarrett, "Bell's theorem: Does quantum mechanics contradict relativity?" *American Journal of Physics* **55,** 696–701 (1987).

[55] A. Shimony, "Controllable and Uncontrollable Non-Locality," *International Symposium on Foundations of Quantum Mechanics* (Tokyo: Physical Society of Japan, 1984), pp. 225–230.

Epilogue

Historians usually divide the history of mankind into three epochs: Antiquity, Middle Ages, and Modern Times. In a similar manner, philosophers distinguish between ancient, medieval, and modern philosophy; physicists speak of ancient, classical (Newtonian), and modern physics. Because the concept of simultaneity, as shown in preceding chapters, is a subject of both philosophical and physical studies, it is tempting to apply a similar chronological trisection to the history of this notion even though it denotes only a (temporal) relation.

This is feasible because, in the history of this notion, there were precisely two temporally separated major conceptual developments that radically revolutionized the meaning of this concept. Consequently each of them defines the end of an epoch and the beginning of a consecutive one.

The first of these two critical events was Roemer's discovery of the finite velocity of light in 1676, a decade before the publication of Newton's *Principia*. This discovery changed fundamentally the ancient conception of simultaneity, for it abolished what was called "the visual simultaneity thesis" (see chapter 4) by implying that events seen simultaneously have in reality not occurred simultaneously unless they had been equidistant from the observer when observed. In other words, and strictly speaking, the world as we see it is an optical illusion.

The first period would then end in the late seventeenth century and be followed by an epoch that lasts until the beginning of the twentieth century, an epoch during which the concept of simultaneity was recognized to raise important problems for philosophical and physical inquiry (Leibniz, Kant, and Poincaré).

The third and last epoch begins in 1905 with the publication of Einstein's ground-breaking paper on the special theory of relativity, which offered an operational definition of the notion of simultaneity, proved that it is a relative (frame-dependent) concept, and initiated the conventionality thesis that is still a matter of dispute.

Philosophers and physicists date the beginnings of their disciplines from the day when Thales of Miletus and his school, or the *physicists*, as Aristotle called them, relegated the Greek mythological gods to the domain of fable and instead explained nature by principles and causes.

In contrast to the history of (Western) philosophy and physics, in the history of the notion of simultaneity it is impossible to determine a precise beginning of the first epoch. For long before the beginning of Greek (or Oriental) philosophy primitive man had an intuitive, albeit imprecise, idea of a global "present" or "now" and thereby of a worldwide simultaneity.[1] It has been even contended that this kind of simultaneity is the prototype of the real simultaneity for it is, as Bergson[2] called it, "perceived and lived" (perçu et vécu). In contrast, it is claimed, the simultaneity defined by modern relativists is based merely on mathematical symbols, is relative, and not experienced by anyone.

In a similar vein a noted philosopher of modern physics claimed that "the classical idea of world-wide instants, containing simultaneously spatially separated events, still haunts the subconscious even of relativistic physicists; though verbally rejected, it manifests itself, like a Freudian symbol, in a certain conservation of language."[3]

In any case, in agreement with the metalinguistic remarks made in the earlier sections in chapter 2, we have to conclude that the beginning of the first epoch of the trisection under discussion is not precisely determinable.

[1]See chapter 1, note 26.

[2]See chapter 8, note 21.

[3]M. Čapec, *The Philosophical Impact of Contemporary Physics* (Princeton: Van Nostrand, 1961), pp. 190–191.

That the third epoch in the history of the concept of simultaneity also is not yet a closed chapter follows already from the fact that the controversy about the conventionality thesis of distant simultaneity has not yet abated and that, from time to time, new ideas and new methods are proposed to tackle outstanding problems. A typical example is Hans C. Ohanian's recent suggestion that the conventionality thesis has so far been widely discussed by physicists and philosophers only in the context of kinematics and that an appropriate application of the laws of dynamics could resolve all ambiguities in synchronization and thus, in particular, the conventionality problem. A nonstandard synchronization, he contended, "introduces pseudoforces into the equations of motion, and these pseudoforces are fingerprints of the nonstandard synchronization, just as the centrifugal and Coriolis pseudoforces are fingerprints of a rotating reference frame." He concluded, therefore, that "in an inertial frame, the nonstandard synchronization is forbidden."[4]

We will not discuss this argumentation in detail but simply point out that his paper was criticized by Albert A. Martinez[5] and Alan Macdonald[6] as unconvincing because the equations that involve what Ohanian calls "pseudoforces," though deviating from their conventional formulation, do not predict any physical differences in the actual material behavior of physical systems. In his reply to these two papers Ohanian claimed that "the essential point of my paper is that Reichenbach's [nonstandard] synchronization gives rise to extra terms in the equations of motion, and conventionalists cannot make this problem go away by disputing matters of terminology."[7]

Another example of a debate about simultaneity that is still a matter of dispute concerns Malament's contention[8] that the standard simultaneity relation is the only nontrivial equivalence relation definable in terms of the causal connectibility relation κ and Sarkar and Stachel's claim published in 1999 that Malament's proof contains an unwarranted assumption.[9] In a profound analysis of this issue Robert Rynasciewicz contested in 2001 that their claim was

[4] H. C. Ohanian, "The role of dynamics in the synchronization problem," *American Journal of Physics* **72,** 141–148 (2004).

[5] A. A. Martinez, "Conventions and inertial reference frames," *American Journal of Physics* **73,** 454–452 (2005).

[6] A. Macdonald, "Comment on 'The role of dynamics in the synchronization problem' by Hans C. Ohanian," ibid., 454–455.

[7] H. C. Ohanian, "Reply to Comment on 'The role of dynamics in the synchronization problem,' " ibid., 456–457.

[8] Chapter 14, note 11.

[9] Chapter 14, note 38.

"based on a misunderstanding of the criteria for the definability of a relation, a misunderstanding that Malament's original treatment helped to foster."

This conclusion, Rynasciewicz continued, does not imply, however, that simultaneity must be an equivalence relation. In addition, it may be questioned whether nonconventionality of temporal relations means their unique definability from causal connectibility. Rynasciewicz claimed that what Malament established is that the only (interesting) equivalence relation definable from κ and O is that of lying on the same hypersurface space–time orthogonal to O. Malament concluded his essay with what he regarded as the most important question in this context.

> Suppose an inertial observer emits a light pulse in all directions. Consider the intersection of the resulting light cone with some subsequent hypersurface orthogonal to the observer. Does causal connectibility (plus O if you like) completely determine the *spatial* geometry of the light pulse on that hypersurface *in the absence of some stipulation as to the one-way velocity of light?* If not (and I urge you to think not), then relative simultaneity *does* involve a conventional component corresponding to a degree of freedom in choosing a (3+1)-dimensional representation of an intrinsically four-dimensional geometry.[10]

In a paper written a few weeks later, "Is Simultaneity conventional despite Malament's result?"[11] Rynasiewicz did not criticize the logico-mathematical contents of Malament's paper as such. Rather, what he criticized was its generally accepted interpretation as a refutation of the claim that the criterion for simultaneity in the special theory of relativity is a matter of convention. For no reason was given why simultaneity has to be relative *only* to the choice of an inertial system. "Is this a postulate of the theory of relativity? Can we put it to test? Is it an *a priori* truth? Or is it rather an artifact of having gotten so used to working with the standard one-way speed of light assumption that we take it entirely for granted and therefore overlook the possibility that the requirement is itself just one of several available conventions, one in fact equivalent to the standard Einstein convention?"[12]

[10] R. Rynasiewicz, "Definition, convention, and simultaneity: Malament's result and its alleged refutation by Sarkar and Stachel," *Philosophy of Science* **68,** S345–S357 (2001).

[11] R. Rynasiewicz, "Is simultaneity conventional despite Malament's result?," paper read at a meeting of the Canadian Society for the History and Philosophy of Science at Université Laval, Quebec City, Canada, 24 May 2001.

[12] Ibid.

Likewise motivated by the above-mentioned essay by Malament, which, as we recall, has been claimed to invalidate the conventionality thesis of simultaneity, Domenico Giulini of the University of Freiburg recently published an important investigation of the problem concerning the uniqueness of certain simultaneity structures is flat space–time. Giulini pointed out that the strategy adopted by Malament and his followers to refute the conventionality thesis consisted of identifying, at first, nonconventionality with uniqueness and then, in proving this uniqueness. But this approach can be challenged by pointing out that "every proof of uniqueness rests upon some hypotheses which the simultaneity relation is supposed to satisfy and which may themselves be regarded as conventional . . . For this reason we concentrate on the question of uniqueness, which seems to be a much better behaved notion about which statements can be made once conditions for simultaneity relations are specified." Giulini's proposal can be summarized as follows. Let Aut_X denote the subgroup of automorphisms in the flat space–time M which preserves (stabilizes) the subset X so that it maps lines to lines. Then "simultaneity relative to X" can be defined as a nontrivial Aut_X-invariant equivalence relation on M, each class of which intersects any physically realizable timelike trajectory in at most one point, a statement to be called "the relative simultaneity definition." Let $ILor_X$ denote the inhomogenous Lorentz group (Poincaré group), where X represents an inertial frame, that is, a foliation of M by timelike straight lines. Then, Giulini demonstrated, standard (or Einstein) simultaneity is the unique simultaneity satisfying the "relative simultaneity definition for $Aut_X = ILor_X$. Giulini concluded his paper with a discussion of how far the arguments presented by Malament and by Sarkar and Stachel have to be modified to be valid.[13]

The preceding review of discussions on the concept of simultaneity, as conceived in the course of the third and last period in our above-mentioned trisection of the history of this notion, dealt with only the more important investigations on this subject. An exhaustive treatment would not only make the book unreasonably long, it would also require expert knowledge of modern algebra, group theory, differential geometry, and the mathematics of modern gauge theory. As we have seen in the preceding pages, the last period in this trisection, starting with Einstein's standard definition of simultaneity, became increasingly more logically and mathematically sophisticated. It is surprising how the conceptual development of a relatively simple notion like si-

[13]D. Giulini, "Uniqueness of Simultaneity," *British Journal for the Philosophy of Science* **52,** 651–670 (2001).

multaneity has reached a stage at which an intelligent treatment of it requires scholarly expertise not generally possessed even by an intelligent reader. Even such intellectual giants as Poincaré or Einstein, who initiated this development, would have hardly imagined such a development.

Yet, despite this unprecedented sophistication, the question of whether the thesis of the conventionality of the concept of distant simultaneity is correct has not yet reached a final or generally accepted satisfactory solution.

It is also thought provoking to note that despite Minkowski's famous statement that "henceforth [i.e. from the beginning of the third period in our trisection] space by itself, and time by itself, are doomed to fade away into mere shadows, and only a kind of union of the two will preserve an independent reality."[14] the following differentiation between space and time is still valid.

While the concept of events occurring at different places in space but at the same moment of time (i.e., distant simultaneity) is the subject of heated discussions, the analogous concept of two events occurring at different moments of time but at the same place in space has hardly, if ever, been given serious attention. In the usual Minkowski space–time diagram (for simplicity, assumed to possess only one spatial dimension, described by a horizontal x axis) the former situation is described by points lying on the same straight line parallel to the x axis; whereas the second situation is described by points lying on the same straight line parallel to the t axis. The very fact, that, despite this geometrical symmetry, only the former situation gave rise to the profound philosophical discussions presented in the book clearly indicates that despite the just-mentioned symmetry, time differs from space.

Furthermore, despite the fact that our present third period in the history of the concept of distant simultaneity can be characterized by Minkowski's statement that "only a union of the two [i.e., space and time] will preserve an independent reality," the notion of "the temporal coincidence of spatially separated events" and not the notion of "the spatial coincidence of temporally separated events" has become the subject of such profound discussions as described in this book.

[14]Chapter 5, note 75.

Index

absolute time, 69, 70, 73
Adler, F., 221, 222
Aenesidemus, 45
Agostinelli, L., 122
Aharoni, Y., 290
Albertus Magnus, 56
Alcock, G. R., 290
Alexander of Aphrodisias, 54, 55
Alexander, A., 149
Alexander, H. G., 74
Allcock, G. R., 290
Al-Nazzam, A. I., 27
Alonso, M., 99
Amerio, A., 221
analogies of experience (Kant), 86
Anderson, A., 283
Anderson, R., 263, 269, 270
Aničin, B. A., 229
Aquinas, T., 52, 53
Aristotle, 24–27, 31, 36, 40, 45, 53, 81
arrow of time, 26
Arthur, R. T. W., 83
Arzeliès, H., 146, 165, 178, 242, 271
Aspect, A., 294
astrology, 21, 49
Augustinus, St., 31, 44, 48, 53, 65
Augustynek, Z., 227
Autolycus, 68
automorphisms, 299
Ayer, A. A., 4

Babovic, V. M., 229
Baensch, H., 234
Ballentine, L. E., 294
Barbour, J. B., 3, 30
Bare, C., 238
Bar-Hillel, Y., 14
Barnes, J., 25, 29
Barrow, L., 69, 70, 71
Barth, G., 234
Bartlett, J., 9
Bartrum, C. O., 181
Basri, S. A., 271, 277
Baumann, J. J., 81
Bayle, P., 28
Beauregard, O. C. de, 288
Beck, A., 113
Bell, J. S., 293, 294
Benedicks, C., 222
Bentley, R., 34
Berber, H., 19
Bergmann, H., 149
Bergmann, P. G., 169
Bergson, H., 149, 155, 156, 162
Berkeley, G., 77, 78
Bernays, P., 153
Berry, G. R., 29
Besso, M., 106
Bilger, H. R., 283
Blasius of Parma, 43, 65
Blum, W., 2
Blumenthal, O., 111
Boethius, 50, 51
Bohm, D., 290, 292, 293
Bol, K., 231
Bonnycastle, J., 13
Borel, A., 105
Born, M., 165, 229
Bose, N., 113
Bostock, D., 40
Bowman, P., 211, 246
Boyd, R., 199
Bradley, J., 64
Bradwardine, Th., 68

Brahee, T., 101
Brennich, H., 267
Bridgman, P. W., 198, 199
Brill, A., 149
Brown, G. B., 227
Brown, J. R., 82, 294
Browne, P. F., 5
Budge, W. E., 8
Bussey, P. J., 294

Čapek, M., 60, 95, 100, 162, 296
Capelle, C., 22
Carnap, R., 14, 192
Cartan, E., 287
Cartwright, M., 6
Cassini, G. D., 63, 67
Cattaneo, C., 284
causal automorphism, 256
causal chain, 193
causal theory of time, 82
causality, 45
cavity microwave resonance method, 231–33
Chang, H., 6, 242
Chang, T., 233
Charlton, W., 74
Ciufolini, I., 271
Clark, R. W., ix, 4
Clarke, S., 74
Clifton, R. K., 236
clocks, 61–63
coexistence, 57
Cohen, I. B., 64
Cohen, J. M., 280, 293
Cohen, R. S., 140
Cohn, E., 145
Cole, T. W., 233
Coleman, R. A., 269
consecutiveness, 83
contemporaneity, 10, 11
conventionalism, 95, 100, 101
conventionality of simultaneity, 104, 135, 171–91, 251–70
Cover, J. A., 82
Cunningham, E., 152

D'Abro, A., 162
Dalibard, J., 294
Danton, A., 193
Davidovič, D. M., 229
Davis, P. C. W., 294
De Ritis, D., 236
Debrock, G., 265
Debs, T. A., 262
Democritos, 23
detemporalization of simultaneity, 97
dichotomy, 24
Dieks, D., 265
Diels, H., 23, 29
dimensionality speculations, 93
Dingle, H., 165
Dingler, H., 114, 115, 241
Diogenes Laertius, 26
direction-dependent synchronization parameter, 187
Dirks, U., 175
Driver, S. R., 18
Duhem, P., 57
Dumbleton, J., 68
Dunoyer, L., 149
duration, 76, 80
Durbin, K. A., 190
Dũrr, H. P., 2

Eagle, A., 223
Earman, J., 79, 200, 250
Eberhard, P. H., 294
Eckoff, W. J., 83
Eddington, A. S., 172, 173
Edwards, P., 162
Edwards, W. F., 251
Egusquiza, I. C., 289
Ehrenfest paradox, 278
Ehrenfest, P., 278
Einstein, A., ix, 2, 9, 106–142, 181, 278, 291
Einstein, M. M., 106
Einstein-Podolsky-Rosen paradox, 295
Ellis, B., 211, 246, 249
embankment thought experiment, 132, 196
Engelmann, F., 290
entangled states, 291
equal passage principle, 253
equivalence class, 202
equivalence relation, 201, 297
Eratosthenes, 19

Erlichson, H., 226, 227
Ernst, F., 9
Essen, L., 227, 229, 231
ET simultaneity, 52
eternity, 39, 50, 51
Euclidean geometry, 99
Euler, L., 82
event simultaneity, 10
event, 9
Eyfferth, M., 92

Fakhry, M., 65
Feenberg, E., 224
Feigl, H., 4
Feynman, R. P., 231
Fick, E., 290
Field, J. H., 146
Fink, E., 14
Finn, E. J., 99
Flaclas, C., 282, 283
Flückinger, M., 122
Fock, V. A., 290
Fölsing, A., 3, 108
Föppl, P., 17
Fraassen, B. C. van, 82, 90, 211
Francis, R., 216, 269
Frank, P., 138, 146, 267
Freeman, K., 23, 51
Frege, F. L. G., 143
Freundlich, E., 186
Friedman, M., 200, 250, 254, 261
Friedman, Y., 146
Fritz, K. von, 24, 26
Froome, K. D., 237
Fung, S. F., 232

Gale, R. G., 24
Galilei, G., 12, 69, 125
Galison, P., 121
galvanometric method, 22
Gardiner, A. H., 8
Gasper, P., 199
Gassendi, P., 61, 72
gauge transformations, 270
Gauquelin, M., 20
Gelfand, I. M., 287
Gemma (Frisius), R., 62
genethialogy, 19
geometrical representation, 25
geometrical symmetry, 302
Gerhardt, I. C., 78
Gerteis, M., 165
Geulinex, A., 66
Ghirardi, G. C., 284
Giannoni, C., 225, 254
Giulini, D., 306
Glymour, C., 79, 200, 250
Goethe, J. W. von, 20
Gofman, Y., 146
Goldberg, S., 105, 169
Goldstein, B. R., 19
Goldstein, H., 6
Gordon-Smith, A. C., 231
Gorini, K., 267
Grimm, C., 10, 150
Grimm, J., 11, 84
Grimm, W., 11, 84
Grunbaum, A., 24, 135, 162, 190, 193, 196, 197, 199, 211, 231, 265
Guccione, S., 236
Gulati, S. P., 234
Gunn, D., 286
Gunn, J. A., 94
Gunter, P. A. Y., 157
Guthrie, W. K. C., 33
Gutzmar, A., 14

Hadley, M. J., 259
Hafele, J. C., 64
Hall, D. B., 72, 73, 164
Hall, M. B., 72, 73
hama, 21–23, 27, 31–33, 35, 36
Handyside, J., 83
Harrison, J., 63
Hartz, G. A., 82
Haubold, H. J., 107
Havas, P., 237
Hawking, S., 31
Hazelett, R., 184, 244
Hebrew, 18
Heisenberg, W., 2
Held, A., 278
Helmont, J. B. van, 60
Heytesbury, W., 68

hieroglyph, 5
Hill, H., 234
Hodgson, S. H., 91
Homer, 21
Hopi, 17
Howard, D., 292
Hoyer, U., 153
Hsieh, K. G., 232
Hussey, E., 34
Huygens, Ch., 62
hypertime, 44

Ibn Roshd (Averroës), 56
Ibn Sina (Avicenna), 55
Ignatowski, W. von, 146, 267
incomprehensibility, 81,
inertial system, definition of, 142–44
Infeld, L., 109
Inomata, A., 294
instantaneousness, 43
intertranslatability, 258
interval simultaneity, 10
intervening velocity, 245–47
intransitivity of nonstandard simultaneity, 212–16
Inwood, M., 41
irreconcilability between light principle and additional rule of velocities, 107
Ishiwara, J., 107
Ives, H. E., 244, 245

Jackson, F., 224
Jacob, M., 221
Jammer, M., 3, 9, 52, 77, 99, 143, 289, 290
Janich, P., 144
Janis, A., 200, 211, 260
Jarrett, J. P., 294
Jeans, J. H., 181
Jeffrey, G. B., 111, 112
Jowett, B., 28

Kacser, C., 113
Kalotas, T. M., 263
Kamlah, A., 174, 175
Kanitscheider, B., 57
Kant, I., 31, 82, 86–89, 91
Kar, K. C., 124
Karakostas, V., 287
Karlov, L., 216, 229
Kauffmann, L. H., 146
Keating, R. E., 164
Keswani, G. H., 153
Kilmister, C. W., 113
Kim, D. S., 153
Kittel, C., 113
Kneale, W. C., 33
Kolen, P., 237
Korte, H., 269
Krallmann, D., 86
Kraus, O., 155
Kretschmann, E., 287
Kretzmann, N., 51
Kuhn, T. S., 3
Kummer, J. W., 271

Lambert, R., 8
Landau, L., 87, 273
Lange, L., 138, 141, 143
Langevin, P., 242
Laplace, P. S. M. de, 105
Larousse, P., 105
Laue, M. von, 1, 138, 144, 171
law of inertia, 99
Lawson, R. W., 133
Lechalas, G., 89, 90
Lechalier, J., 28
Lee, A. R., 267
Leibniz, G. W., 66, 78, 80, 84
Leighton, R. B., 231
Lémeray, M., 244
Leplin, J., 216
Levy-Leblond, J. M., 267
Liceti, F., 126
Lichtenberg, D. B., 281
Liebowitz, B., 231
Lifshitz, E., 271
light principle, 122
light signal chain, 277
linearity principle, 253
Lippincott, K., 61
local time, 104
Locke, J., 75, 76
Lodge, O. J., 283
Loemker, L. E., 79

logical priority of simultaneity over time, 76, 110
Long, J. D., 234
López, J. A., 166
Lorentz, H. A., 103, 141
Lovejoy, A. O., 183–85
Lukoschek, B., 145

MacCorquodale, K., 246
Macdonald, A., 297
Macdonald, A. L., 218
Machamer, P. K., 82, 250
Maier, A., 57
Majorana, Q., 182
Malament, D., 255, 256, 259, 298
Malebranche, N., 65
Mallinckrodt, A. J., 170
Mansuri, R., 229
Marder, L., 65, 242
Marinov, S., 233
Maritain, J., 162
Martin, H. A., 86
Martinez, A. A., 297
Massey, G. J., 200
Mayato, R. S., 289
McCausland, I., 141
McGilvary, E. B., 244
McGuire, J. E., 82
McLaughlin, R., 4
McMullin, E., 5
McTaggart, J. M. E., 29
Meehl, P. E., 246
Mehlberg, H., 90
Meitner, L., 127
Mellin, H. J., 153, 155, 182
Mermin, N. D., 146
Metz, A., 136, 157, 162, 243
Michel, F. C., 282
Migne, J. P., 65
Miller, A. I., 105, 113, 204
Milnos, R. A., 287
Minguzzi, E., 270
Minkowski, H., 10, 91, 119, 301
Misner, C. W., 285
Mittelstaedt, P., 254, 267, 280
Mohr, R. D., 33
Møller, C., 129, 165, 232, 237, 284
More, H., 69
Morgenbesser, S., 193
Morrand, E., 162
Moser, S., 40
Moses, H. E., 280, 283
Muga, J. G., 289
Mühlhölzer, F., 266

Naber, G. L., 10
Nelson, A., 167
Nerlich, G., 4, 269
Newton, I., 15, 72–74
Newton, R. G., 281
Nissim-Sabat, C., 237
Noonan, T. W., 143
Nordenson, H., 135
Nordmann, Ch., 158, 162
Northrop, F. S. C., 161, 162
Norton, J. D., 147, 264
now, 14, 15, 26, 35, 38–40, 41, 208

occasionalism, 65, 66
Ohanian, H. C., 297
Øhrstrøm, P., 224, 225
Okruhlik, K., 82
omnipresence, 74
Ono, Y. A., 107
operational definition, 49
operationism, 198
Ortroy, F. van, 62
Otis, A. S., 234
Otto, E., 9, 10
Owen, G. E. L., 29

Pagel, W., 60
Pais, A., 142
Palagyi, M., 76
Pargetter, R., 224
Parish, L., 166
Parmenides, 28
Patrizi, F., 58
Paty, M., 155
Pauli matrices, 286
Pauli, W., 128
Pearsell, J., 12
Penrose, R., 288
Perret, D., 122

Perrett, W., 112
Petkov, V., 238
Pfarr, J., 144
Photius, 45
Pihl, M., 228
Planck, M., 1
Plato, 31–35
Plotinus of Lycopolis, 50
plurality of times, 54, 57
Podlaha, M., 142, 234
Podolsky, B., 291
Pogo, A., 62
Poincaré, H., 98, 100–5, 240
Porphyry, 50
Poser, H., 175
Post, E. J., 283
principle of coexistence (Kant), 87
principle of relativity, 103
principle of the succession of time (Kant), 88
process, 10
Prokhovnik, K. S. J., 248, 262
Putnam, H., 238, 239

Quinn, P. L., 214, 259

rapidity, 243
Rapier, P. M., 233
reality of time, 149
Rechenberg, H., 2
Redhead, M., 200, 262
reductio ad absurdum, 28
Reiche, F., 3
Reichenbach, H., 5, 13, 14, 71, 79, 127, 174, 177, 182, 183, 193, 197, 206, 228
Reignier, J., 105
relational theory of time, 78
relativity of simultaneity, 116–18
Rescher, N., 82, 200, 268
Richer, J., 67
Rietdijk, C. W., 239
Rimini, A., 294
Rindler, W., 268, 288
Rizzi, G., 284
Robb, A. A., 153, 255
Robertson, H. P., 143
Roemer, O., 64, 226
Roemer's method, 226–80
Roger, G., 294
Rosen, N., 290
Rosenblum, A., 283
Rossi, B., 164
Roth, L. M., 294
Rothe, H., 146, 267
round-trip axiom, 207
round-trip light principle (Winnie), 253
Rowland, D. R., 169
Roxburgh, I. W., 215
Ruderfer, M., 232
Ruebenbauer, K., 234
Rumpf, H., 289
Russell, B., 23, 34, 79
Rynasiewicz, R., 73, 297, 298, 299

Sagnac effect, 283
Sagnac, G., 283
Saha, M. N., 113
Salecker, H., 290
Salmon, M. H., 264
Salmon, W. C., 24, 175, 211, 238, 249, 250
Sands, M., 231
Sapir, E., 15
Sarkar, S., 263
Sauter, J., 122
Scherr, R. E., 167
Scheurer, P. B., 265
Schilpp, P. A., 14
Schlick, M., 158, 159, 186
Schofield, M., 30
Schopenhauer, A., 66, 89
Schrödinger, E., 268
Schwartz, H. M., 113, 123, 129, 268
Schweizer, A., 221
Scott-Iversen, P. A., 187, 197, 228
Scribner, Ch., 105, 113
Sears, F. W., 129, 191
Seelig, C., 145
Seeliger, H., 141
sem, 11
Sexl, R. U., 229
Sextus Empiricus, 19, 21, 45
Shaffer, P. S., 167
shaft synchronization, 222–26
Shakespeare, W., 9
Shamos, M. H., 113

Shapiro, Z. Y., 287
Shea, J. H., 229
Shimony, A., 294
Sigwart, C., 81, 93
Silberstein, L., 206
Simplicius, 24, 33, 43, 44
simul, 11, 12, 48, 51
singleness of time, 53
Sjödin, T., 234
Smith, Q., 10, 15
Solovine, M., 106
Sommerfeld, A., 1, 228
Sorabji, R., 29, 32
space, 60
spatial measurements, 4
Speziali, P., 106
spinors, 287
Spirtes, P. L., 257, 258
Stachel, J., 79, 200, 250, 263, 278
stadium, 27
standard signal synchrony, 180
Stedman, G. E., 263, 269, 270, 283
Stein, H., 261
Stein, L., 65, 66
Stiegler, K., 153
Stier, H. E., 9
Stolakis, G., 235
Strasser, H., 153
Strato of Lampsacus, 43
Strauss, M., 144, 252
Stump, L., 51
Suarez, F., 59, 60
Suppes, P., 175
Süssmann, G., 267
Swineshead, R., 68
symmetry of simultaneity, 201–19
synchronization procedures, 104

T simultaneity, 52
Talmon, S., 18
Tartaglia, A., 284
Telesio, B., 58
Terletzki, Y. P., 267
Thirring, H., 183
Thomson, J., 98
Thorne, K. S., 285
time, 33, 60, 61, 77, 78, 80
Tipler, P. A., 170
Tolman, E. C., 246
topological simultaneity, 105
Toraldo di Francua, G., ix
Torr, D. G., 237
Torretti, R., 39, 143, 166, 178, 225
Townsend, B., 226, 237
train-embankment thought experiment, 132, 133
transitivity of simultaneity, 201–19
Treber, M., 283
Trendelenburg, A., 91, 92
Trout, J. D., 199
Trumble, B., 12
Turnbull, H. W., 74
Turnbull, R. G., 82, 250
Turner, D., 184, 244
twin paradox, 70, 165, 262
twin rod experiment, 128

Ungar, A., 254
Urmson, J. O., 33

Vaihinger, H., 155
Valentiner, G., 150
Varičak, V., 128
Velleius, G., 32
velocity of light, 63
Vescovini, G. F., 65
Vetharaniam, I., 263, 286
visual simultaneity thesis, 55, 64
Vlastos, G., 24
Vokos, S., 167
vortex tensor, 284

Wahrig, G., 9
Waismann, F., 4
Wald, R. M., 285
Wartofsky, M. W., 190
Weber, T., 294
Weingard, R., 238
Weisse, C. H., 92
Wertheimer, M., 121, 202
Weyl, H., 175, 239
Wheeler, J. A., 271, 285, 290, 294
Whitehead, A. N., 159, 161
Whitrow, G. J., 7, 290

Whorf, B. L., 17
Wiechert, E., 151
Wien, W., 221
Wigner, E. P., 290
William of Ockham, 56, 57
Williams, L. P., 113
Winnie, J. A., 79, 129, 190, 250, 253
Winternitz, J., 242
Wittgenstein, L., 14
Wolters, G., 115
world point, 10
Wroblewski, A., 64

Yasui, E., 107

Zangari, M., 285
Zeller, E., 40
Zeno of Elea, 23
Zielinski, Th., 18
Zurek, W. H., 291, 294